China in the Global Economy

WITHDRAWN

Environment, Water Resources and Agricultural Policies

LESSONS FROM CHINA AND OECD COUNTRIES

DATE DUE

ILL#27285977	
DUE 03/15/2007	
JAN 15 2008	

ORGANISATION FOR ECONOMIC CO-OPERATION AND DEVELOPMENT

10174867

S
494.5
.W3
E69
2006

ORGANISATION FOR ECONOMIC CO-OPERATION AND DEVELOPMENT

The OECD is a unique forum where the governments of 30 democracies work together to address the economic, social and environmental challenges of globalisation. The OECD is also at the forefront of efforts to understand and to help governments respond to new developments and concerns, such as corporate governance, the information economy and the challenges of an ageing population. The Organisation provides a setting where governments can compare policy experiences, seek answers to common problems, identify good practice and work to co-ordinate domestic and international policies.

The OECD member countries are: Australia, Austria, Belgium, Canada, the Czech Republic, Denmark, Finland, France, Germany, Greece, Hungary, Iceland, Ireland, Italy, Japan, Korea, Luxembourg, Mexico, the Netherlands, New Zealand, Norway, Poland, Portugal, the Slovak Republic, Spain, Sweden, Switzerland, Turkey, the United Kingdom and the United States. The Commission of the European Communities takes part in the work of the OECD.

OECD Publishing disseminates widely the results of the Organisation's statistics gathering and research on economic, social and environmental issues, as well as the conventions, guidelines and standards agreed by its members.

This work is published on the responsibility of the Secretary-General of the OECD. The opinions expressed and arguments employed herein do not necessarily reflect the official views of the Organisation or of the governments of its member countries.

© OECD 2006

No reproduction, copy, transmission or translation of this publication may be made without written permission. Applications should be sent to OECD Publishing: *rights@oecd.org* or by fax (33-1) 45 24 99 30. Permission to photocopy a portion of this work should be addressed to the Centre français d'exploitation du droit de copie (CFC), 20, rue des Grands-Augustins, 75006 Paris, France, fax (33-1) 46 34 67 19, *contact@cfcopies.com* or (for US only) to Copyright Clearance Center (CCC), 222 Rosewood Drive Danvers, MA 01923, USA, fax (978) 646 8600, *info@copyright.com*.

FOREWORD

The *Workshop on Environment, Resources and Agricultural Policies* in China, held in Beijing on 19-21 June 2006, marked a decade of co-operation between the Chinese Ministry of Agriculture and OECD's Directorate for Food, Agriculture and Fisheries. During this period seven workshops were jointly organised, each time on a new theme.

The theme of agro-environmental issues with a special focus on water was identified in the *OECD Review of Agricultural Policies in China* (2005) as one of several urgent issues to be examined and was given priority by the Chinese Ministry of Agriculture. This issue also falls within the framework of work on China by the OECD's Environment Directorate, in particular the preparation of the *Environmental Performance Review of China* currently underway and which involves close co-operation between the OECD Agriculture and Environment Directorates.

The overall aim of this Workshop was to conduct a dialogue between Chinese authorities and OECD countries with a view to outlining policy options to address agri-environmental policies in China.

The Workshop attracted over 60 participants including representatives from 14 OECD member and non-member countries; Chinese policy makers (from the Ministry of Agriculture, State Council, Development Research Center of the State Council, National Development and Reform Commission, and Office of the Leading Group for Finance and Economy); academics and researchers from China and OECD countries; extensive representation from the business community (including Syngenta, International Fertilizer Industry Association, Unilever, Nestlé, and BASF); and NGO representation from IUCN-World Conservation Union, as well as representatives from the FAO and the World Bank. The meeting was opened and closed by Mr. Fan Xiaojian, Vice Minister in the Ministry of Agriculture, who has assured that the results of the workshop will be reflected in the policy making process in China.

The Executive Summary captures the essence of the discussion and main conclusions of the Workshop, as well as highlights of each paper presented in the three sessions. Each paper is preceded by an abstract to orient the reader. The reader will find a wealth of material relating to key issues in the interface between agriculture and environment, policy design and practice, and challenges for the future.

AVANT-PROPOS

L'Atelier sur l'environnement, les ressources et la politique agricole en Chine tenu à Beijing du 19 au 21 juin 2006 a marqué dix ans de coopération entre le ministère chinois de l'Agriculture et la Direction de l'alimentation, de l'agriculture et des pêcheries de l'OCDE. Cette décennie a vu le déroulement de sept ateliers, qui ont à chaque fois été organisés conjointement sur un thème différent.

L'Examen OCDE des politiques agricoles de la Chine (2005) avait mis en évidence plusieurs dossiers nécessitant un examen urgent, parmi lesquels celui des questions agro-environnementales, et plus particulièrement le problème de l'eau, auquel le ministère chinois de l'Agriculture a donné la priorité. Ce thème fait également partie des travaux que la Direction de l'environnement de l'OCDE consacre à la Chine, et notamment de la préparation de *L'Examen des performances environnementales de la Chine,* qui est actuellement en cours et donne lieu à une coopération étroite entre les Directions de l'agriculture et de l'environnement.

Cet atelier visait globalement à permettre un dialogue entre les autorités chinoises et les pays de l'OCDE, l'objectif étant de cerner les choix s'offrant à la Chine pour élaborer sa politique agro-environnementale.

L'atelier a réuni plus de 60 participants, parmi lesquels des représentants de 14 pays membres et non membres de l'OCDE, des hauts fonctionnaires chinois (du ministère de l'Agriculture, du Conseil d'Etat, du Centre de recherche sur le développement du Conseil des Affaires d'Etat, de la Commission d'Etat pour le développement et la réforme, ainsi que du cabinet du Groupe dirigeant chargé des finances et de l'économie), d'universitaires et de chercheurs chinois et de pays de l'OCDE, d'un nombre important de représentants du monde des affaires (en particulier : Syngenta, International Fertilizer Industry Association, Unilever, Nestlé et BASF), des représentants d'ONG appartenant à l'Union mondiale pour la nature (UICN), ainsi que des représentants de la FAO et de la Banque mondiale. M. Fan Xiaojian, vice ministre de l'Agriculture, qui a ouvert et clôturé la réunion, a déclaré que les conclusions de l'atelier seraient prises en compte dans le processus d'élaboration des politiques chinoises.

Le résumé reprend l'essentiel du débat et des principales conclusions de l'atelier, ainsi que les principaux essentiels de chacune des communications présentées au cours des trois sessions. Chaque communication est précédée d'un résumé destiné à orienter le lecteur, qui trouvera dans cette publication d'abondantes informations sur les enjeux clés que pose l'interface entre agriculture et environnement, conception et exécution des politiques, et défis de demain.

ACKNOWLEDGEMENTS

These proceedings bring together papers from the *Workshop on Environment, Resources and Agricultural Policies in China*, held in Beijing, on 19-21 June 2006. For the OECD, Andrzej Kwieciński was responsible for the design and organisation of the Workshop with assistance from Anita Lari. Li Na prepared and organised the Workshop on behalf of the Ministry of Agriculture, China. These proceedings were edited by Brad Gilmour and Andrzej Kwieciński. Special thanks are extended to all those who provided papers and contributed to the success of the discussion. Anita Lari assembled and formatted the final publication with assistance from Michèle Patterson and Claude Nenert.

Table of contents

EXECUTIVE SUMMARY

Background

China's leaders face the daunting challenge of feeding the world's largest population with limited resources. China's per capita endowment of arable land falls between Germany and Chile. Its endowment of water resources is extremely low and badly distributed, with the North China Plain having one of the lowest per capita endowments in the world.

Historically, increased food production to satisfy China's large and growing population was achieved by expanding the area of land cultivated. When this was no longer possible, increases in output were achieved by increasing land productivity through higher water, fertiliser, pesticide and mechanical inputs. However, the effects of some of these input-intensive agronomic practices are starting to take their toll. Nutrient and pesticide loading of land and watercourses are on the increase; desertification, soil erosion and land degradation contribute to declines in productivity. Pollution of some rivers and watercourses has become so serious that the water is no longer fit for irrigation, let alone human consumption. Thus water pollution is contributing to water scarcity as well as increasing the costs of provision of water for agriculture. Water scarcity is of particular concern as irrigated agriculture accounts for over 70% of national consumption, but only about 40 to 60% of water flowing through irrigation systems is used effectively.

Like China, many OECD countries are searching for ways and means of better aligning resource use with societal interests and environmental sustainability. The Workshop on which this volume is based was motivated by the idea that sharing know-how and experiences would bring mutual benefits and insights as well as be of assistance in better anticipating and responding to stresses rather than simply reacting to them. Sponsored by the OECD and China's Ministry of Agriculture, the Workshop was held on 19-21 June 2006 in Beijing and involved specialists representing agricultural, environmental, agri-food and water industry interests from government, academia, the private sector, and international and non-governmental organisations. The main objectives of this workshop involved exploring questions in the following three areas:

- How does one balance the pressing need to expand food production for a growing population with the desire to improve rural well-being, protect the environment, and assure a sustainable future, while continuing to meet international trade commitments and obligations?
- What is the best way to ensure greater harmony between individual behaviour and societal priorities? Is it desirable to shift from institutional administrative decrees to market and incentive-based approaches? Will this better ensure that polluters and others producing negative externalities pay while providers of environmental and societal benefits are rewarded?

- What would be a good policy mix in China with respect to taxes, levies, payments, regulations, entitlements, market-based, voluntary measures, and other means of involving stakeholders?

Priorities and perspectives

Participants found it useful to reflect on the criteria on which policies are based. What is the main policy goal? What types of practices are put in place to achieve that goal? What are the long-term ramifications?

- ***Agricultural production targets.*** In this instance, authorities provide as many resources as are required to meet production targets. This is exemplified by China's current policy environment where authorities strive to maintain a high level of self-reliance for staple grains.
- ***Agricultural-use productivity.*** Authorities price resources like water to ensure maximum efficacy, or "crop per drop."
- ***Resource-use efficiency.*** Resources are priced to reflect their economic scarcity values.
- ***Sustainability criteria***. Resources are priced to reflect market and non-market (economic, social, environmental) values, costs and benefits.

It is also important to think about who and what to target in order to achieve positive changes. The ability to monitor and measure performance is a critical issue; it is impossible to enforce laws, guidelines and regulations intended to discipline behaviour or results where performance itself is difficult to measure.

- ***People?*** How might those who own natural resources or ecosystems or those who practice progressive environmental stewardship be encouraged or rewarded? How might those who pollute or otherwise contribute to environmental degradation be penalised?
- ***Practices or behaviours?*** Are there incentives that can be put in place *ex ante* to encourage progressive environmental stewardship which delivers agri-environmental benefits, while concurrently penalising methods with adverse effects.
- ***State of the resource?*** In instances where *ex ante* policy intervention is difficult, it may be possible to undertake periodical *ex post* environmental audits which can be used to assess the state of the natural resource at any point in time and reward those who have maintained or improved the natural resource. This could be based on the physical state of the resource (*e.g.* grassland, wetlands, forested area, etc.) and its relative value to society, or by applying indicators such as nutrient and pesticide loading, soil erosion, organic matter, etc.

Main messages

Clearly defined land and water rights and markets promote good resource stewardship. A clear correspondence between authority over and responsibility for resources and their use appears to be the most reliable way of ensuring good stewardship over the environment and natural resources. For water, it may be useful to divide authority and responsibility and commensurate rights into three areas: the right to extract, the right to allocate and deliver, and the right to use. For land, there are considerations like cultivation rights, sub-surface mineral rights, the right to sell or transfer land, and the right to convert land to other uses. Farmers, resource owners, and service providers tend not to make long term investments that are environmentally progressive if their rights are unclear or insecure. Tenure insecurity also appears to contribute to excessive nutrient and pesticide loading because returning migrants apply inputs "all at once" in an effort to manage their time and the risks of a crop shortfall.

Infrastructure, institutions, entitlements, obligations and incentive systems need to be aligned. The South to North Water Transfer project will transfer 40 million cubic metres of water annually from China's south to its water-deficit north. At the same time, under-pricing of water has led to water being treated as an abundant resource in many northern localities. This has led to the "virtual" southbound transfer of water in the form of water-intensive agri-food products that exceed the northbound volumes of the South to North Water Transfer project. Obviously, some effort to correct incentives for water use and establish an efficient water allocation method (EWAM) is warranted. Australia's experience with institutional, entitlement, and pricing reform is helpful. Pricing and institutional reforms ensured the commercial viability of government-linked enterprises. Institutional and entitlement reforms better defined the business relationship between those allocating water and those using it. Entitlement and pricing reforms helped to better define and determine 'product' and 'price'. All these refinements allowed the roles of regulators, resource managers, infrastructure operators, and service providers to be more clearly delineated in addition to allowing for the separation of water property rights from land title and tenure issues.

Providing cheap access to scarce resources like water and land is counter-productive in practice. Water is a particularly scarce resource that is not fully reflected in the price paid by farmers or consumers. Under-valued resources are typically either rationed or over-exploited. Theory and experience show that, when resources or goods are under-valued and rationed, queuing occurs and some citizens obtain preferential access to resources over others. Resource allocations under such circumstances do not typically reflect societal priorities. When a resource is under-valued and over-exploited it may deteriorate or even reach a point of collapse. When scarce resources like water and land are under-valued, they may be treated as though they are in abundance – exacerbating strains on the resource. In essence, consumers today are receiving a subsidy from future generations in that the fees they pay for water, land and other resource are too low. Ironically, foreign consumers are also receiving benefits from China's water subsidies through the export of water-intensive goods.

Agriculture is a leading source of non-point source water pollution and land degradation. This stems in part from efforts to promote commercial fertilisers and pesticides. In China, researchers, extension workers and input suppliers appear to be biased in favour of application levels of 25% or more above what is required. Such behaviour has been observed in OECD as well, although not to the same degree. Since nutrient and pesticide loading is equivalent to farmers throwing money down nature's drain, such behaviour could probably be remedied through improved extension. China may wish to examine the European Community's "license to produce" approach where farmers are rewarded through the market and regulatory system for good stewardship and penalised for misconduct. With growing intensification in livestock operations, excessive loading from effluents is also a concern. China's efforts to develop "circular agriculture" may partly address such problems. Domestic "effluent trading" based on land availability and appropriate manure application can provide an additional management mechanism.

Concerns regarding transparency and the danger of policy capture can be addressed by harnessing market forces to send signals regarding resource scarcity and society's preferences. Administrative and institutional "bricks and mortar" mechanisms can be captured, with vested interests developing that may attempt to increase their longevity or their stature by perpetuating problems rather than resolving them. Dealing with farmers and rural citizens in an inclusive manner rather than in a top-down fashion may also lead to better targeting, transparency and accountability.

Reform takes time, involves on-going adjustment, and needs to be inclusive in nature. Because of differences across agri-ecological zones, societies, cultures and economies, there is no magic recipe. Current efforts to develop a planning framework that will identify the differences between good and bad resource stewardship and how to go about encouraging the former is a positive initiative as it is desirable to anticipate rather than react to crises. But one size does not fit all and a mixture of incentives and disciplinary measures is often warranted. Europe's experience has been that the road map may require occasional modifications, as lessons are learned and as individuals and sectors change their behaviour. Reform should be inclusive in nature, drawing on the expertise and input of ecologists, scientists and social scientists alike. Even more important, it must incorporate the input and interests of its clients – farmers and rural citizens who use the resources in question and who are concurrently the primary stewards of such resources. Plans that ignore the circumstances and incentives facing farmers and rural communities will remain just that – plans with a low probability of becoming truly operational.

China possesses the prerequisite expertise and institutions to deal with the challenges faced. However, authority and responsibility are not always well aligned to achieve the best results. There is widespread recognition of the need for greater use of market-based instruments, such as better pricing structures and tradable permits, accompanied by government regulations, as well as more inclusive co-operative efforts among resource users.

Highlights of the papers

In **Session 1**: Agri-environmental situation and policies in China, Tang Renjian discussed ***China's plan for "Building of a New Socialist Countryside"***. The plan's major focus is how to continue maintaining a high degree of self reliance while improving the well-being of China's farmers and dealing with pressing environmental issues such as the loss of arable land, declining water tables, and pollution arising from agricultural production and rural industries. Tang emphasised that – given China's resources - higher consumption and production targets can only be met through more efficient use of land, water, and other resources. Tang also touched on how better management, use and treatment of agricultural biomass could help reduce dependence on fossil fuels and reduce pollution in the countryside. In general, improving the efficiency of use for agricultural inputs and resources has the potential to concurrently increase production, lower input costs, and halt environmental degradation. The challenge is to implement incentives and policies which encourage such results; the underpinnings of the 2006 Number 1 Document and the 11th five-year plan provide a good starting point.

Simon Spooner surveys the current situation with respect to ***China's water resources and their management***. Overall, China has only about one-quarter of the world's average water endowment per capita. In the northern and western parts of the country, this figure is much lower at roughly one-tenth. When combined with stresses caused by inequitable distribution, over-extraction, saline intrusion and pollution from point and non-point sources, China's hydrologic and aquatic ecosystems have suffered serious degradation and – unless dealt with constructively – will place limits on future economic development and prospects for improving the livelihoods of the population. Current engineering efforts will partly address the water deficit situation by transferring water from the Yangtze River basin and its tributaries to river basins in northern China. But such undertakings will still not meet all of the requirements for ecological recovery, let alone support future economic growth. Improved incentive systems will be critical to future prospects. Major policy changes underway include: strengthening basin management; defining and protecting water sources zones; defining and managing trans-boundary water pollution; planning economic development and exploitation to better suit water resource conditions; improving the efficiency of water use in agriculture; investing to provide 70% wastewater treatment in urban areas by 2010. Water allocation entitlements are also being implemented on a provincial and sub-provincial basis.

While the "polluter pays" approach has been mandated since the promulgation of the 1979 Environmental Protection Law, it is unevenly applied. In many instances, discretion lies with local governments as to whether to charge a levy or not and, if charged, how much. Levies paid often reflect bargaining between the polluting firm and the local administration as much as the gravity of the infraction. Surveys suggest that state-linked firms pay lower rates than privately owned firms and that levy rates for pollution tend to be positively linked to firm profitability. User charges for water and other resources are increasing in both their incidence and level, but they still typically fall well below their economic value or opportunity cost, to say nothing of their societal and ecological value. Newly emerging, basin and community-based water associations show promise in balancing irrigation needs and consumptive uses with available water supplies and are a promising vehicle for arriving at appropriate user charges and delivery mechanisms.

Deforestation, soil erosion, land degradation and desertification are among the most serious ecological challenges China faces. In this context, **Han Jun** discusses the universal problem of how to achieve long term ecological improvement while dealing with the immediate needs of improving the well-being of rural citizens. Most countries share this challenge, but China's case is particularly daunting considering its natural resource endowment and the large number of rural citizens living in very difficult circumstances. For China, integrated ecosystem management will require: comprehensive cross-sector planning and co-ordination; building and drawing more on human capital; extending and protecting land users' legal rights; improving the targeting of poverty reduction policies; promoting inclusive, community-driven and locally appropriate land management technologies.

While a number of government agencies and bureaux play a role in promoting conservation practices, there is little in the way of co-ordination or follow-up. Each agency or line department manages investment and activities according to its own understanding, leading to a patchwork quilt where efforts are rare in some areas and redundant in others. Further, central agencies are often more concerned with the disbursement of funds then with the wise allocation and management of these funds. There is little in the way of *ex post* evaluation of efficacy. Decision-making, although fragmented across agencies, remains highly centralised within agencies, often resulting in inappropriate local practices. Greater effort is spent on packaging and approving new projects than on ensuring that existing projects are well delivered. Since funds for conservation projects are typically transferred from the central to the local level in the form of grants, diversion of funds is not uncommon. Han advocates greater involvement of rural citizens in both the design and implementation of conservation projects. This would improve design and targeting in view of local conditions and make officials more accountable, thereby reducing the extent to which funds can be diverted.

Ma Xiaohe and Fang Songhai note that China faces daunting, although not insurmountable, challenges with respect to its water resources. Its endowment of water resources per capita is low and badly distributed. Water supplies are relatively static, but demand for urban, industrial and residential uses is rising steadily. Current farming practices, if they are to meet the needs of China's growing population, will also place increasing demand on irrigation capacity. In addition to problems with water shortages, the north also faces problems with water quality: the Hai, Liao, Huai and Huang river basins all have high pollution indices which render them unsafe for drinking and, in many instances, unsafe for irrigation. Given these stark realities, current practices are unsustainable and several suggestions to remedy the situation are made. These include raising water charges to more appropriate levels, improving incentives for adopting water saving irrigation practices, increasing support to water conservation facilities, promoting cyclical irrigation methods, supporting research in water-saving techniques and drought tolerant crops, and taking stronger measures to combat water pollution. Ma and Fang suggest that ***correcting incentives should be a higher priority than reallocating water resources via the South to North water transfer project.***

Agri-environmental Resource Management in OECD Countries was the main focus of **Session 2.** Collectively, the agricultural sectors of OECD countries have managed to achieve unprecedented growth since the middle of the 20th century. They have done this with fewer workers and slightly less land, but using more water, chemicals, and machinery, coming at a cost to the environment and natural resources. While most OECD countries indicate that they attach a high priority to environmental issues as they relate to agriculture, only a small share of support is actually directed at agri-environmental concerns. Improvements have been made recently, but the lion's share of support remains linked to production and input use – support types that are less transfer efficient and less environmentally friendly.

Wilfrid Legg suggests that the underlying objective of sustainable natural resource management policies is to contribute to the overall well-being by ***optimising economic efficiency and net environmental benefits*** within the context of a country's societal values. This means taking into consideration both commercial and non-commercial resource uses, on-site and off-site effects, and balancing the interests of current and future generations. While the natural resources associated with agriculture – land, water, and ecosystems – are largely renewable resources, they can be depleted or degraded if not managed responsibly. The challenge for those involved in policy design and implementation is to align incentives to optimise both economic efficiency and environmental benefits when the latter in particular are not easily captured within the market mechanism. The experience of OECD – both positive and negative – is that policy can play a critical role in the sustainable exploitation of natural resources associated with the agri-food industry. Several areas are worthy of note: developing property rights and markets; reforming policies that provide production-linked support; reducing resource degradation and enhancing ecological service provision; dealing with information shortfalls; addressing distributional implications of resource management policies; managing publicly owned natural capital. A key underlying principle is to establish a correspondence between incentives and the actual scarcity value of a good, service or resource to society.

Market mechanisms are now a central element in the allocation of water resources in Australia. **Seamus Parker** discusses how ***water trading*** was introduced and describes how it has evolved to meet the needs of existing users, new users and society. Although focussing on enabling market forces to play a role, it is emphasised that high level intergovernmental agreements at the state level are necessary and that these are underpinned by financial incentives to elicit the responses desired from market participants. Prior to reform, water license arrangements were dealt with on an *ad hoc* basis, without considering the cumulative effects of licenses in place. New uses proved particularly problematic as water entitlements were tied to land ownership and the only way to obtain a water license was to buy land. In making decisions regarding the granting of licenses, little attention was paid to how much water was actually required to sustain the ecosystem. Conflicts over water use grew in number and intensity and, when disputes did arise, highly adversarial approaches were employed to address what really were problems of planning and priorities.

Reforms were needed in three key areas: institutions, entitlements, and pricing areas. Pricing and institutional reforms combined to ensure the commercial viability of government-linked enterprises. Institutional and entitlement reform combined to better define the business relationship between those allocating the water and those using it. Entitlement reform and pricing reform combined to better define and co-determine "product" and "price". These refinements allowed the roles and responsibilities of regulators, resource managers, infrastructure operators and service providers to be more clearly defined and separated. They also separated water property rights from land title and tenure issues. Over time, as rights and entitlements became clearer, trading arrangements for water entitlements evolved – concurrently reducing conflict over use rights and allowing pricing to more accurately reflect cost and volumes used. Parker provides a roadmap on how Australia has developed an operational, catchment- and-market-based system for trading water entitlements which reflects society's priorities, while allowing resources to be allocated more efficiently than if administrative means or government fiat were employed.

In the Netherlands, agriculture intensified dramatically between 1950 and 1985. This came at a cost, however, with an increased loading of the environment with nutrients, pesticides, and other by-products of agricultural production. **Peter van Boheemen** outlines the measures that governments, farmers and the agri-food industry have taken ***to reduce the adverse impacts agricultural production has had on the environment***. He describes the extent and effects of excessive nutrient loading over time, particularly phosphorus and nitrogen, followed by a discussion of water quality trends over time. The two-pronged Dutch approach to reducing nutrient emissions is then examined: controlling manure production and discharge and improving on-farm mineral management. In the mid-1980s, quotas were placed on livestock to limit their numbers. In 1987, standards on manure application were introduced where farmers with large surpluses of manure were required to transport it to farms and areas with a manure shortfall. Alternatively, farmers could purchase additional land or reduce their livestock numbers.

A market contracting system was in place in the Netherlands between 2002 and 2005 for dealing with excess livestock manure and effluents, but there is less need today for such a system as the nutrient load is more balanced. Since 1990, regulatory controls on crop protection products have also become more stringent, with limits on application levels, proximity to watercourses, application in the presence of wind, among other things. New systems of "license to deliver" agri-food products based on proper environmental, food safety and other criterion are now in place. Contravening the regulations can have serious consequences, including both administrative fines and criminal justice. However, a key reason for their success is that through identification and licensing the new system allows the market mechanism to reward farmers with progressive husbandry practices, while concurrently penalising those who have less than stellar records.

In the United States, most current issues regarding water availability and water quality in agriculture have arisen due to increasing competition for limited resources and growing public awareness of environmental issues. **Dennis Wichelns** suggests that public officials must resolve two primary issues. First, the best ways to allocate limited water supplies must be determined in an era of increasing demands for agricultural, residential, urban, industrial and environmental uses. Second, ***the best means to define and achieve state and national quality objectives without unduly impinging on growth and development must also be determined***. A particularly challenging set of water management issues exists in the states serviced by the Ogalla Aquifer. There is a serious overdraft of groundwater. The pressure on groundwater supplies is unlikely to abate as demands for growing traditional crops, bio-mass crops for ethanol, and ecosystem water are all increasing while states sharing the aquifer pursue their different goals, employ different policies, and have no co-ordinated plan or shared priorities.

In other jurisdictions, increasing competition for water supplies and greater focus on the environmental impacts of agriculture motivated both farmers and entrepreneurs in other industries to improve water management practices. These improvements have enabled both agriculture and industries to continue expanding, while using roughly the same volume of water used in 1985. Different states continue to experiment with approaches to ensure that fees for water and other natural resources truly reflect their scarcity value. As in China, non-point source pollution is a critical issue, and the public is becoming more aware of its ramifications for the quality of drinking water, environmental amenities, wildlife habitat, fisheries, wetlands, and other ecosystem services. Wichelns observes that the policy and planning process is unwieldy, time-consuming and costly, but that the failure to engage *ex ante* will result in even more unwieldy, time-consuming and costly measures to clean-up *ex post*. Wichelns concludes that innovation, tenacity and political courage are needed to ensure wise resource use.

To police and discipline behaviour, one needs to be able to monitor and measure it. **Kevin Parris** discusses how ***improvements in environmental indicators*** and in the way we monitor and evaluate agri-environmental policies ***can better inform policy design and implementation for sustainable resource use***. OECD uses a "Driving Force-State-Response" (DSR) framework to analyse agri-environmental linkages. Simply summarised, the DSR approach attempts to answer the following questions, among others. What is causing environmental conditions in agricultural areas to change? What effect are these changes having on the condition of the environment in agricultural areas, with possible linkages elsewhere? What actions can be taken to respond to (detrimental) changes in the state of the environment in agricultural areas?

Although a daunting task, considerable effort is made to keep the key agri-environmental indicators used in the DSR framework policy relevant, easy to interpret, analytically sound, and measurable. Indicators are collected in three areas where agriculture plays a role: protecting the stock of natural resources impacted by agriculture; reducing agri-environmental pollution; improving agri-environmental management practices and resource use. While linkages between policies and production impacts have been reasonably well quantified empirically, the same cannot be said for agri-environmental linkages. Assessments and understanding of the links between policies and incentives, actions, biological and ecological responses, and remedial measures are still at the formative stage. OECD is developing a Stylized Agri-Environmental Policy Impact Model to capture differences in agri-environmental systems of OECD member states and to better understand cause-effect linkages through integrated economic and bio-physical modelling. OECD also employs its AGLINK partial equilibrium model of world agriculture to examine the effects of removing agricultural subsidies and the application of taxes and levies on agri-chemical use. Due to difficulties in gathering appropriate data and establishing policy counterfactuals and clear links between actions and outcomes, such models cannot substitute for an *ex post* analysis of the impact of a specific policy. However, they can provide guidance regarding progressive stewardship behaviour as well as how to avoid calamities. Parris observes that societies and policy makers are increasingly recognising the environmental costs and benefits associated with agricultural activities and makes a number of suggestions on how to improve our analytical framework and models to assist in making informed choices.

In Session 3: Policy Options for China, Chen Mengshan provides us with an ***informative history of fertiliser use in China***, with insights regarding the present situation and problems. The ancient texts Chen quotes show both the trade-offs and the complementarity that can occur when making agri-environmental choices, with the Zhou Li texts advocating that marshland grasses be killed to plant wheat, while Shi Jing suggests that (when) the weeds decay, the millet grows luxuriantly. In one instance, wetlands and aquatic habitat would be destroyed in order to make way for wheat production while, in another instance, the productive contribution of decomposing organic matter and green manure is recognised. Even 3 000 years ago, the role that organic matter and manure played in sustaining and augmenting soil fertility and buffering and releasing both water and soil nutrients was well recognised, if not fully understood.

China began to embrace commercial chemical fertilisers in the early 1980s, with the intent of raising aggregate production to meet the nation's growing food requirements. In the late 1980s and early 1990s, greater attention was paid to more targeted, high efficiency and high quality products. Over the past several years, awareness of ecology, safety and stewardship has come to the fore. In spite of increased cognisance, however, there remains a much higher proportion of single ingredient and low-grade fertilisers used than high grade or complex fertilisers, with nitrogen and phosphorus typically being applied in excess of requirements. New breakthroughs in controlled release fertilisers have been realised, but their use is not yet widespread. A number of new standards and handbooks for applying fertiliser and other agricultural inputs have been updated and released in recent years, drawing on the latest scientific knowledge. However, soil testing and the scientific application of fertiliser lag scientific knowledge somewhat and work is now underway to extend their scientific application across the country.

Jeff McNeely provides a synopsis of the global Millennium Ecosystem Assessment, drawing out implications in the areas of bio-diversity conservation and water system management in particular. McNeely points out that the human race is part of the global ecosystem and that we depend on it for our very survival. McNeely expands on ***the concept of ecosystem services and the idea that those who provide the services and goods (often treated as a public good in the past) deserve to be compensated*** when they manage ecosystems to deliver more of such services to others. It is recognised that farmers know better than anyone else that a healthy, resilient farm is essential for a productive and profitable ecosystem. Basing the conservation of ecosystem services on economic incentives recognises farmers' capabilities and role in land and resource stewardship. For China, the challenge lies with aligning private incentives with public interest in a way that the three goals of sustainable agriculture (food security; employment and income generation; and resource conservation and environmental protection) can be realised.

McNeely explores operational questions related to the valuation of "non-market value" ecosystem services. By posing questions and reflecting on the nature of the responses, we begin to imagine how markets can be harnessed to value services and goods that were previously unvalued or under-valued in the marketplace. What specific services does an ecosystem provide? Who are the primary recipients of such services? How important are such services? How do the recipients use them? What would happen to these services if the ecosystem were managed differently? How might one capture the value of such services? McNeely emphasises the importance of establishing strong, clearly articulated property rights and legal frameworks to enable ecosystem markets to develop. He also shares his thoughts regarding the care that must be taken to ensure that the poorest members in society are not disenfranchised by efforts to protect the environment. McNeely closes by noting that initiatives that place additional burdens on society's poorest members will not be sustainable.

Ke Bingsheng suggests that a country with the size and population of China must take food security as its principal objective and that such security – for the most part – must come from domestic sources. However, Ke notes that constraints on resource availability are increasingly evident and that China is already past the point where it can increase its draw on available natural resources. Hence, the only way to meet China's food security objectives in a sustainable way is to protect existing resources while concurrently raising efficiency and productivity with respect to their use. Ke then goes on to ***discuss ways and means to: protect existing cultivable land; protect water resources; increase land productivity; enhance technical innovation; limit supply-chain losses; improve agricultural and rural infrastructure***. Ke observes that farmers are likely to make long term investments in productivity-enhancing capital and land improvements if they have more secure tenure arrangements. Moves toward pricing resources like water and land according to their scarcity value and on the basis of all of the non-market and market-based services they render will also encourage greater efficiency and productivity in their use. Ke also notes some of the challenges and asymmetries with regard to land. When land is to be used for urban or industrial uses, most of the benefit stream can be captured privately and, therefore, the private value in the market place will be much the same as its total societal value. With land that is classified for resource industry or agricultural uses, a large part of the benefit stream cannot be captured privately and, therefore, the private value for land zoned or classified as agricultural will be much less than its total societal value.

Tang Huajun and Yin Changbin explore the ***lessons and possible gains from embracing a "Circular System" approach to agricultural development***. Simply put, such an approach attempts to maximise the economic and societal "footprint" of the sector while concurrently minimising the ecological "footprint" of its agricultural production. By viewing the wastes of one economic activity as resources for use in another productive activity, utilisation rates and efficiency can be boosted while achieving a parallel reduction in pollutants. This concept has some parallels with Michael Porter's clustering and agglomeration concept to reap the rewards of economic complementarity across enterprises and firms. However, this approach extends the ideas of complementarity, clustering and agglomeration across both the economic commercial dimension and the unvalued or under-valued ecological-environmental dimension.

Jeff Au uses case studies to discuss ***the crop protection industry's role in supporting agricultural development***. Au is candid in his views regarding the challenges of extending new technologies to some 200 million poor farm households across a vast territory, most with small land holdings and low education levels. Many of the cases examined relate to no-till or low till cultivation practices, combined with timely pesticide application. Trade-offs between the benefits of improved soil tilth, improved fertility, lower erosion, higher organic matter and better water retention characteristics must be weighed against higher needs for timely pesticide applications and downstream effects on water, bio-diversity and ecosystems.

Au observes that, although generally not well educated, farmers make the best decisions they can in order to feed and clothe their families. Given the limitations of their education, however, complex instructions must be avoided and follow-up support is required over an extended period of time. One needs to think not just of the new technology but also the appropriate model for extending it in order to be successful. Sample plots and turnkey demonstrator farmers are found to be critical to successful extension. The importance of having inclusive farmer engagement if progressive practices are to be adopted is also emphasised. Au emphasises that it is necessary to demonstrate the economic dividends of adoption to farmers if new technologies are to be embraced.

Crop insurance has been advocated as a policy choice to support farmers' incomes and agricultural production, while remaining consistent with China's WTO obligations. **Zhong Funing, Ning Manxiu and Xing Li** undertake an empirical examination of ***crop insurance's influence on agrochemical use*** in Xinjiang. Beginning with scientifically underpinned hypotheses regarding whether particular inputs augment production or reduce the likelihood of a poor crop or catastrophic outcome, Zhong *et al.* test their hypotheses and make a number of important observations. Zhong *et al.* find that full-time farmers are more likely to carry crop insurance, possibly because they are more reliant on farming for their well being. Farmers who are relatively specialised were also more likely to carry insurance as were those less able to spread risk across different production alternatives. Older, more established farmers are less likely to buy crop insurance. This may be related to older farmers' real or perceived ability to manage risk. It may also be related to their debt level and degree of diversification. Finally, farmers with crop insurance applied more fertiliser and agri-film while applying less pesticide than farmers who did not have crop insurance. This makes sense as fertiliser and film increase yields, while pesticides prevent crop loss and yield shortfalls.

Huang Jikun and his colleagues examine ***the application of commercial fertilisers and pesticides*** in China. Application rates have approximately doubled since 1980 and China now ranks fourth in the world in the intensity of fertiliser use, after Japan, Korea and the Netherlands. Pesticide use has increased almost three-fold. While chemicals have played an important role in increasing agricultural production, they can also increase production costs, increase the risk of certain food quality and food safety problems, as well as contribute to environmental pollution. Several recent studies have shown that chemical fertilisers are now over-applied at rates between 20 and 50%. For pesticides, the over application rate appears to be even higher, at between 40 and 55%. This does not seem to make sense as over-application of such inputs effectively amounts to throwing money away. It can also have serious environmental and health-related consequences. Are farmers aware of what they are doing?

In research currently underway, the Centre for Chinese Agricultural Policy has found that there were no significant yield shortfalls on plots where fertiliser use was reduced by 25 to 35%. Farmers taking part in this research were stunned when they saw the results. A number of hypotheses may explain this. It may be that some farmers know what they are doing and that the over application of inputs is part of a risk management strategy. There is some circumstantial evidence that tenure and migration issues play a role in the pattern of excess application of commercial inputs. When migrant workers return home, they often apply inputs "all at once" rather than in optimal amounts at critical times in the growing cycle because the time they have during their home visit is limited. But the study also shows that there is even more evidence that the government, scientific community, plant breeders, extension agents, and input suppliers have convinced farmers that "if a little bit is good, a lot is better". While these findings are tentative, they suggest that ***incentives within the existing research, extension education, and agricultural input suppliers need to be re-examined***. Is the information that farmers are receiving credible? Do their sources of information have vested interests in ensuring higher input use and in reaching sales or output targets that are not necessarily to the benefit of farmers or the environment? In the coming few years, efforts will be made to educate a broader group of farmers about the merits of reducing input use, both in terms of their own income and in terms of environmental damage.

RÉSUMÉ

Généralités

Nourrir la plus forte population du monde avec des ressources limitées – tel est l'audacieux défi auquel sont confrontés les dirigeants chinois. Le capital naturel par habitant de la Chine se situe entre celui de l'Allemagne et celui du Chili pour ce qui concerne les terres arables, tandis que pour les ressources en eau, il est extrêmement faible et mal réparti, la plaine du Nord étant à cet égard l'une des moins bien dotée du monde.

Pour répondre aux besoins croissants d'une population de plus en plus nombreuse, la Chine a jusqu'à présent réussi à augmenter sa production agricole en étendant les superficies cultivées. Lorsque cette solution s'est révélée insuffisante, l'accroissement de la production a été obtenu en augmentant la productivité des terres grâce à une consommation plus forte d'eau, d'engrais, de pesticides et d'intrants mécaniques, mais les effets négatifs de certaines de ces pratiques agronomiques commencent à se faire sentir. Ainsi, la charge des sols et des cours d'eau en éléments nutritifs et en pesticides est en augmentation, tandis que la désertification, l'érosion des sols et la dégradation des terres contribuent à faire chuter la productivité. Aujourd'hui, certains cours d'eau et rivières sont tellement pollués que leur eau n'est utilisable ni pour l'irrigation, ni - a fortiori- pour la consommation humaine. La pollution de l'eau contribue donc à sa raréfaction et à l'augmentation du coût de l'approvisionnement en eau du secteur agricole. La rareté de l'eau est d'autant plus préoccupante que l'agriculture irriguée absorbe plus de 70% de la consommation nationale d'eau, mais que 40 à 60% seulement de l'eau destinée à l'irrigation est utilisée de manière efficace.

Comme la Chine, de nombreux pays de l'OCDE cherchent actuellement à définir les voies et moyens permettant de mieux concilier, d'une part, l'exploitation des ressources et, d'autre part, les intérêts sociétaux et la durabilité de l'environnement. A la base, cet atelier repose sur l'idée que partager savoir-faire et expériences apporterait des avantages et éclairages mutuels et permettrait de mieux anticiper les contraintes et y répondre plutôt que se contenter d'y réagir. Organisé sous les auspices de l'OCDE et du ministère chinois de l'Agriculture, l'atelier qui donne lieu à la présente publication s'est déroulé du 19 au 21 juin 2006 à Beijing, où il a réuni des spécialistes des sphères gouvernementales, universitaires et privées, ainsi que des organisations internationales et non gouvernementales, qui représentaient les intérêts des secteurs agricole, environnemental, agroalimentaire et de l'eau. Il s'agissait notamment de mener une réflexion sur diverses questions portant sur les trois thèmes suivants :

- Comment concilier l'impératif d'augmentation de la production agricole face à l'accroissement démographique et le souhait d'améliorer le bien-être en milieu rural, de protéger l'environnement et de garantir un futur viable, tout en continuant à remplir les engagements et obligations commerciales internationales du pays ?

- Comment harmoniser au mieux comportements individuels et priorités de la collectivité ? Est-il souhaitable d'abandonner les décrets administratifs institutionnels au profit d'approches marchandes ou incitatives ? Dans quelle mesure ces choix seront-ils mieux à même de faire supporter les coûts par les pollueurs et autres producteurs d'externalités négatives et d'assurer une rémunération à ceux qui procurent des avantages environnementaux et sociétaux ?
- Comment s'articuleraient en Chine les instruments fiscaux, les redevances, les paiements, les réglementations, les droits, les mesures volontaires fondées sur le marché et d'autres dispositifs cherchant à impliquer les acteurs concernés ?

Priorités et perspectives

Les participants ont jugé utile de réfléchir aux critères pris en compte pour l'élaboration des politiques : quel est l'objectif premier de l'action publique ? Quels sont les types de pratiques mis en place pour atteindre cet objectif ? Quelles en sont les incidences à long terme ?

- ***Objectifs de production agricole.*** En l'occurrence, les autorités apportent toutes les ressources nécessaires pour atteindre les objectifs de production. Le cadre d'action en vigueur en est un bon exemple, puisque les autorités chinoises s'efforcent de maintenir un taux élevé d'autosuffisance en céréales de base.
- ***Productivité des usages agricoles.*** Les autorités fixent le prix des ressources, telles que l'eau, de manière à assurer une efficacité maximum (« more crop per drop », c'est-à-dire une amélioration de la productivité agricole de l'eau).
- ***Efficience de l'exploitation des ressources.*** Le prix fixé pour les différentes ressources reflète leur valeur de rareté.
- ***Critères de durabilité.*** Le prix des ressources est déterminé de telle sorte qu'il rende compte de leurs valeurs, de leurs coûts et de leurs avantages (économiques, sociaux et environnementaux) marchands et non marchands.

Si l'on veut obtenir des changements positifs, il importe également de réfléchir aux bénéficiaires et aux objectifs que l'on souhaite cibler. Il est en particulier essentiel d'être à même de suivre et mesurer les performances, car il est autrement impossible de faire appliquer les lois, lignes directrices et réglementations destinées à maîtriser les comportements ou les résultats.

- ***Au niveau des individus.*** Comment encourager ou rémunérer les détenteurs de ressources naturelles ou d'écosystèmes ou ceux qui adoptent des pratiques modernes de protection de l'environnement ? Comment pénaliser ceux qui sont responsables de pollutions ou contribuent d'une manière ou d'une autre à la dégradation de l'environnement ?
- ***Au niveau des pratiques et des comportements.*** Quelles mesures incitatives ex ante peut-on mettre en place pour encourager une protection innovante de l'environnement qui apporte des avantages agro-environnementaux et sanctionner parallèlement les méthodes ayant des conséquences préjudiciables ?

- *Etat de la ressource.* Lorsqu'il est difficile d'intervenir préalablement, il est peut-être possible de réaliser a posteriori des audits environnementaux périodiques, lesquels peuvent être utilisés pour évaluer l'état de la ressource naturelle au temps t et rémunérer ceux qui l'ont entretenue ou améliorée. Ces évaluations pourraient se fonder sur l'état physique de la ressource (par exemple : pâturages, zones humides, superficies boisées) et de sa valeur relative pour la collectivité, ou sur l'application d'indicateurs tels que la charge en éléments nutritifs et en pesticides, l'érosion du sol, le taux de matière organique, etc.

Principaux messages dégagés

L'existence de droits et de marchés clairement définis pour l'eau et les terres favorise une bonne protection de ces ressources. A cet égard, en effet, il apparaît que la meilleure solution pour protéger l'environnement et les ressources naturelles consiste à établir une correspondance explicite entre, d'une part, les pouvoirs et responsabilités et, d'autre part, l'exploitation des ressources. Dans le cas de l'eau, il peut être judicieux de répartir les pouvoirs et les responsabilités et de subdiviser les droits en trois catégories : droit de prélèvement, droit de répartition et de distribution, et droit d'usage. En ce qui concerne les terres, il importe de prendre en considération les droits de cultiver, les droits sur le sous-sol, le droit de vendre ou de transférer des terres, et le droit de convertir des terres à d'autres usages. Si leurs droits ne sont pas précisément définis ou garantis, les agriculteurs, les détenteurs de ressources et les prestataires de services hésiteront à consentir des investissements de long terme améliorant l'environnement. L'insécurité foncière semble également jouer un rôle dans l'augmentation excessive de la charge en éléments nutritifs et en pesticides car, lorsque des migrants reviennent chez eux, ils appliquent les intrants « d'un seul coup » pour tenter de gagner du temps et d'éviter les risques de mauvaises récoltes.

Il est impératif d'articuler infrastructures, institutions, droits, obligations et systèmes incitatifs. Le projet de transfert des eaux Sud-Nord a pour ambition de détourner un volume annuel de 40 millions de mètres cubes d'eau du Sud de la Chine vers le Nord fortement déficitaire en eau. Par ailleurs, comme le prix de l'eau est peu élevé, de nombreuses localités du Nord de la Chine l'utilisent sans compter, d'où l'existence d'un transfert d'eau « virtuel » en direction du Sud sous la forme de produits agroalimentaires, dont la production nécessite d'importantes quantités d'eau. Or, ce transfert dépasse les volumes d'eau qui seront détournés dans le cadre du projet Sud-Nord. Dès lors, il est logique de chercher à réajuster les incitations à la consommation d'eau et à mettre en place une méthode efficace d'affectation de l'eau. A cet égard, il est intéressant d'examiner les acquis de l'Australie en matière de réforme des institutions, des droits et de la tarification. S'agissant des institutions et de la tarification de l'eau, les réformes ont assuré la viabilité commerciale des entreprises liées aux administrations publiques. Les réformes des institutions et des droits ont permis de mieux définir les relations commerciales entre les structures chargées de la répartition des eaux et celles qui les exploitent. Enfin, les réformes des droits et de la tarification ont amené à mieux déterminer les notions de « produit » et de « prix ». Ces diverses améliorations ont permis non seulement de préciser le rôle respectif des réglementeurs, des gestionnaires de ressources, des responsables d'infrastructures et des prestataires de services, mais aussi de bien séparer les droits de propriété sur l'eau des questions de régimes et de titres fonciers.

Fournir un accès bon marché à des ressources rares telles que l'eau et les terres est en définitive contreproductif. L'eau est une ressource d'une très grande rareté, laquelle n'est pas entièrement reflétée dans le prix acquitté par les agriculteurs ou les consommateurs. De façon générale, les ressources insuffisamment valorisées sont soit rationnées, soit surexploitées. En théorie comme en pratique, on sait bien que la sous-évaluation et le rationnement de ressources ou de biens entraînent la création d'un système de files d'attente, avec pour corollaire l'octroi d'un accès préférentiel à certains au détriment des autres. Dans ces conditions, l'allocation des ressources ne reflète généralement pas les priorités de la collectivité. D'autre part, dès lors qu'une ressource est sous-évaluée et surexploitée, elle peut se détériorer ou même venir à manquer. Comme des ressources rares telles que l'eau et les terres sont sous-évaluées, on risque de considérer qu'elles sont abondantes – ce qui exacerbe les contraintes auxquelles ces ressources sont soumises. Ce qu'il faut comprendre, c'est qu'aujourd'hui, les consommateurs chinois reçoivent une subvention des générations futures, puisque les redevances qu'ils versent sur l'eau, les terres ou d'autres ressources sont très faibles. Comble d'ironie, les consommateurs étrangers profitent eux aussi du subventionnement de l'eau par la Chine à travers les exportations de biens exigeant une forte consommation d'eau.

L'agriculture est l'une des principales sources de pollution diffuse de l'eau et de dégradation des terres. Cette situation est en partie imputable aux efforts déployés pour encourager l'utilisation d'engrais et de pesticides commerciaux. Il semble qu'en Chine, les chercheurs, les vulgarisateurs et les fournisseurs d'intrants ont tendance à conseiller des niveaux d'apport ou d'application supérieurs de 25% ou plus aux doses nécessaires. Ce comportement a également été observé dans les pays de l'OCDE, mais il n'est pas aussi répandu. Etant donné que la charge en éléments nutritifs et pesticides qui en résulte constitue pour les agriculteurs un gaspillage pur et simple, on pourrait probablement remédier à cette situation en améliorant les services de vulgarisation. La Chine pourrait souhaiter étudier l'approche adoptée par la Communauté européenne, qui passe par la délivrance d'une « autorisation de production », ce qui permet de rémunérer les agriculteurs via les mécanismes réglementaires et le marché lorsqu'ils adoptent des pratiques respectueuses de l'environnement et de les pénaliser dans le cas contraire. Avec l'intensification croissante des élevages, se pose également le problème de la pollution de l'eau et des sols imputable aux effluents. Ces différentes difficultés pourraient en partie être résolues grâce aux efforts déployés par la Chine pour mettre en place une « agriculture circulaire ». Par ailleurs, un « système national d'échange d'effluents », reposant sur la disponibilité des terres et un épandage adéquat pourrait compléter le mécanisme de gestion.

Face aux problèmes de transparence et au danger d'appropriation de l'action publique, la solution consiste peut-être à canaliser les forces du marché de telle sorte qu'elles envoient des signaux concernant la rareté des ressources et les préférences de la population. En effet, les mécanismes administratifs et institutionnels de base peuvent être détournés au profit des groupes d'intérêt qui se développent et qui peuvent chercher à se pérenniser ou à renforcer leur position en entretenant les problèmes plutôt qu'en les résolvant. Si les autorités publiques associent les agriculteurs et les habitants des zones rurales à l'élaboration des règles au lieu de les leur imposer, elles devraient également pouvoir améliorer le ciblage et la transparence des mesures, ainsi que le sens des responsabilités.

La réalisation d'une réforme prend du temps, exige un ajustement permanent et impose un mode participatif. Compte tenu des disparités entre les zones agro-écologiques, les sociétés, les cultures et les économies, les solutions clés en main n'existent pas. Dans la mesure où il est souhaitable d'anticiper les crises plutôt que d'y réagir, les efforts actuellement déployés pour élaborer un cadre de travail permettant de recenser les différences entre les bonnes et les mauvaises pratiques de protection des ressources et déterminer comment encourager l'adoption des premières constituent une démarche constructive. Il n'existe cependant pas de formule standard, et il est donc souvent justifié d'associer des mesures incitatives et des mesures disciplinaires. L'expérience européenne a montré que le calendrier de mise en œuvre des réformes peut nécessiter quelques révisions, lesquelles sont opérées ça et là en fonction des acquis et de l'évolution du comportement des individus et des secteurs. La réforme devrait être menée sur le mode participatif, autrement dit elle devrait s'appuyer sur les compétences et les contributions des écologistes, des scientifiques et des experts en sciences sociales. Point plus important encore : la réforme doit tenir compte des points de vue et intérêts de ses principaux acteurs, à savoir les agriculteurs et les habitants des zones rurales qui exploitent les ressources en question et qui assurent conjointement les premières mesures de protection de ces ressources. Tout plan ne tenant pas compte de ces éléments, ni des incitations auxquelles sont confrontés les agriculteurs et les populations rurales, aura peu de chances d'être véritablement opérationnel.

La Chine possède d'ores et déjà les compétences et les institutions lui permettant de relever ces défis. Toutefois, la répartition des pouvoirs et responsabilités est telle qu'elle ne permet pas toujours d'obtenir des résultats optimaux. On s'accorde généralement à reconnaître la nécessité d'avoir plus souvent recours à des instruments de marché – structures des prix améliorées et permis échangeables – accompagnés de réglementations publiques, ainsi que d'actions coopératives associant davantage les usagers des ressources.

Principaux points des communications présentées

Session 1 : Contexte agro-environnemental et action publique en Chine. **M. Tang Renjian** présente le ***Projet d'instauration d'une nouvelle campagne socialiste***, qui a pour finalité première de déterminer comment maintenir un degré élevé d'autosuffisance tout en améliorant le bien-être des agriculteurs chinois et en remédiant aux problèmes environnementaux les plus urgents, comme la perte de terres arables, la baisse des nappes phréatiques et la pollution due aux productions agricoles et aux industries rurales. M. Tang fait observer que, compte tenu des ressources dont dispose la Chine, les objectifs d'accroissement de la consommation et de la production ne pourront être atteints que grâce à une meilleure efficacité de l'utilisation des terres, de l'eau et des autres ressources. Il évoque également la possibilité de réduire la dépendance du pays vis-à-vis des combustibles fossiles, ainsi que la pollution des campagnes, en améliorant la gestion, l'exploitation et le traitement de la biomasse agricole. De façon générale, augmenter l'efficacité d'utilisation des intrants et des ressources agricoles devrait permettre non seulement d'accroître la production, mais également d'abaisser le coût des intrants et de stopper la dégradation de l'environnement. Il faut donc mettre en œuvre des mesures incitatives et des politiques allant dans ce sens et, à cet égard, les principes définis dans le Document N° 1 publié en 2006, ainsi que dans le 11ème plan quinquennal, constituent une bonne base de départ.

M. Simon Spooner fait le point sur les ***ressources en eau de la Chine et leur gestion.*** Globalement, la dotation en eau par habitant de la Chine s'élève à environ un quart seulement de la moyenne mondiale. Dans les parties septentrionale et occidentale du pays, elle est même beaucoup plus faible, puisqu'elle représente globalement un dixième de la dotation mondiale moyenne. En outre, si l'on prend en compte les stress causés par une répartition inégale de la ressource, sa surexploitation, les remontées salines, ainsi que la pollution ponctuelle et diffuse, on constate que les écosystèmes aquatiques et hydrologiques de la Chine se sont considérablement dégradés et que – à moins que des mesures constructives ne soient prises – cela constituera un frein au développement et aux perspectives économiques qui permettraient d'améliorer les moyens de subsistance de la population. Les aménagements hydrauliques en cours remédieront en partie au déficit hydrique de la Chine du Nord grâce au transfert d'importants volumes d'eau du Yang Tsé Kiang et de ses affluents vers les bassins septentrionaux. Néanmoins, ils demeureront insuffisants pour réaliser une véritable restauration écologique, a fortiori pour soutenir la croissance économique de demain. L'amélioration des systèmes incitatifs sera déterminante pour l'avenir. L'action publique connaît d'importantes évolutions, notamment le renforcement de la gestion des bassins ; la définition et la protection des zones de captage ; la définition et la gestion de la pollution des eaux transfrontières ; la planification d'un développement économique et d'une exploitation mieux adaptés à l'état des ressources en eau ; l'amélioration de l'efficacité d'utilisation de l'eau par le secteur agricole ; les investissements dans le traitement des eaux usées, afin que 70 % des eaux urbaines soient traités d'ici 2010. Par ailleurs, des droits de répartition sont mis en place à l'échelle des provinces et des niveaux inférieurs d'administration.

Bien que l'approche « pollueur payeur » soit obligatoire depuis la promulgation de la loi de 1979 sur la protection de l'environnement, elle est appliquée de manière inégale. Très souvent, en effet, il est difficile de savoir si les autorités locales prélèvent ou non une taxe et, dans l'affirmative, quel est son montant, lequel reflète souvent le résultat des négociations entre l'entreprise polluante et l'administration locale tout autant que la gravité de l'infraction. Selon certaines études, les entreprises d'Etat acquittent des taux plus faibles que les entreprises privées, et il existe généralement une corrélation positive entre le taux appliqué et la rentabilité de l'entreprise. Les redevances sur l'eau et d'autres ressources versées par les usagers se multiplient et sont de plus en plus élevées, mais elles demeurent généralement très inférieures à leur valeur économique ou à leur coût d'opportunité, pour ne rien dire de leurs valeurs sociétale et écologique. De grands espoirs sont mis dans les toutes nouvelles associations locales de bassin, qui devraient permettre de trouver un équilibre entre besoins d'irrigation et consommation en fonction des volumes d'eau disponibles et arriver à déterminer des redevances d'usage et des mécanismes de distribution adéquats.

La déforestation, l'érosion des sols, la dégradation des terres et la désertification figurent parmi les défis écologiques les plus sérieux que la Chine ait à relever. Dans ce contexte, **M. Han Jun** évoque le problème universel que pose le double objectif d'amélioration à long terme de l'environnement et d'amélioration immédiate du bien-être des populations rurales. La plupart des pays se trouvent confrontés à ce défi, mais le cas de la Chine est particulièrement délicat en raison de sa dotation en ressources naturelles et du nombre important de ruraux vivant dans des conditions très difficiles. Pour la Chine, une gestion intégrée des écosystèmes impliquera une planification et une coordination horizontales globales ; le développement d'un capital humain et sa valorisation ; l'extension et la protection des droits d'usage sur les terres ; l'amélioration du ciblage des politiques de lutte contre la pauvreté ; la promotion des techniques de gestion des terres de nature participative et adaptées au contexte local.

Bien qu'un certain nombre d'organismes et bureaux officiels s'attachent à promouvoir des pratiques de conservation, il n'y a guère de coordination ou de suivi. Chaque organisme ou ministère opérationnel gère les investissements et les activités à sa façon, d'où une situation disparate, certaines zones bénéficiant de très nombreuses actions et d'autres pratiquement d'aucune. De plus, les organismes centraux se préoccupent souvent davantage du montant des fonds à verser que de leurs bonne affectation et gestion. Il n'existe pratiquement aucune évaluation ex post de l'efficacité des mesures prises. Le processus décisionnel, bien que réparti entre différents organismes, demeure très centralisé au sein de ceux-ci, ce qui conduit souvent à des pratiques locales inappropriées. Le montage de nouveaux projets et leur approbation retiennent davantage l'attention que la vérification de leur bonne exécution. Comme les crédits destinés à financer les projets de conservation sont généralement transférés de l'administration centrale aux autorités locales sous la forme de subventions, il n'est pas rare que ces fonds soient détournés. M. Han plaide en faveur d'une plus grande participation des populations rurales à la conception et à la mise en œuvre des projets de conservation, ce qui permettrait une élaboration et un ciblage plus adaptés aux conditions locales et imposerait aux fonctionnaires une plus grande transparence, réduisant ainsi les possibilités de détournement de fonds.

MM. Ma Xiaohe et **Fang Songhai** font observer que les défis que posent à la Chine les ressources en eau sont considérables, mais qu'ils ne sont pas pour autant insurmontables. La dotation en eau par habitant de la Chine est faible et mal répartie. L'offre d'eau est relativement statique, mais la demande urbaine, industrielle et résidentielle ne cesse de croître. Par ailleurs, les pratiques agricoles actuelles entraîneront une augmentation de la demande d'eau d'irrigation si l'agriculture chinoise veut pouvoir répondre aux besoins d'une population de plus en plus nombreuse. Outre les problèmes de pénurie d'eau, le Nord de la Chine rencontre également des problèmes de qualité de l'eau : les bassins des rivières Hai, Liao, Huai et Huang affichent tous des indices de pollution élevés, qui rendent leur eau impropre à la consommation et, très souvent, inutilisable pour l'irrigation. Ces réalités indiscutables font que les pratiques actuelles ne sont pas durables, et plusieurs suggestions sont faites pour remédier à cette situation. Il s'agit en l'occurrence de relever les redevances sur l'eau à des niveaux plus réalistes, d'améliorer les mesures incitant à adopter des pratiques d'irrigation économes en eau, d'accroître le soutien aux installations de protection de l'eau, de promouvoir des méthodes d'irrigation utilisant de l'eau recyclée, de financer des recherches pour mettre au point des techniques d'économie d'eau et des cultures tolérantes à la sécheresse, ainsi que de prendre des mesures plus strictes de lutte contre la pollution de l'eau. Selon MM. Ma et Fang, ***il est plus urgent de revoir les mesures incitatives que de réaffecter les ressources en eau dans le cadre du projet de transfert d'eau Sud-Nord.***

La **session 2** portait essentiellement sur le thème de la **gestion des ressources agro-environnementales dans les pays de l'OCDE.** Globalement, les secteurs agricoles des pays de l'OCDE sont parvenus à obtenir une croissance sans précédent depuis le milieu du XXème siècle. Pour ce faire, ils ont diminué la main-d'œuvre agricole et réduit légèrement les superficies exploitées, mais ils ont utilisé davantage d'eau, de produits chimiques et de machines, ce qui s'est fait au détriment de l'environnement et des ressources naturelles. Bien que, dans leur majorité, les pays de l'OCDE indiquent qu'ils attachent une priorité élevée aux questions d'environnement liées à l'agriculture, seule une faible part du soutien concerne effectivement les aspects agro-environnementaux. Récemment, la situation s'est améliorée, mais l'essentiel des mesures de soutien demeure lié à la production et à l'utilisation d'intrants – c'est-à-dire des formes de soutien qui présentent une moindre efficacité de transfert et sont moins favorables à l'environnement.

M. Wilfrid Legg indique que l'objectif sous-jacent des politiques de gestion durable est de contribuer au bien-être global en ***optimisant l'efficience économique et les***

avantages environnementaux nets en fonction des valeurs sociétales du pays considéré. Autrement dit, il s'agit de prendre en considération les usages marchands et non marchands des ressources, ainsi que les effets locaux et à plus longue distance, et de trouver un juste équilibre entre les intérêts des générations actuelles et ceux des générations futures. Même si les ressources naturelles sur lesquelles se fonde l'agriculture, à savoir les terres, l'eau et les écosystèmes, sont en grande partie des ressources renouvelables, elles peuvent s'appauvrir ou se détériorer si elles ne sont pas gérées de manière responsable. Les acteurs impliqués dans l'élaboration et la mise en œuvre des politiques ont donc pour tâche d'harmoniser les mesures incitatives de manière à optimiser à la fois l'efficience économique et les avantages environnementaux dès lors que ces derniers, en particulier, ne sont pas facilement pris en compte par le marché. L'expérience, aussi bien positive que négative, acquise par les pays de l'OCDE montre que l'action publique peut jouer un rôle décisif dans l'exploitation durable des ressources naturelles dont le secteur agroalimentaire est tributaire. Plusieurs domaines d'action méritent d'être mis en avant : développer les droits de propriété et les marchés ; réformer les politiques apportant un soutien lié à la production ; limiter la dégradation des ressources et renforcer la fourniture de services écologiques ; remédier aux déficits d'information ; régler les problèmes de répartition induits par les politiques de gestion des ressources ; et, enfin, gérer le capital naturel relevant du domaine public. Un principe de base sous-tend ces différents axes de travail, à savoir établir une correspondance entre les mesures incitatives et la valeur de rareté effective d'un bien, d'un service ou d'une ressource pour la collectivité.

Les mécanismes de marché sont actuellement au cœur du système d'affectation des ressources en eau en Australie. **M. Seamus Parker** retrace l'historique de la mise en place d'un ***marché de l'eau*** et décrit de quelle manière celui-ci a évolué pour répondre aux besoins des usagers anciens et nouveaux, ainsi que de la collectivité. Même si l'objectif premier consiste à favoriser le jeu des forces du marché, il n'en demeure pas moins qu'il est indispensable de conclure des accords intergouvernementaux au niveau des Etats et que ceux-ci sont fondés sur des incitations financières destinées à amener les intervenants sur ce marché à avoir les comportements souhaités. Avant la réforme, les accords d'exploitation étaient conclus au cas par cas sans que soient pris en compte les effets cumulatifs avec les autorisations déjà délivrées. Les nouveaux usages de l'eau ont posé des problèmes particulièrement épineux dans la mesure où les droits sur l'eau étaient liés à la propriété des terres et que le seul moyen d'obtenir une autorisation d'exploitation consistait à acheter des terres. Lorsqu'il s'est agi d'arrêter les règles d'octroi des autorisations, il n'a guère été tenu compte des volumes d'eau nécessaires à l'entretien de l'écosystème. Les conflits portant sur les usages de l'eau se sont multipliés et intensifiés, et lorsque de véritables différends surgissaient, les approches employées pour résoudre des questions relevant en réalité de problèmes de planification et de priorités étaient extrêmement contradictoires.

Trois grands domaines exigeaient des réformes : les institutions, les droits et la tarification. En ce qui concerne les institutions et la tarification de l'eau, les réformes ont assuré la viabilité commerciale des entreprises relevant d'une administration publique. Les réformes des institutions et des droits ont permis de mieux définir les relations commerciales entre les structures chargées de la répartition des eaux et celles qui les exploitent. Enfin, les réformes des droits et de la tarification ont amené à mieux déterminer l'une par rapport à l'autre les notions de « produit » et de « prix ». Ces diverses améliorations ont permis de préciser le rôle respectif des réglementeurs, des gestionnaires de ressources, des responsables d'infrastructures et des prestataires de services, et de bien séparer les droits de propriété sur l'eau des questions de régimes et de titres fonciers. Les marchés de l'eau ont évolué à mesure que la définition des droits de propriété et des droits sur l'eau se précisait – avec pour conséquences la diminution des conflits concernant les droits d'usage et une tarification reflétant mieux les coûts et les volumes consommés. M. Parker présente le calendrier adopté par l'Australie pour la mise en place d'un marché de l'eau opérationnel reposant à la fois sur les bassins versants et les mécanismes de l'économie, lequel rend compte des priorités de la collectivité tout en autorisant une affectation des ressources plus efficace que si l'on avait eu recours à des procédures administratives ou à des autorisations officielles.

Aux Pays-Bas, l'intensification de l'agriculture entre 1950 et 1985 a été spectaculaire, mais elle s'est opérée au détriment de l'environnement, où la charge en éléments nutritifs, pesticides et autres sous-produits de la production agricole a augmenté. M. **Peter van Boheemen** décrit les mesures prises par les pouvoirs publics, les agriculteurs et le secteur agroalimentaire ***pour réduire les impacts négatifs de la production agricole sur l'environnement.*** Après avoir évoqué le niveau et les effets à long terme d'une charge excessive en éléments nutritifs, en particulier le phosphore et l'azote, il présente un bilan de l'évolution de la qualité des eaux au fil des ans. Il fait ensuite le point sur la double approche adoptée par les Pays-Bas pour réduire ses émissions d'éléments nutritifs : maîtriser la production et les rejets d'effluents d'élevage, et améliorer la gestion des éléments minéraux sur les exploitations. Au milieu des années 80, des quotas ont été institués pour limiter les effectifs de bétail. En 1987, des normes sur l'épandage des effluents d'élevage ont imposé aux agriculteurs produisant d'importants excédents de les acheminer vers des exploitations ou des zones déficitaires. Les agriculteurs avaient en outre la possibilité d'opter pour l'achat de terres supplémentaires ou la réduction de leur cheptel.

Entre 2002 et 2005, les Pays-Bas ont mis en place un système de maîtrise du marché destiné à limiter les excédents de lisiers et d'effluents d'élevage, mais celui-ci se justifie moins aujourd'hui dans la mesure où la charge en éléments nutritifs est globalement plus équilibrée. Depuis 1990, les contrôles réglementaires des produits phytosanitaires se sont durcis, notamment en ce qui concerne les doses limites, la proximité de cours d'eau, ou les traitements effectués en présence de vent. Ont par ailleurs été mis en place des systèmes « d'autorisation de distribuer » des produits agroalimentaires dont la production répond à des critères tels qu'un niveau satisfaisant de sécurité environnementale et alimentaire. Tout infraction à ces réglementations peut avoir de graves conséquences, notamment des sanctions administratives et pénales. Néanmoins, le succès de ces mesures s'explique en particulier par le fait que, grâce aux mesures d'identification et d'autorisation, ce nouveau système permet une rémunération par le marché des agriculteurs ayant adopté des pratiques d'élevage innovantes, tout en pénalisant ceux dont les résultats sont plus que médiocres.

Aux Etats-Unis, la concurrence de plus en plus vive dont font l'objet des ressources limitées et la sensibilisation croissante de la population aux questions d'environnement sont à l'origine de la plupart des problèmes qui se posent aujourd'hui concernant la disponibilité et la qualité de l'eau en agriculture. Selon M. **Dennis Wichelns,** les autorités publiques doivent relever deux grands défis. En premier lieu, il est impératif de déterminer les meilleures modalités d'affectation des ressources en eau désormais limitées, alors même que la demande d'eau pour des usages agricoles, résidentiels, urbains, industriels et environnementaux ne cesse de croître. En second lieu, ***il faut également déterminer quels sont les meilleurs moyens pour définir et atteindre les objectifs de qualité au niveau des Etats et au niveau national sans obérer outre mesure***

la croissance et le développement. La gestion de l'eau pose des problèmes particulièrement aigus dans les Etats où opère la société Ogalla Aquifer. En effet, la surexploitation des eaux souterraines y est très importante et n'est sans doute pas près de diminuer étant donné que la demande en eau ne cesse d'augmenter, que ce soit pour des cultures traditionnelles, des cultures destinées à la production de biomasse pour la fabrication de l'éthanol ou de l'eau nécessaire aux écosystèmes, et que, parallèlement, les Etats partageant cet aquifère poursuivent des buts différents, appliquent des politiques différentes et n'ont défini ni plan coordonné, ni priorités communes.

Dans d'autres Etats, la concurrence de plus en plus grande vis-à-vis des ressources en eau et l'attention croissante portée aux effets de l'agriculture sur l'environnement ont incité les agriculteurs et les industriels d'autres secteurs à améliorer leurs pratiques de gestion de l'eau. Non seulement les améliorations apportées ont permis à l'agriculture et aux secteurs concernés de poursuivre leur croissance, mais elles ont aussi permis de stabiliser à peu près le volume d'eau utilisé à son niveau de 1985. Plusieurs Etats expérimentent actuellement différentes méthodes de fixation des redevances sur l'eau et d'autres ressources naturelles, afin de déterminer celle qui sera la mieux à même de refléter véritablement leur valeur de rareté. Comme en Chine, la pollution diffuse est particulièrement préoccupante, et le grand public est désormais bien conscient de ses liens avec la qualité de l'eau de consommation, des aménités environnementales, de l'habitat de la faune et de la flore sauvages, des pêches, des zones humides et d'autres services écosystémiques. M. Wichelns fait observer que le processus d'élaboration de l'action publique et de planification est lourd, chronophage et coûteux, mais que si rien n'est fait aujourd'hui, les mesures de réhabilitation qui devront être prises ultérieurement seront encore plus lourdes, chronophages et coûteuses. Dès lors, seuls l'innovation, la ténacité et le courage politique peuvent conduire à une exploitation prévoyante des ressources.

Si l'on veut maîtriser et contrôler les comportements, il faut tout d'abord être en mesure de les observer et de les mesurer. Dans son intervention, M. **Kevin Parris** montre comment ***l'amélioration des indicateurs environnementaux*** et des procédures utilisées par l'OCDE pour suivre et évaluer les politiques agro-environnementales ***peut permettre de mieux éclairer le processus d'élaboration et de mise en œuvre de mesures propices à une exploitation durable des ressources.*** Pour analyser les relations entre agriculture et environnement, l'OCDE fait appel à un modèle « causes agissantes – états – réponse », dont l'application doit permettre de répondre, entre autres, aux questions suivantes : quelles sont les causes à l'origine des modifications de l'environnement dans les zones agricoles ? Quels sont les effets de ces modifications sur l'état de l'environnement dans les zones agricoles et, éventuellement, dans d'autres zones ? Quelles mesures peut-on prendre pour tenir compte les évolutions (préjudiciables) de l'état de l'environnement dans les zones agricoles ?

Même s'il s'agit d'une véritable gageure, des efforts considérables sont déployés pour faire en sorte que les principaux indicateurs agro-environnementaux utilisés pour le modèle causes agissantes – états – réponse demeure pertinent, d'interprétation aisée, fiable au plan analytique, et mesurable. Les indicateurs collectés concernent trois secteurs dans lesquels l'agriculture joue un rôle : protéger le stock de ressources naturelles sur lesquelles l'agriculture influe ; réduire la pollution agro-environnementale ; améliorer les pratiques de gestion agro-environnementales et l'exploitation des ressources. Si les liens entre les politiques et leurs effets sur la production ont été relativement bien quantifiés empiriquement, ce n'est pas le cas des relations agro-environnementales. L'évaluation et la compréhension des liens entre les politiques et les mesures incitatives, les actions, les réponses biologiques et écologiques, et les mesures correctives n'en sont qu'à leurs premiers balbutiements. L'OCDE met actuellement au point un modèle simplifié d'impact des politiques agro-environnementales (SAPIM), qui doit permettre de rendre compte des disparités entre les systèmes agro-environnementaux des pays membres de l'OCDE et de mieux appréhender les liens de causalité existants grâce à une modélisation économique et biophysique intégrée. Par ailleurs, l'OCDE a recours à son modèle d'équilibre partiel de l'agriculture mondiale, AGLINK, pour étudier les effets de la suppression des subventions agricoles et de l'application de taxes et prélèvements divers sur l'utilisation de produits agrochimiques. Compte tenu des difficultés posées par la collecte des données appropriées et l'établissement de scénarios contrefactuels et de liens clairs entre actions et résultats, ces modèles ne sauraient se substituer à une analyse ex post de l'impact d'une politique donnée. Néanmoins, ils peuvent fournir des orientations quant à la définition d'un comportement écologique et aux moyens d'éviter les catastrophes. M. Parris indique que les populations comme les décideurs publics prennent de plus en plus en compte les coûts et avantages environnementaux des activités agricoles, puis présente différentes suggestions pour améliorer le cadre analytique et les modèles mis au point par l'OCDE dans le but de mieux éclairer les choix publics.

Lors de **la session 3 consacrée aux moyens d'action envisageables pour la Chine, M. Chen Mengshan** ***retrace l'histoire de l'utilisation des engrais en Chine,*** en développant plus particulièrement son propos sur la situation et les problèmes actuels. Les premiers textes cités montrent quels peuvent être les arbitrages à opérer dans ce domaine, mais aussi la complémentarité des choix agro-environnementaux. Ainsi, Zhou Li préconise d'éliminer les graminées des prairies humides pour pouvoir cultiver le blé, tandis que Shi Jing estime que la décomposition des mauvaises herbes favorise une croissance luxuriante du millet. Dans le premier cas, les zones humides et les habitats aquatiques seraient détruits au profit de la production de blé, tandis que dans l'autre, l'intérêt de la décomposition de la matière organique et des engrais verts pour l'amélioration de la production est pris en compte. Il y a 3 000 ans déjà, la matière organique et le fumier jouaient un rôle dans l'entretien et l'amélioration de la fertilité du sol et de son pouvoir tampon, et même si ce phénomène demeurait en partie inexpliqué, on avait fort bien compris toute l'importance de la circulation de l'eau et des éléments nutritifs du sol.

La Chine a commencé à adopter les engrais chimiques commerciaux au début des années 1980, dans le but affiché d'augmenter la production globale pour répondre à l'accroissement des besoins du pays. A la fin des années 1980 et au début des années 1990, on s'est davantage préoccupé de disposer de produits de qualité, à la fois plus ciblés et plus efficaces, puis l'on a pris conscience au cours de ces quelques dernières années des questions d'écologie, de sécurité et de protection de l'environnement. Les connaissances acquises ne suffisent cependant pas, puisque l'on continue d'utiliser beaucoup plus d'engrais simples de peu de qualité que d'engrais complexes ou de grande qualité, l'azote et le phosphore étant généralement appliqués à des doses supérieures aux besoins des cultures. De nouveaux progrès ont été réalisés dans le domaine des engrais à libération lente, mais l'utilisation de ceux-ci demeure marginale. Au cours de ces dernières années, on a mis à jour et diffusé un certain nombre de nouvelles normes et de nouveaux manuels sur l'application des engrais et d'autres intrants agricoles en s'appuyant sur les toutes dernières recherches scientifiques. Néanmoins, tels qu'ils sont pratiqués aujourd'hui, l'analyse des sols et l'apport rationnel des engrais ne tiennent pas véritablement compte des connaissances scientifiques, et l'on prépare actuellement le lancement d'une campagne nationale de sensibilisation sur ce sujet.

M. Jeff McNeely présente ensuite une synthèse du Bilan du Millénaire relatif aux écosystèmes et en dégage les enseignements, en particulier pour la conservation de la biodiversité et la gestion des systèmes hydriques. Il fait remarquer que dans la mesure où la race humaine fait partie de l'écosystème mondial, notre propre survie dépend de ce dernier. Il évoque longuement ***le concept de services écosystémiques et l'idée selon laquelle ceux qui fournissent ces biens et services (qui ont souvent été considérés jusqu'ici comme un bien d'intérêt public) méritent une rémunération*** dès lors qu'ils gèrent les écosystèmes en vue d'accroître les services fournis à la collectivité. Or, personne n'ignore que les agriculteurs sont mieux placés que quiconque pour savoir que la productivité et la rentabilité d'un écosystème dépend essentiellement de l'existence d'exploitations saines et adaptables. En fondant la préservation des services écosystémiques sur des incitations économiques, on reconnaît ipso facto les qualifications des agriculteurs et le rôle qu'ils jouent dans la protection des terres et des ressources. Dans le cas de la Chine, l'enjeu réside dans la mise en adéquation des incitations privées et de l'intérêt public, de manière que les trois objectifs de l'agriculture durable (sécurité alimentaire, emploi et génération de revenus, et conservation des ressources et protection de l'environnement) puissent être atteints.

M. McNeely passe en revue les aspects opérationnels de l'évaluation des services écosystémiques à « valeur non marchande ». En se posant des questions et en réfléchissant à la nature des réponses à y apporter, on commence à imaginer de quelle manière les marchés peuvent être mobilisés pour valoriser les services et les biens auxquels, jusqu'ici, le marché n'attribuait aucune valeur ou la sous-estimait. Quels sont les services spécifiques fournis par un écosystème ? Quels sont les principaux bénéficiaires de ces services ? Quelle est l'importance de ces services ? Quels usages en font leurs bénéficiaires ? Que deviendraient ces services si l'écosystème était géré différemment ? Comment peut-on s'approprier la valeur de ces services ? M. McNeely souligne l'importance de l'instauration de droits de propriété et de cadres juridiques solides et clairement définis pour permettre aux marchés des écosystèmes de se développer. Il fait également part aux participants de son souci de veiller à ce que les membres les plus pauvres de la collectivité ne soient pas exclus des efforts de protection de l'environnement. En conclusion, il fait observer que les actions faisant supporter une charge supplémentaire à ses membres les plus défavorisés ne sauraient être viables.

Selon **M. Ke Bingsheng,** un pays ayant la superficie et la population de la Chine doit avoir pour priorité première sa sécurité alimentaire, et cette sécurité doit provenir, pour l'essentiel, de sources intérieures. Toutefois, M. Ke indique que les contraintes pesant sur la disponibilité des ressources se font de plus en plus ressentir et que la Chine a déjà augmenté ses prélèvements sur les ressources naturelles disponibles au-delà de ce qu'il est possible de supporter. De ce fait, la seule manière de satisfaire aux objectifs de sécurité alimentaire de la Chine de manière durable consiste à protéger les ressources existantes tout en accroissant l'efficacité et la productivité de leur exploitation. M. Ke évoque ensuite ***les voies et moyens susceptibles de permettre de protéger les terres cultivables existantes, de préserver les ressources en eau, d'accroître la productivité des terres, de renforcer l'innovation technique, de limiter les pertes intervenant tout au long de la chaîne d'approvisionnement et d'améliorer les infrastructures agricoles et rurales.*** Il fait remarquer que les agriculteurs consentiront probablement des investissements de long terme dans des équipements ou des terres en vue d'améliorer leur productivité s'ils bénéficient de régimes fonciers plus fiables. Par ailleurs, la fixation du prix de ressources telles que l'eau et les terres en fonction de leur valeur de rareté et de l'ensemble des services marchands et non marchands qu'elles fournissent encouragera l'amélioration de l'efficience et de la productivité de leur exploitation. Enfin, M. Ke évoque certains des enjeux fonciers et des asymétries existant dans ce domaine. Lorsque les terres doivent servir au développement urbain ou industriel, la majeure partie des avantages qui en découlent vont à des acteurs privés et, par conséquent, leur valeur de marché sera assez proche de leur valeur sociétale globale. Par contre, lorsque les terres sont affectées à des usages agricoles ou au secteur des ressources, une part non négligeable des avantages qui en résultent ne peut être récupérée par le secteur privé et, de ce fait, la valeur de marché des terres dites agricoles dans les zonages ou classifications sera beaucoup moins élevée que leur valeur société globale.

MM. Tang Huajun et Yin Changbin ***passent en revue les enseignements tirés de l'adoption d'une approche « en système fermé » du développement agricole, et les gains que cette approche est susceptible d'apporter.*** Pour résumer, cette approche cherche à maximiser « l'empreinte » économique et sociétale du secteur tout en minimisant « l'empreinte » écologique de sa production agricole. Si l'on considère les déchets d'une activité économique comme des ressources pour une autre activité productive, on peut alors augmenter leur taux d'utilisation et leur efficience en réduisant parallèlement les polluants. Ce concept évoque les notions de grappes d'activités et d'agglomérations d'entreprises de Michael Porter, qui permettent de tirer parti de la complémentarité économique entre différentes activités et entreprises. Néanmoins, cette approche étend les notions de complémentarité, de grappe et d'agglomération non seulement à l'aspect commercial, mais également à l'aspect écologique/environnemental, dont la valeur est sous-évaluée ou n'est pas prise en compte.

M. Jeff Au se fonde sur différentes études de cas pour examiner ***le rôle de l'industrie phytosanitaire dans le développement agricole.*** M. Au ne se cache pas la difficulté que présente la diffusion de nouvelles technologies auprès de quelque 200 millions de ménages agricoles pauvres répartis sur un vaste territoire et dont la plupart ne possède que de petites superficies exploitables et ont un faible niveau d'instruction. Dans la plupart des cas évoqués, il s'agit de pratiques d'absence de travail du sol ou de travail minimum du sol, combinées à une application raisonnée de pesticides. Il s'agit donc d'arbitrer entre, d'une part, les avantages d'une amélioration de la structure et de la fertilité du sol, de la diminution de l'érosion, de l'augmentation du taux de matière organique et des capacités de rétention de l'eau et, d'autre part, l'urgence d'adopter des traitements raisonnés dont les effets positifs se feront sentir en aval sur l'eau, la biodiversité et les écosystèmes.

Selon M. Au, bien que les agriculteurs n'aient généralement pas un niveau élevé d'instruction, ils prennent les décisions les mieux adaptées à leur objectif, à savoir nourrir et habiller leur famille. Cela dit, il faut éviter de leur donner des instructions complexes, et il importera de leur apporter un soutien suivi pendant une période relativement longue. Par ailleurs, il ne faut pas se contenter de réfléchir à la nouvelle technique à implanter, mais également au modèle qui permettra de la faire adopter. Il est par conséquent essentiel de mettre en place des parcelles témoins et des exploitations pilotes clés en main. M. Au souligne également l'importance d'associer les agriculteurs à cette démarche si l'on veut qu'ils adoptent des pratiques plus performantes, ainsi que de leur apporter la preuve des avantages économiques qu'ils retireraient de l'application de nouvelles technologies.

Parmi les choix publics qui permettraient de soutenir les revenus et la production agricoles, on a préconisé l'assurance récolte, qui présenterait l'avantage de demeurer compatible avec les obligations de la Chine vis-à-vis de l'OMC. **MM Zhong Funing,**

Ning Manxiu et Xing Li présentent une étude économétrique de ***l'influence de l'assurance récolte sur l'utilisation de produits agrochimiques*** dans la province du Xinjiang. Ils exposent tout d'abord différentes hypothèses qui, étayées par des données scientifiques, doivent permettre de déterminer si des intrants particuliers augmentent la production ou réduisent la probabilité d'une récolte mauvaise ou catastrophique, puis présentent les résultats des tests de ces hypothèses et font plusieurs observations importantes. Selon eux, la probabilité que les agriculteurs à temps plein contractent une assurance récolte est plus grande, sans doute parce que leur bien-être dépend essentiellement de l'agriculture. C'est également le cas des agriculteurs relativement spécialisés, ainsi que de ceux dont les risques peuvent moins facilement être répartis sur plusieurs productions. En revanche, les agriculteurs établis depuis longtemps seront moins attirés par l'assurance récolte, peut-être en raison de leur capacité, réelle ou supposée, à gérer les risques, mais aussi de leur niveau d'endettement et de leur degré de diversification. Enfin, les agriculteurs ayant contracté une assurance récolte utilisent des quantités plus importantes d'engrais et de film plastique que ceux sans assurance récolte, mais ils limitent davantage les traitements pesticides. Ce comportement est logique, puisque les engrais et les films plastiques permettent d'augmenter les rendements, tandis que les pesticides permettent d'éviter les pertes à la récolte et les chutes de rendement.

L'intervention de **M. Huang Jikun** et de ses collègues porte sur ***l'application d'engrais et de pesticides commerciaux*** en Chine. Les doses utilisées ont à peu près doublé depuis 1980, la Chine se situant aujourd'hui au 4ème rang mondial du classement par intensité de consommation de l'engrais, derrière le Japon, la Corée et les Pays-Bas. Quant à la consommation de pesticides, elle a pratiquement été multipliée par trois. S'il est vrai que les produits chimiques ont joué un rôle important dans la progression de la production agricole, ils peuvent aussi accroître les coûts de production, augmenter le risque de survenue de certains problèmes de sécurité et de qualité des aliments, et contribuer à la pollution l'environnement. Plusieurs études récentes montrent que les apports actuels d'engrais chimiques sont excessifs (supérieurs de 20 à 50% aux doses préconisées). En ce qui concerne les pesticides, les excès seraient encore plus considérables, avec des doses supérieures de 40 à 55% au niveau recommandé. Ce comportement est illogique, parce que ces excès reviennent effectivement à jeter de l'argent par les fenêtres, mais également parce qu'ils peuvent avoir de graves conséquences pour l'environnement et la santé. Mais les agriculteurs sont-ils conscients de ce qu'ils font ?

Des travaux de recherche menés actuellement au Centre de la politique agricole chinoise ont montré qu'on n'observait aucune chute significative de rendement sur les parcelles où les apports d'engrais avaient été réduits de 25 à 35%, ce qui n'a pas manqué de surprendre les agriculteurs ayant pris part à ces études. Plusieurs hypothèses peuvent être avancées à ce sujet. Il se peut que certains agriculteurs sachent effectivement ce qu'ils font et que l'apport excessif d'intrants fasse partie de leur stratégie de gestion des risques. Par ailleurs, selon certaines données, le régime foncier et les problèmes de migration peuvent jouer un rôle dans ce recours excessif à des intrants commerciaux. En effet, lorsque des migrants reviennent chez eux, ils préfèrent souvent appliquer les intrants « d'un seul coup » plutôt que de les répartir de manière optimale à diverses périodes critiques du cycle de développement des végétaux, car leur séjour leur laisse peu de temps. Néanmoins, il ressort également de cette étude que selon des données beaucoup plus nombreuses encore, les autorités publiques, la communauté scientifique, les sélectionneurs, les agents de vulgarisation et les fournisseurs d'intrants ont convaincu les agriculteurs que « si de faibles apports ont des effets positifs, des apports élevés ont des effets encore plus favorables ». Certes, ces conclusions demandent à être confirmées, mais elles suggèrent toutefois que ***les incitations données actuellement par les organismes de recherche, le système de vulgarisation et les fournisseurs d'intrants agricoles doivent être réexaminées.*** Les informations reçues par les agriculteurs sont-elles crédibles ? Les sources d'information proviennent-elles de groupes d'intérêt cherchant à augmenter la consommation d'intrants et à parvenir à des objectifs de vente ou de production qui ne sont pas nécessairement à l'avantage des agriculteurs ou de l'environnement ? Dans les années à venir, des efforts seront déployés pour dispenser plus largement aux agriculteurs une formation destinée à leur faire comprendre l'intérêt de diminuer leur consommation d'intrants, à la fois pour leur propre revenu et pour épargner l'environnement

PART I.

AGRI-ENVIRONMENTAL SITUATION AND POLICIES IN CHINA: PRACTICE AND OUTCOMES

Chapter 1.

THE NEW SOCIALIST COUNTRYSIDE AND ITS IMPLICATIONS FOR CHINA'S AGRICULTURE AND NATURAL RESOURCES

Editor's summary of the speech by Tang Renjian
Leading Group for Finance and the Economy

The 2006 State Council/CCCP "Number 1 Document" emphasises the "Building of a New Socialist Countryside". This year's document is particularly noteworthy as it sets the direction for China's 11th 5 Year Plan and makes significant commitments to China's agri-food sector and rural industries. While it also includes provisions to boost spending on health care, education and social security, the major focus is on improving the well-being of China's farmers and redressing the imbalance in living conditions between urban and rural citizens.

China is moving from placing burdens on agriculture to modest support, although support levels remain low in comparison to most OECD countries. We have placed an emphasis on improving rural infrastructure, lowering taxes, improving access to credit, improving incomes and containing input costs. The government remains concerned with maintaining a high level of "self sufficiency" with respect to staple grains, but we are also aware of the need to improve farmers' incomes and well-being. Resource and environmental stewardship are also high priorities, including challenges relating to the loss of arable land, declining water tables, and pollution arising from agricultural production.

There are three main elements of the new plan's agri-rural component I'd like to discuss.

First, it is obvious that – to meet the needs of a growing population and rising living standards – agricultural production must be maintained and augmented even further. Given China's fixed resource base, this must come about through the more efficient use of land, water and other resources. Our success in this area will affect the degree to which China can maintain its self-reliance and improve the well-being of 200 million rural families.

A second element is to make better use of biomass in both energy conservation and in energy creation. It may be possible to use cellulose production (*e.g.* rice stalks) to produce bio-energy. Since such crop residues are often burned anyway, this would concurrently reduce China's reliance on fossil fuels and reduce unnecessary pollution in the countryside.

The third challenge is to reduce waste discharge and improve the efficiency of use for various agricultural inputs, including water, land, livestock effluent, crop residues and purchased inputs. If successful, this will concurrently reduce waste and lower soil and water contamination. Benchmarking with other countries suggests we can make significant progress in this area.

If successful in dealing with these three areas of emphasis, we will actually be doing "more with less". Moreover, pollution and stresses on the environment will actually decline if such reforms are undertaken intelligently. If we pay closer attention to the incentives we are giving, we can not only halt the deterioration in China's natural bio-capital but we can begin rebuilding it – maintaining natural capital for future generations. In addition, China might be able to partially capture the benefits of such improved husbandry practices by promoting bio-tourism, providing another possible source of employment and income for China's rural citizens.

Redressing the imbalance between urban and rural areas remains a major policy preoccupation in China. However, in pursuing these policy priorities, China faces many formidable challenges in such areas as water scarcity, fixed land and soil resources, land tenure, rural-urban migration, and infrastructure – to name a few. The major components to "Building a New Socialist Countryside" provide us with many of the tools needed to meet those challenges.

Chapter 2.

SELECTED ASPECTS OF WATER MANAGEMENT IN CHINA: CONDITIONS, POLICY RESPONSES AND FUTURE TRENDS[1]

Simon Spooner
Consultant to the OECD's Environment Directorate

Abstract

Although China has a large water resource in absolute terms, it's availability is low on a per capita basis and unevenly distributed both geographically and seasonally. Overall China has about one quarter of the world's average water per capita, but this falls to one tenth in northern and western areas. In almost all areas of China, water courses, lakes and groundwater are severely polluted due to agricultural pollution, and industrial and household discharges. Severe over utilisation and pollution have resulted in scarcity, inequitable distribution and degradation of aquatic ecosystems that have constrained economic development in addition to posing a major threat to the health and livelihoods of the population. In using 70% of abstracted water and returning less than a third of this water to watercourses, agriculture places by far the greatest stress on water resources.

This paper reviews the state of water resources in China in terms of available quantities and quality. It also discusses the pressures resulting from agriculture practices. It also presents key elements of the regulatory and institutional framework for water management and the application of various policy instruments.

Introduction

China faces serious problems with its water resources. In many parts of China, and especially in the developed areas of the north, water resources are under enormous pressure from over utilisation and pollution. This has become a major constraint to economic development and severely impacts upon the people's health and quality of life. It is estimated that this is costing the equivalent of USD 14 to 27 billion per year in lost economic activity[2]. Around 400 of the 600 major cities suffer from water shortages and 70% of water courses are severely polluted[3]. There is a critical shortage of 40 billion m^3 per year of water in the north of China.

The Chinese leadership is aware of this possible crisis and called for action to ensure that both the quantity and quality of water resources is improved if the nation is to maintain economic growth, food security, the health of the people and the health of the ecosystems that support the society. "Water saving before water transfer, pollution

1. The views expressed in this paper are those of the author and do not necessarily reflect those of the Chinese authorities, the OECD or its member countries.

2. A broad range of cost impacts exist based on estimates.

3. Statements of Chinese officials and official website publications.

control before water delivery and environmental protection before water consumption"[4] have been the key principles set out by the General Secretary, Hu Jintao[5] and Premier Wen Jiabao. The Minister for Water Resources (MWR)[6] and the Administrator of the State Environmental Protection Administration (SEPA) have also reiterated these directions and principles. Strenuous efforts are being made to implement changes, yet this remains a great challenge.

State of water resources

Surface water availability

The climate of China is divided into 3 main zones, generally south of the Yangtze River is warm and wet with rainfall of 800 to 2 000 mm throughout the year but concentrated in monsoons from April to September. The north western part of China is arid with less than 200 mm rain per year. The central, northern and north eastern part of China has 400-600 mm rain per year of which between 70 and 90% falls in just 3 months during the summer. High summer temperatures throughout China contribute to high levels of evaporation. The uneven distribution of rainfall through the year and large variation year to year lead to frequent episodes of flood and drought. The main rivers generally flow from the west to the East while most of the recent industrial development and wealth creation is in the East of the country. Thus the West of the country is a source of primary resources and ecological services while the East, suffering from excessive pollution, is increasingly relocating the most polluting industries back to the West.

The total available water resource in China is quite large at around 2 400 km^3 in 2004[7], however for 1.3 billion people this works out at just 1 850 m^3 per capita[8], this is less than a quarter of the world average. Water use per capita is 427 m^3 which is relatively low but the overall rate of utilisation of water resources is quite high at 23%, (Figure 1). Domestic consumption per capita is 212 l/h/d in urban areas, this compares with 150 l/h/d in UK and 427 l/h/d in USA. Water consumption in rural areas is 68 l/h/d.

4. Direct quote from Wang 2003, very similar to quote in Hu 2004 (allowing for translation).

5. For example, speech by Hu Jintao at "Population Resource and Environment work conference", 10 March 2004.

6. For example, speech by MWR minister Wang Shucheng at 2003 World Water conference in Kyoto, Japan.

7. Ministry of Water Resources (MWR): Water resources Bulletin 2004, note this is 13% lower than in 2003.

8. 1 850 m^3 calculated for 2004 – but the figure quoted in most official publications is 2 200 m^3/capita as average –2003 was 2 162 m^3/capita.

Figure 1. Freshwater use, early 2000s[1]

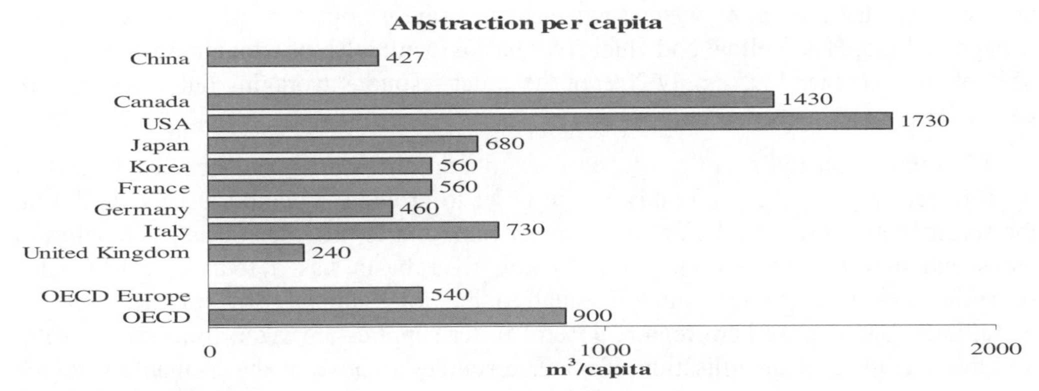

1. Or latest available year.
Source: OECD Environment Directorate.

Figure 2 illustrates the regional pattern of water resources and water use in China over the last 25 years. It shows the high reliance on groundwater resources in Northern China, the greater resource availability in Southern China and the increasing proportion of domestic and industrial water use compared to agricultural use in the developed parts of China. Not all used water is "consumed" by evaporation, much wastewater is returned to the rivers or aquifer: in 2003 the rate of water consumption was 68% for agricultural irrigation, 24% for industries, 24% for urban domestic uses, and 88% for rural domestic uses.

Figure 2. Water resource distribution, use by source, consumption by sector and exploitation index

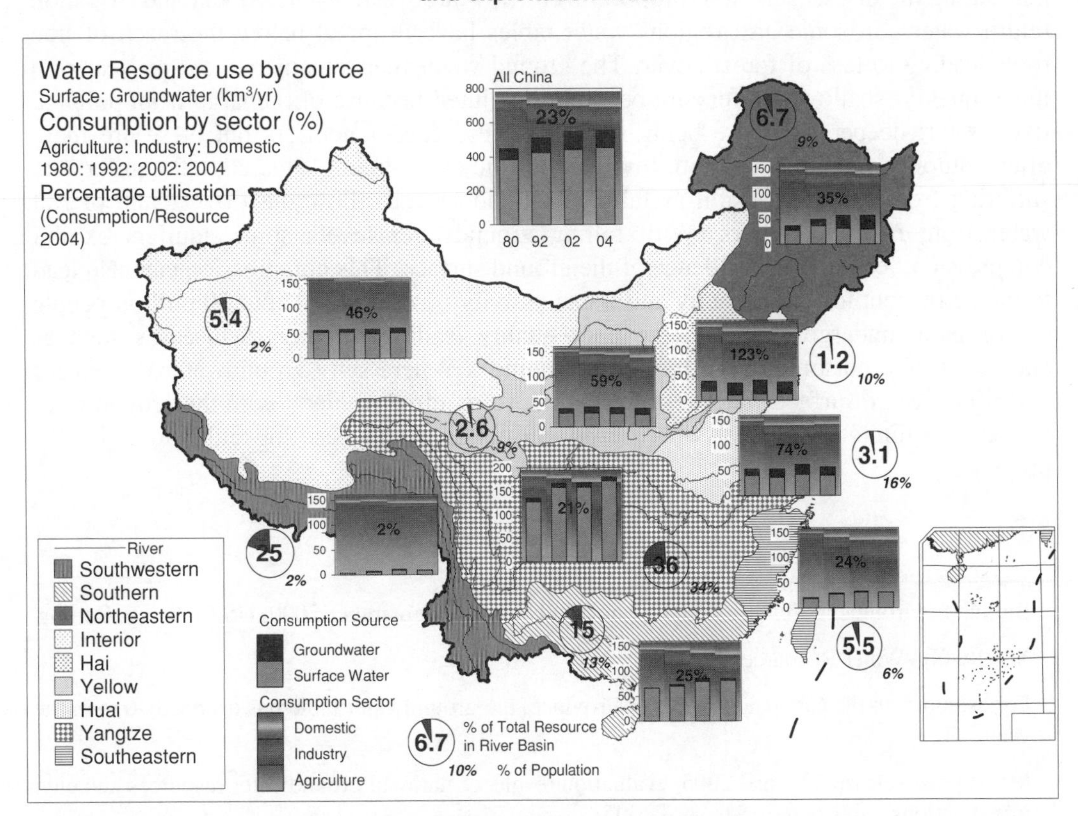

Source: Report on water resources in China, MWR, 2003.

The water resource situation is made considerably worse by the uneven geographical and seasonal distribution of water such that the heavily populated northern watersheds (Songhua, Liao, Hai, Yellow and Huai river basins) with 44% of China's population and 65% of cultivated land have only 20% of the water resource averaging out at just 575 m^3 per capita.

With most rain falling in a few summer months these areas are heavily dependent upon reservoir storage and groundwater in order to provide irrigation, especially during the spring planting season (before the summer rain) and to provide consistent supplies to cities and industry. For example the Yellow river basin has a total of 3 100 dams providing reservoir storage roughly equal to the total annual flow in the river[9] of 58 billion m^3. In the northern region, 43% of water supplies are taken from groundwater and the overall level of utilisation of water resources is 56% of the available resource consumed, this rises to more than 120% in the Hai River Basin, these figures indicate severe and unsustainable water stress.

Groundwater availability

Groundwater levels are monitored by the Ministry of Water resources (MWR) with 20 000 monitoring stations nationally. In many areas groundwater is being abstracted much faster than it can be replenished and in the northern region water tables are falling year on year in the majority of areas with monitoring identifying 72 000 km^2 with sinking water levels[10]. In some areas of the northern plain water tables have fallen by 70 to 100 m[11] and in some western areas by more than a hundred meters over the last decade.

With falling water tables rivers and lakes have been seeping away and drying up such that during the dry season large proportions of flow are wastewater effluent and irrigation return waters. In some arid regions water tables have dropped below the reach of tree roots leading to loss of forest cover. The ground water that remains is year on year of a lower quality; shallow aquifers are becoming polluted because of recharge from polluted river water; deeper wells are being sunk to access deeper aquifers but these are more often naturally contaminated with fluoride, arsenic and mineral salts; coastal aquifers are suffering from saline intrusion as falling land side levels allow the flow inland of salt water from marine aquifers. With falling groundwater levels some aquifers exhibit compression, leading to subsidence of the ground surface. This compression can also lead to the permanent loss of capacity and transmissivity of aquifers. About 70 million people are drinking underground water of poor quality leading directly to diseases such as chronic arsenic poisoning and fluoride poisoning[12]. Where salty groundwater is used for irrigation, scarce surface water resources are also required to blend with the groundwater to reduce salinity to a level where crop germination and seedling cultivation is still possible.

---

9. Beijing Environment, Science and Technology Update, 1 September 2000, US Embassy, Beijing.

10. MWR 2004 Water Resources Bulletin.

11. For example, in the Cangzhou in Hebei Province, the groundwater table has dropped 70 m in the past ten years.

12. MWR press release 22 April 2005, evaluation results of national groundwater resources and environment investigations published on 21 April 2005.

Trends and responses on water availability

There has been major progress in the management of water resources in some of the inland catchments. Over exploitation of the Yellow River was so extreme that by 1997 it did not flow as to the sea on 226 days of the year. Now water rights management and water transfer tunnels and storage reservoirs have resulted in the yellow river maintaining flows and not running dry for 6 years. The Tarim River in Xinjiang is an inland river 1 400 km long terminating in a system of ephemeral inland lakes. As a result of successful implementation of water rights management, including allocation of water quotas to the environment[13] and water saving agriculture methods, the 363 km lower section of the Tarim River and the Hei River are now flowing after 20 years of being dry due to over abstraction in the upper reaches[14]. The livelihoods of farmers in these regions are greatly enhanced.

To address the emerging water resource crisis in Northern China the nation is undertaking an enormous investment in a project to transfer 40 billion m^3 a year from the Yangtze Basin to the north China plain by 2010. This should increase to 60 billion m^3 in the following years to 2020. However this will still not meet all the needs for economic growth and ecological recovery.

In addition to large scale supply side engineering solutions to the water resources crisis China is considering developing an enormous number of small scale local watershed management projects to improve the retention and utilisation of available rainwater sources in rural areas by collecting runoff in bunds and small ponds for local use and aquifer recharge. Urban rainwater water can be collected and, together with recycled wastewater treatment effluent, used for irrigation of urban green areas and irrigation of peri-urban horticultural farming.

More suggestions for water saving are set out in the China Water Conservation Technology Policy Outline[15]. However the local benefits of such schemes must be balanced against possible reduced flows downstream and the hazard of open ponds as disease vectors.

Surface water quality

Rivers

The main indicator of surface water quality is the comparison of monitored data with the water quality standards set in legislation[16]. The standards cover 20 to 30 determinants with the principal ones being COD, BOD, ammonia, petroleum and phenol compounds for rivers. For lakes total nitrogen and total phosphorous are also critical. Figure 3 shows an analysis of data on each of the main river basins available from State Environmental

13. Tarim Basin 2 Project, World Bank project report, 27 June 2005.
14. Taitema lake now fills up to a maximal area of 200 km^2 after being dry for 30 years.
15. China Water Conservation Technology Policy Outline, 21 April 2005, NDRC, MOST, MWR, MOA, MOC joint publication.
16. Surface water quality standards, GB3838-88, GHZB 1-99, and GB3838-2002.

Protection Administration (SEPA) from 1996 to 2004[17]. It also presents the percentage of the monitored river under each grade. It has to be noted that the changes in standards brought some step changes in 2000 and 2002 at which times the number and locations of the monitored river reaches were also revised. In some cases data is not presented for all of the Grades individually but groups a number of grades together – in the charts this has been represented by the lowest grade in the group.

Figure 3 shows that the water quality in the northern and eastern areas of the country is much worse than in southern and western districts. There would appear to have been a deteriorating trend in many areas up to 2000 or 2002 but since 2002 most areas show three years of a strongly improving trend, apart from the Songhua basin in the far Northeast which appears to be continuing to deteriorate. Though water resources are much more abundant in the South of the country pollution and poor water quality is still a major constraint to the usability of these resources.

Figure 3. Water quality in China

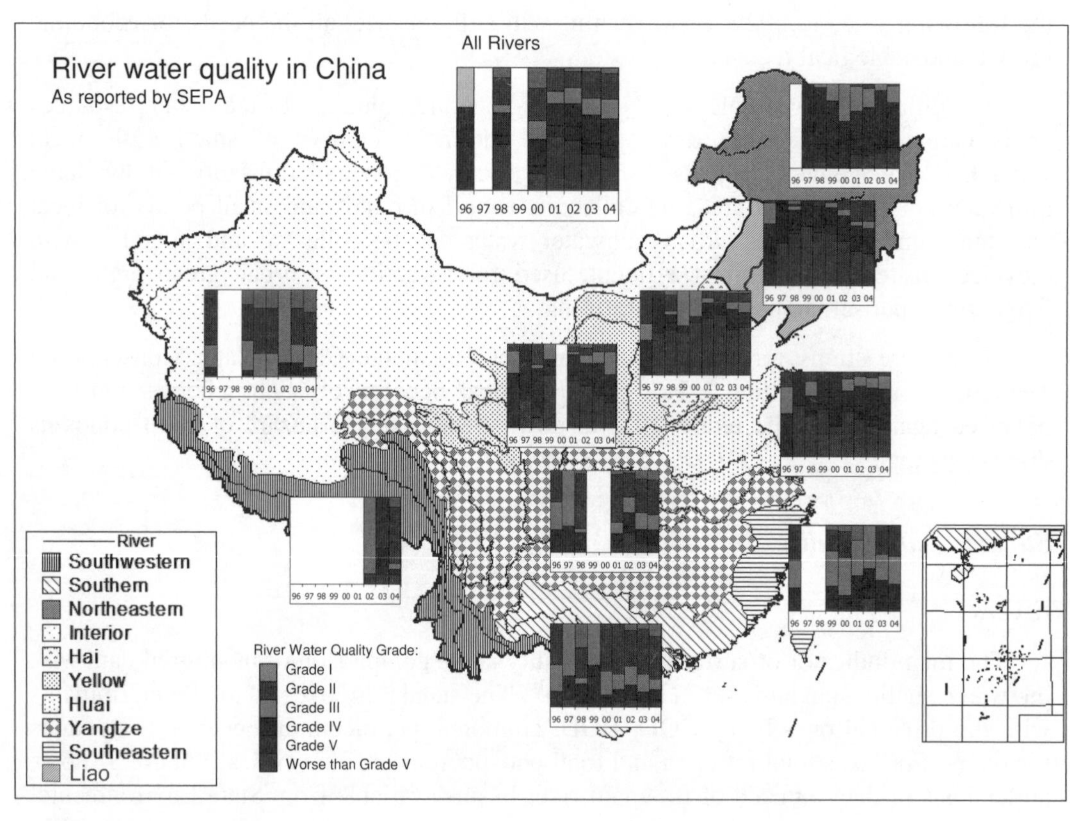

Source: SEPA data.

The 2004 SEPA data are based on reports from 488 monitoring stations around China[18]. The MWR also monitor river water quality and report, in their 2004 Annual

17. From SEPA State of the Environment, 1996, 1997, 1998, 1999, 2000, 2001, 2002, 2003, 2004 English and Chinese language versions – where there are discrepancies then the Chinese language version has been taken.

18. SEPA State of the Environment 2004, Note that there was a major change in numbers of reported monitoring points with around 750 points used nationwide between 1999 and 2002 and around

Water Resources Bulletin[19], a summary of more than 3 200 monitoring stations nationwide. Table 1 shows a comparison of the 2004 Water quality results for all stations for SEPA and MWR together with the 229 provincial level stations of MWR.

Table 1. Comparison of the 2004 water quality results for all stations for SEPA and MWR

Source	Reaches	% in each Grade					
		I	II	III	IV	V	<V
MWR	>3 200 in 1 300 rivers	6.3	27.2	25.9	12.8	6	21.8
SEPA	412 in 7 main river basins	4.6	20.9	16.3	21.6	8.7	27.9
MWR	229 Provincial		34.5		31		34.5

Table 1 illustrates how summary reporting is influenced by choice of river reaches reported and data source. The remoter water source regions of a river basin tend to be better quality thus the MWR data including more of these indicates better overall quality. The SEPA data for major rivers and major tributaries paints a less encouraging picture and finally the MWR data selecting only the major provincial level stations in the main rivers paints a worse picture still of overall river water quality.

Lakes

Both SEPA and MWR report data on the quality of lakes and reservoirs. Table 2 shows the percentage of the number of lakes monitored that meet the defined grades. In most cases the reason for failure to achieve a grade was because of excesses of the nutrients nitrogen and phosphorous, in a few cases failure was also due to excess COD.

Table 2. Data on quality of lakes, 2004

Source		Number	% in Grade		
			I-III	IV / V	<V
SEPA	Lakes	17	18	29	53
SEPA	Reservoirs	10	40	50	10
MWR	Lakes	50	36	26	38
MWR	Reservoirs	332	82	14	4

There does not appear to be a significant trend of improvement in general lake quality in the last few years. For the three key national lakes – Dianchi next to Kunming in Yunnan; Chao, in Central Anhui; and Tai in southern Jiangsu – the water quality data, as presented in Figure 4, do not yet show a consistent trend of improvement despite considerable effort and investment over many years.

400 points from 2003 onwards. No data on number of points before 1999. Before 1997 quality worse than grade V was reported as grade V.

19. MWR 2004 Annual Water Resources Bulletin.

Figure 4. Trends in water quality of lakes Dianchi, Chao and Tai

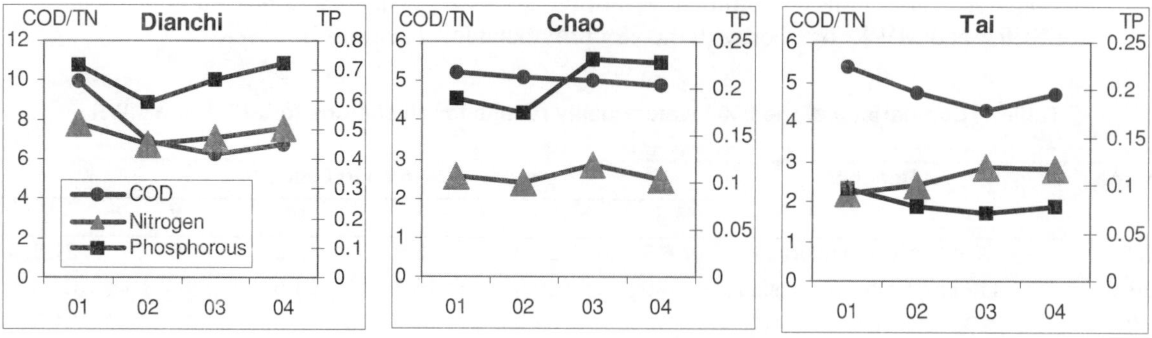

Source: SEPA. Note the change in scales for Lake Dianchi, which is far more polluted than Lakes Chao and Tai.

Groundwater quality

Of the total national groundwater resources, 63% are usable as drinking water without treatment, 17% can be used for drinking water after appropriate treatment, 12% are unsuitable for drinking water but can be used as industrial and agricultural water sources, and 8% can be used as industrial water only after special treatment[20].

Water pollution by agriculture

Agricultural chemicals are widely applied in Chinese agriculture (Figure 5). Both fertiliser production and use are major sources of river water pollution, excess nitrate and ammonium can also be a significant source of NOx air pollution as a result of soil microbial processes. Dust blowing off the land is an additional source of phosphorous nutrient pollution of lakes and the sea. Production and distribution of pesticides and fertilisers are heavily subsidised by the government as an incentive to achieve grain production targets while the efficiency of their use is relatively low. Also there is a lack of understanding among many farmers of the hazards associated with pesticides. Evidence shows that farmers' demand for agricultural chemical inputs is relatively unresponsive to price changes. Additionally, the price signal transferred from farmers to final consumers is very weak, so that farmers would ultimately bear a big part of the burden of increased prices.

20. SEPA State of the Environment 2004. Detailed or regional public data on groundwater quality are not available.

Figure 5. Agricultural inputs, early 2000s

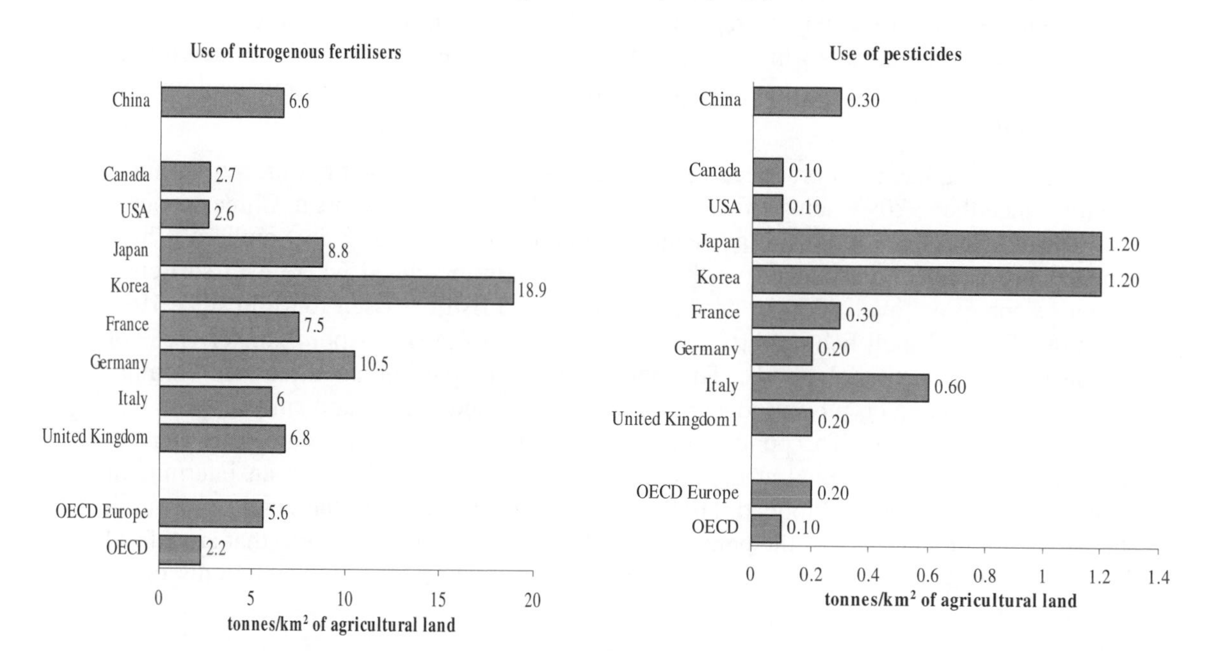

1. Great Britain only.

Source: FAO (2004), FAOSTAT data; OECD Environment Directorate, SEPA "Report on state of the environment in China 2003".

Farmers make decisions on fertiliser use largely based on their own estimations. Much of the application is by hand without mechanical dosing controls. Only a very few farmers have access to training on fertiliser and pesticide application. There is large scope for technical advice and education to play in improving fertiliser use. Policy set out in 11th Five Year Programme is to promote better use of fertiliser by applying them together with training in soil testing to scientifically identify dose requirements. However, the 11th FYP also proposes to increase subsidies on fertiliser and diesel to promote higher productivity.

Details of land use planning influence the amount of polluted runoff entering watercourses. Incorporating buffer areas along watercourses can improve the ecosystems of rivers and lakes and prevent arable and livestock runoff from directly entering the water.

Increasingly consumers are willing to pay for high quality agriculture products. Organic farm products are gaining a market both domestically and especially for export markets where they can attract a significant price premium. Organic farming is also seen as a way of improving the image of an area, attracting foreign investment and providing a more attractive way of life.

A major area of concern for agricultural pollution is the rapidly growing intensive livestock industry. The amount of livestock farming has increased dramatically, since 1990 the production of pork has doubled; beef has increased fivefold; sheep and poultry fourfold; egg output is 3.5 times higher and dairy milk production has increased fivefold, mostly in the last 5 years. Most of this increase has been achieved by switching from livestock free ranging or being in individual pens to large scale intensive livestock facilities. Regulations require effluent management for these, but not all facilities comply with these regulations or operate the effluent management systems effectively. Ironically, effluent control has been cited as one of the drivers towards centralised livestock facilities, in order to keep animals and their waste a distance from human dwellings and water sources. Livestock may be regarded as a diffuse pollution source when outside, or as a rural point source when in sheds and farmyards. Discharges from these facilities to smaller upstream catchments can have a devastating impact on water quality, ecology and sanitary safety.

Fish farming has been practiced in China for thousands of years but has grown most rapidly since the 1970s when artificial breeding methods were developed, China accounts for more than 60% of world aquaculture production[21]. Starting in southern China aquaculture has been introduced northwards and is now practiced where ever sufficient water resources are available; the production of farmed fish has risen fourfold since 1990 such that fish and shell fish production now equals that of pork[22]. Approximately 70% of production is in artificial ponds. The ponds can form part of an integrated farming system, reprocessing crop by-products, algae and fungi and human and animal waste into valuable animal protein. The ponds require a reasonably high quality source of water but are only emptied for renewal at most once a year so are normally only an intermittent source of pollution unless flooded. High levels of eutrophication in the pond water are a benefit to the productivity of the ponds by accelerating the growth of algae that are a food source for the fish; it is common practice to fit mechanical aerators to maintain dissolved oxygen levels.

In coastal areas marine aquaculture of fish and shellfish is also developing rapidly. In these areas eutrophication is a major threat to production. Recent "Red Tides" in the Bohai and East China have caused great economic losses because the marine algal blooms include cyano-bacteria which produce toxins, even if these do not directly kill the fish stock they render them unfit for consumption and they must be destroyed. The construction of marine aquaculture facilities and the subsequent constant emissions from them results in loss and damage of natural coastal habitats, many of which may be vital in the life cycle of open marine harvested fish species.

As an overall source of non-point pollution the contribution by agriculture practices in China is not well understood. If pollution control is to be implemented based on achieving water quality objectives by rational controls on all discharges and abstractions then the contribution of different agricultural land uses will need to be calculated. These difficulties in understanding and managing farm pollution are certainly not unique to China; a major feature of the European water framework directive is the monitoring, research and institutional structures required to bridge this knowledge gap. Hopefully China, the EU and other OECD countries will be able to learn together how to address such a problem. New management methods and institutions may be required as farming practices become more industrialised and so start to cause pollution on even larger scale.

Flood and drought management

Flooding

China is one of the countries most hit by severe flood disasters. Fast socio-economic development and high population density mean that the consequences of flooding in flood-prone areas have substantially increased. Since 1990, national average losses

21. FAO, Review of the state of world aquaculture, 1997.

22. Ministry of Agriculture website data, 2004.

resulted from floods amount to RMB 110 billion each year, equivalent to 1% of the national GDP in the same period[23]. Since 1906 more than 6.5 million people have died in China's floods.

Flood risks are mitigated in three ways: i) infrastructure such as dams or sacrificial flood plains to reduce the development of flood waves along rivers; ii) levees and flood walls to keep flood waters contained in the river channel and protect assets; iii) building flood resilient infrastructure, warning systems and emergency plans so that when floods do occur the damage is minimised and lives are not put at risk.

China now has more than 85 000 dams with flood protection functions[24] (normally combined with water resource and power generation). More than 465 million people in China live behind flood protection systems. The Three Gorges dam is designed to help protect the heavily populated Yangtze plains from potentially catastrophic flooding. When complete in 2009 it will raise protection on the downstream plain from 1 in 10 years to 1 in 100 years. The last major period of flooding was in 1998, more than 1 000 people died and damage cost hundreds of billions of Yuan. Many thousands of kilometres of flood embankments have been constructed along major rivers around China and the older embankments are being raised and reinforced to provide higher levels of protection – typically 1 in 100 years flood protection for major urban areas and 1 in 50 years for peri-urban and rural areas. The Yellow River, which was historically subject to frequent devastating floods, has not flooded for more than 50 years partly as a result of flood prevention measures.

In recent years large areas of flood plains that had been drained for agriculture and protected from flood have been re-designated as flood plains and protection removed so that they can once more act as flood storage so reduce flood risks downstream. Where agriculture still continues in these areas the farmers are made fully aware of the risks and warning and evacuation plans are put in place. Compensation schemes are being developed. Flood warning monitoring systems are being extended and flood insurance is being introduced.

Many areas remain still at risk of flooding as China's climate, with rain concentrated in a few months, makes flash flooding very common. The south and East of the country are subject to typhoons in the summer. These cause extensive damage but far less loss of life now than they used to because of development planning to avoid construction of unsuitable buildings in vulnerable locations and timely evacuation procedures.

Droughts

Historically, droughts are second only to floods as the natural disaster resulting in deaths in China, having claimed some 3.5 million lives since 1906. The death toll due to drought in the last few decades is dramatically reduced with national management of storage and distribution of grain. However economic losses due to drought occur frequently and, as water resources become more and more stressed, so areas become more vulnerable to climatic changes.

23. Speech by H.E. Dr. Jiao Yong, Vice Minister of Water Resources, People's Republic of China, (22 November 2005 - New Delhi).

24. Total storage 518 billion m^3, Including around 400 large reservoirs storing 63% and 2 700 medium reservoirs storing 14% and the remaining 23% in more than 80 thousand small reservoirs (2 003 MWR water resources in China report).

Soil erosion, deforestation, overgrazing and other factors have lead to desertification in areas of China, especially in the north and west. These processes increase drought risks. Strong measures are being taken by the government to mitigate desertification factors and protect key population and farming regions.

Chinese authorities are obliged to prepare drought management plans for their administrative regions. The plans for water rights management and abstraction permitting are being developed to incorporate mechanisms to fairly restrict and redistribute water resources under drought conditions to ensure that vulnerable users do not lose out disproportionately to large industrial and urban users, as can be the case at present.

Main policy and legislative instruments for water management

The process of delivering the policy objectives is threefold, based on development of i) setting targets and investment planning through the five year development plans, ii) legal framework and regulatory instruments, iii) economic instruments and iv) institutional framework.

Environmental planning

The Chinese environmental authorities have developed Five-Year Environment Plans to correspond with the Five-Year Social and Economic Development Plans. They are supported by sectoral five year plans, including for water management in key rivers and lakes hazardous waste management. The Five-Year Environment Plans have also constituted the basis for sub-national governments and their EPBs to prepare their own five-year environment plans.

The Eighth Five-Year Environment Plan (1991-1995) included various provisions to strengthen environmental management. The Ninth Five-Year Plan for National Environmental Protection issued in 1996 acknowledged worsening ecological destruction and called, explicitly for the first time, for establishing environmental management and legislative systems and curbing pollution. The provisions of the plan were supported by the State Council Decision on Problems Regarding Environmental Protection issued in August 1996. The Decision set forth 10 mechanisms for achieving plan's environmental objectives, many of which are still in force. These measures included, *inter alia*, the "33221" programme which designated three rivers: Huai, Hai and Liao, three lakes: Chao, Diao and Tai, two air pollution control zones: Acid Rain control zone and sulfur dioxide (SO_2) control zone, and one municipality: Beijing, as key environmental control areas. The Decision also called for stricter inspection of firm pollution control through the "three synchronizations"[25] system. China subsequently began a number of environmental initiatives, as for example, "one control and double attainments" programme[26].

The Tenth Five-Year Plan for National Environmental Protection, announced in 2001, aimed at further improvement of the quality of the environment in cities, rural towns, and particularly in large and medium-sized cities by 2005. It set new targets based on progress in the previous period in various areas including for '33221' programme, the Three Gorges Reservoir Area, the Bohai Sea, environmental protection in urban and rural areas and nature protection. Key targets are presented in Table 3.

The 10th 5 Year Plan also envisaged a number of institutional and regulatory measures, such as strengthening decision-making system for integration of environmental matters into economic development, and capacity of environmental management institutions, strengthening environmental regulations and the use of incentive based instruments and strengthening enforcement towards enterprises responsible for severe pollution that damages public health. The programme also called for enhancing research and development efforts for environmental protection, promoting environmental goods and services industry and promoting public participation in environmental policies. Finally the plan provided that the total demand for investment in environmental protection would achieve RMB 700 billion accounting for 1.3% of GDP, and about 3.6% of the total fixed investment in China. The Plan identified a list of 1 137 projects (know as China Green Project Plan – Phase II) most of which have been addressed in some way. It was estimated that these projects would require CHY 262 billion, about 37% of the estimated financial requirement for the 10th environmental 5 Year Plan.

25. The system of "three synchronizations" (also called the "three Simultaneous Steps") that was introduced with the 1989 Environmental Protection Law requires that: *i)* the design; *ii)* the construction; and *iii)* the operation of a new industrial enterprise (or an existing factory expanding or changing its operations) must be synchronised with the design, construction and operation of an appropriate pollution treatment facility. Once the construction of the project is completed, inspection and approval by environmental administrations are required (for large projects, or in case of a dispute at the local level, the approval has to be confirmed by the national level authority). If project operations begin without the approval from the local EPB, the owner of the project can be sanctioned. In many instances though, the sanctions have not been applied and there are many departures from the above-mentioned procedures, especially by many TVEs.

26. The "one control"(一控) required that the total emission loads of major pollutants in all regions nationwide should be kept within nationally specified levels (1995 standards). The "double attainments"(双达标) required that emissions from all industrial sources nationwide should meet both national and local standards by the target date.

Table 3. Examples of environmental targets and indicators in the Tenth Five Year Economic and Social Development Plan (2001-2005)

Issue	Specific targets and indicators
Environmental protection in river basins	Total elimination of the state-controlled sections with the lowest quality (grade V) of water in major rivers.
Urban environmental protection	Meeting relevant water quality national standard by all centralised potable water sources in urban areas. Achieving 45% rate of centralised sewage treatment in urban areas with the daily treatment capacity of 40 m tonnes; Increase the rate of urban waste treatment to 50% (with an annual treatment capacity of 55 m tonnes - or 150 kt/day). Increasing the number of cities targeted for pollution control to 100 from 47 and reaching in all of them applicable standards for air, water and noise
Total pollutant discharge control	Reduction of the total COD discharge to the levels of 12.5 million tonnes.
Nature protection	Increasing the total number of nature reserves to 1 200 with the area of 11.2 million hectares. Increasing the percentage of land designated as nature reserves from 10 to 13% of the total land area. Achieving the rate of forestry coverage to 18.2% of the total land area.

Source: SEPA.

The 11th Five-Year Environmental Plan (2006-2010) is under preparation at the time of writing but the overall Social and Economic Plan identified some environmental targets. Improving resource efficiency, water conservation and pollution control are key aspects of the "scientific concept" of development and have been incorporated into the 11th Five year plan[27]. Detailed guidance on techniques for water conservation to be incorporated to the development of the five year plan were published in the China Water Conservation Technology Policy Outline[28]. However, some of the targets in the 11th Five Year Social and Economic Development Programme (Box 1), *e.g.* on reducing pollution loads and energy efficiency, do not seem sufficiently ambitious in relation to the overall policy objectives of Chinese leaders.

Box 1. Environmentally-related targets in the 11th Five Year Social and Economic Development Programme (2006-2010)

- Total discharge quantity of major pollutants reduced by 10%.
- Water consumption per unit of industrial value-added to be reduced by 30%.
- Water for irrigation in agriculture maintained at current levels.
- Area of farmland to be retained at 120 million hectares.
- Forest coverage to reach 20%.

At the end of 2005, and in the wake of the Songhua river chemical spill, the Central Committee of the National People's Congress judged that the contradictions between socio-economic development and environment had become more obvious and that environmental and other natural resources had "seriously restricted" economic development. Shortly after, the State Council adopted a Decision on "Implementing the

27. The 11th Five-Year Plan: Targets, Paths and Policy Orientation, 23 March 2006, Ma Kai NDRC.

28. China Water Conservation Technology Policy Outline, 21 April 2005, NDRC, MOST, MWR, MOA, MOC joint publication.

Scientific Concept of Development and Strengthening Environmental Protection" (Decree 39 of 3 December 2005). The State Council explained that the rationale for this action was the need "to place environmental protection in a more strategic position." The Decision presents reasons why greater emphasis should be placed on environmental protection (to realise faster and "better" growth; to foster economic activities related to environmental protection and the related employment opportunities; to improve the ethical basis of Chinese society; to protect human health and the quality of life; and to maintain the resource base for present and future generations); identifies a range of environmental priorities; and proposes ways in which the environmental policy and institutional framework should be strengthened, including a number of new measures to strengthen the implementation of environmental policy. SEPA and the Ministry of Supervision are directed to jointly review implementation of the Decision and to report to the State Council annually. With respect to water the Decision sets the key objectives as:

- Strengthening drainage basin management.
- Defining and protecting water sources zones.
- Defining and managing trans-provincial water pollution.
- Planning economic development to suit water resource conditions.
- Building water saving agriculture.
- Investing to provide 70% wastewater treatment in urban areas by 2010.

Legal framework and regulatory instruments

The key laws related to water management include:

- Environmental Protection Law (1989) which provides obligation on government and industry to protect the environment, control discharges and monitor environmental quality.
- Water Law[29] (2002) which is a comprehensive update of the original 1988 Water Law. It sets a framework for water management including, integrated water resource management of river basins based rational assessments of flows in rivers, lakes and groundwater with quantity and quality management coordinated between institutional organisations with obligations to collect and share data with each other and with other stakeholders. Water is identified as an economic commodity with provision for water resources fees and water supply pricing to recover the cost of services and administration. Government organisations at all levels are required to encourage the establishment of and coordinate activities with stakeholder groups such as Water User Associations, who will have defined rights of access to water resources through a water permit system, ownership of water infrastructure and mechanisms for conflict resolution. A system for defining river reach functional use zoning is put forward and detailed in secondary legislation. The law aims to support social change as means of water demand management and to require integration of water resource and economic development planning. The law supports existing and upcoming secondary legislation at national and provincial levels.

29. Water Law produced primarily by Ministry of Water Resources.

- Law on Prevention and Control of Water Pollution[30] (1996) sets obligations on the institutions at different levels of government to manage pollution discharges to the aquatic environment. It also provides the framework for industrial and municipal discharge pollution controls and levies. Detailed standards for discharges of listed pollutants from different industry types are defined in the 1996 Regulations on Integrated Wastewater Discharge Standards. Standards for ambient water quality are set in the 2002 Regulation on Environmental Quality Standards for Surface Water.
- Law on Flood Control (1997) which sets framework for flood management, guarding against and mitigating damage done by floods and water-logging, preserving the safety of people's lives and property, and ensuring flood defence is integrated to the national planning process.

The laws are implemented together with secondary legislation and regulation and supported by the State Council decrees. The above laws are enacted in each province through incorporation of the provisions into the provincial laws. This process can result in some regional variations on interpretation and implementation. The Laws do not generally set clear timeframes for implementation, these timeframes are dependent on the political will for institutional reform and the timeframes set in the national and provincial five year plans.

Though China has a relatively comprehensive high level water environmental framework there are many conflicts, ambiguities and institutional overlaps in the implementation of the laws and secondary regulations. This is especially apparent in the critical area of pollution discharge control where there are both concentration based standards for individual discharges and total load based standards for all discharges in a region but these two methods are not clearly coordinated. Discharge standards also do not integrate in any specified way with the ambient water quality standards and objectives.

Water quality objectives for each major reach of a river are defined on the basis of five categories of chemical quality related to suitability of water for use as a drinking source, recreational resource, aquatic ecosystem support or source for industrial or basic irrigation water. Both the Environmental Protection Bureaus and the Ministry of Water Resources have responsibility for defining the objectives for each water body and for monitoring compliance with objectives, reporting ambient quality and planning interventions. They generally each carry out these activities separately based on slightly different guidelines.

Annual summary statistics of water quality data are published each year by each of the agencies and made freely available on their websites in both Chinese and English. These reports aggregate the data to river basin levels and national summaries with provincial summaries for some factors.[31] However, unlike in most OECD countries, the underlying processed data for each monitoring point are not readily available. These may normally only be obtained by formal application and security approval or upon payment of substantial fees. This is a barrier to transparency in environmental management and limits stakeholders' ability to understand and plan their interaction with the environment and with other users of environmental services.

30. This law was produced and implemented primarily by SEPA.

31. As used in the compilation of this report.

The Chinese Minister for Water Resources, in his speech to the 2003 World Water Forum in Kyoto, stated: "*…Access to clean safe water is a basic right of human beings and foundation of good health…*". Through the 2002 Water Law and pilot initiatives in the Yellow river, Tarim basin and elsewhere authorities in China are developing their system for clarifying and administering water rights. The aim is to formalise and stimulate entrepreneurial activity in water management whist adding monitoring and regulation to promote efficiency, responsibility to other users, economic sustainability and avoid free riders. The approach and principles are described by the Vice Minister of Water Resources in 2005:

"Regarding to water shortage problem, establishment of a water-saving society in China is considered as the ultimate measure to resolve conflict between water supply and demand. To do so, we are developing a mechanism of water saving, in terms of allocation of water right in each river basin to deferent local authorities, in condition that the basic environmental flow must be reserved. Maximum water use availability will be calculated for provinces in the river basin, which is considered as a "macro-control indicator" to confine the upper limit of water use of the provinces. Within the upper limitation of water availability, provinces can further allocate water right to different areas, industries, and water users so as to the initial water rights can be specified. In addition, a "water consumption index system" will also be established as a water saving indicator to measure water saving level of a water user such as an irrigation system or a manufactory. In this way, we hope to enhance the awareness of water saving and protect water resources from adverse utilisation in the whole society, which will help us to achieve sustainable use of water."[32]

Many pilots of water rights management schemes are being undertaken around China, for example, with notable success by the Yellow River Basin Authority and Tarim Basin Authority and now for further dissemination through the MWR led Water Resources Demand Management Assistance Project.[33] These draw on international experience but adjust to the particular conditions of China to establish an active and flexible system.

Economic instruments

Pollution charges

Charging polluters for their emissions and discharges, which is the basis of the "polluter pays" principle, was introduced in China with the promulgation of the 1979 trial Environmental Protection Law. The system was also stipulated in others laws, including the Law of Water Pollution Prevention and Treatment.

Pollution charges in China links an economic incentive for pollution reduction with sanctions in case of non-compliance. A nationally-uniform set of water pollution discharge standards and levies was designed by the State Council in 1982 and revised in 2003. Originally only discharges that exceeded pollution concentration standards were subject to a fee. Currently levy formula incorporates both concentration and volume, as it calculates a pollutant-specific discharge factor based on both total load of pollutants in waste water discharge and the degree to which each pollutant concentration exceeds the

32. Speech by H.E. Dr. Jiao Yong, Vice Minister of Water Resources, People's Republic of China, (22 November 2005 - New Delhi).

33. See www.wrdmap.com.

standard. Charges are levied on 18 pollutants specified by national discharge standards. The charge rate and the discharge factor are set by central government. And concentration standards are set jointly by national and local governments.

Polluters are required to register with local environmental authorities and report increased discharges. Rebates are possible when pollution reductions are verified by environmental authorities. Although the EPBs issue notices with the amounts of due discharge fees, the amount is generally negotiated rather than calculated using formulas detailed in regulations. The levy can be reduced or even eliminated at the discretion of local regulators after appropriate inspections. The levy may also be postponed if the polluter cannot afford to pay it, although reductions or exemptions are not allowed in such cases. Such discretion introduces considerable variation in regional enforcement practices.

The current pollution charge system is comparatively mature and effective though some problems still exist. In particular the fees are still by a large margin lower than the cost of pollution reduction, despite the recent increases of the rates. In many cases the actual levy paid by a firm is the result of bargaining between the administration and the firm. Some surveys show that state-owned enterprises pay lower effective rates than privately owned and that levy rates are positively related to firm profitability. In addition the collection rate of the fees is still low, estimated to be on average at only less than 50% of the charges imposed (varying between 10% in Western provinces to 80% in coastal areas), which reduced their incentive effect. Raising pollution fees would not only spur more pollution abatement, but also become a driver for a local environmental goods industry.

In 2004, the total revenue amount collected from pollution levy was RMB 9.42 billion from nearly 740 000 enterprises. This included RMB 3.05 billion collected from levies on exceeding concentration standards and the rest from other penalties, charges for emissions of SO_2 and levies on wastewater volume discharges.

Even though the revenue from collected charges is now transferred to the Ministry of Finance, which submits expenditure proposals to the National People's Congress, the resources are still earmarked for environmental improvement. However, the revenue is no longer partially used to defray the running expenses of the inspection agency, but now is directed towards: i) general environmental protection; ii) purchase of monitoring equipment; and iii) new technology. Of total revenue from the fees, 10% is transferred to the central government and 90% remains at the sub-national level. The revenue is use in a form of a subsidy or loan interest-rate support for pollution control projects. SEPA and the Ministry of Finance allocate funds on the basis of proposals provided by the provinces. In 2005, about RMB 0.8 billion of the pollution levy revenue was allocated of which 48% was used for regional/river basin level comprehensive pollution control projects, 23% for demonstration or pilot of new technologies for pollution control and the remaining 29% for capacity building and strengthening for environmental monitoring. As the scheme started operating in mid 2003, there has been no comprehensive analysis on the use of pollution levy revenue.

User charges

Farmers are charged a water abstraction fee, typically in the range of 0.02 to 0.25 RMB/m^3. In many cases, however, the water use for agriculture purposes is free of charge. In major river basins such as the Yellow and Yangtze the water resource fees for abstractions from the main rivers are managed by the river basin management authorities. However the tributaries and catchment areas generally still fall under the administration of municipal or county level authorities. With the abolition of agricultural taxes in 2005, there may be more flexibility to implement more effective irrigation fees. It is the case in many irrigation districts the fees charged to the farmers are much less than the cost of providing the water. Most irrigation supplies are not volumetrically metered and management systems are vulnerable to abuse of commons with those taking more than their share benefiting without sanction. The charge system and its management (including fee collection and allocation of water rights) is currently being reformed and strengthened under the implementation of the 2002 Water Law.

The Chinese authorities also plan to establish a comprehensive ecological compensation system which would include setting up tradable quotas for water pollution to encourage inter-regional, inter-industrial and upstream-downstream trading. This is expected to increase the flow of funds from heavily populated water using areas and industries to remoter areas with a water surplus. Pilot projects in Zhejiang as well as Ningxia and Inner Mongolia have already proved successful and will be replicated in other regions.

In rural areas water user associations (WUA) may be formed to manage the irrigation and water supply on behalf of the community. They can provide a means to revive district irrigation schemes. They are able to take ownership of the assets and to set, collect and spend water resource fees. WUAs are being established more widely with the implementation of the 2002 Water Law. However, during their establishment the WUA's have to overcome the obstacle of having to charge and pay realistic prices for water, likely to be much higher than past arrangements. Significant government support for investment, education and training is required to overcome the barriers to establishing effective water saving rural communities but examples from the Tarim Basin and Yellow River Basins have demonstrated that this can be successfully achieved leading to greatly enhanced livelihoods[34].

Institutional structures

The administrative structure of functional departments in China is one of double subordination with vertical management and guidance from the line ministry but horizontal management of finance and administration from the corresponding level of local government. Figure 6 illustrates this for the Ministry of Water resources. Corresponding structures exist for the SEPA and its provincial administration (Environmental Protection Bureaux – EPBs) as well as other ministries.

In some regions at municipal levels the management of water supply and water resources may be carried out by a Water Affairs Bureau as a united authority between Water Resources and Construction Ministry Bureaux[35]. For the seven major trans-provincial river basins, authorities to coordinate basin management have been appointed by MWR with representative offices at provincial and state levels. Smaller intra-province river basins may have coordinating basin authorities at provincial levels. These basin

34. See Tarim Basin II Project, World Bank project report, 27 June 2005.

35. Especially in larger cities such as Beijing and Shanghai. In medium sized cities management is most often by construction bureau and in rural areas and smaller cities by an MWR based water affairs bureau.

management authorities have the duty of resolving conflicts between upstream and downstream users in different administrative jurisdictions.

Figure 6. The vertical and horizontal structure of the Ministry of Water resources and its supporting and influencing institutions

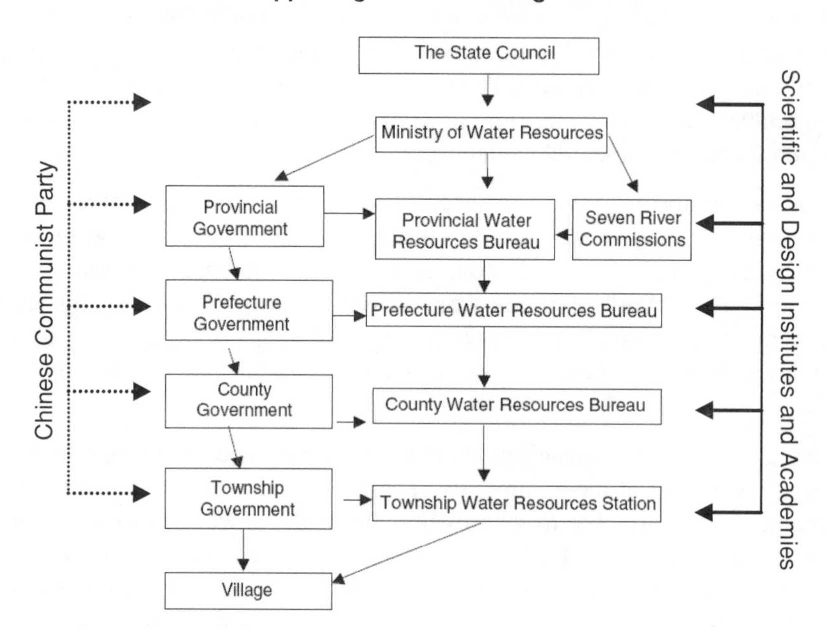

At each administrative level scientific and engineering institutes provide monitoring, surveying, research, policy development and engineering design these "undertaking units" are mostly now nominally independent companies. Policy priorities, performance assessments and career progression of officials against specified targets are affected by the cadre responsibility system which is managed by the Communist Party. This system strongly influences the priorities in administrative decisions, assessments are currently mostly based on targets for economic growth, population management and public order, the system is partly, but far from entirely, quantitative.

Future water demand trends in China

In the late 1970s and early 1980s, it was the countryside that saw the greatest benefits first, with rapidly rising incomes and opportunities. But as continuing development has become more focused on manufacturing and service industries the development is led by the cities and incomes in rural areas lag far behind. Addressing this imbalance is a priority area for the 11th Five Year Programme period.

One of the most significant transitions in Chinese history is currently underway, a huge population migration from the countryside to the cities. This will have far reaching implications for water resource use and pollution. This migration is currently happening at about 1.5% of the rural population permanently relocating to cities, 20 million people each year. In addition a much larger number of people from rural areas seek temporary work in the towns and cities. 30 to 60% of rural families have one or more members working away. Though some women or whole families travel in most cases it is men that work away. Overall this has lead to the feminisation of large areas of the countryside.

Even when families move away they are obliged to keep farming their contracted land[36]. If left fallow then they will be subject to fines then forfeit the land to the state. Therefore they must subcontract the farming to other families, thus, though the statistic of land per household may remain constant, the land actively farmed by each household may be much larger and is increasing year on year.

Registration requirements and strong population controls and migration controls have limited the migration process but it is progressing inevitably forwards. By 2030 it is expected that probably 60% of the population will be urban. This process will greatly change the balance of pressures on water resources and will also affect the methods of agricultural production as fewer people will be expected to produce more food. These trends could potentially be beneficial as urban residents and industrial activities use only a fraction of the water resources of farming and the remaining farmers should have available a greater share of land and capital to invest in improved water efficient farming methods and associated transport and market infrastructure. However there is also the risk of increased conflict between urban and rural communities over access to water resources.

Table 4 summarises projections of water demand as of 2000. The projected negative growth rate in agricultural water consumption is attributed to continuously improving irrigation technology. The projected growth rate decrease in industry is primarily due to improved manufacturing technologies.

Table 4. Projections of water demand (as of 2000)

Year	Agriculture			Industry			City and town			Total (billion m^3)
	Amount (billion m^3)	Growth Rate	Pro-portion	Amount (billion m^3)	Growth Rate	Pro-portion	Amount (billion m^3)	Growth Rate	Pro-portion	
2010	465	–0.13%	78%	93	3.6%	16%	27	2.7%	4.6%	585
2030	453	–0.43%	65%	190	3.0%	28%	46	2.4%	7%	689
2050	416		50%		344	41%	73		9%	833

Source: Chinese Academy of Science, "Analysis of Water Resource Demand and Supply in the First Half of the 21st Century", China Water Resources (January 2000). Reported in: Water Supply and Wastewater Treatment Market in China, US Department of Commerce, International Trade Administration, 2005

Conclusions

As water resources are fundamental to economic development and social health, they should be managed in the context of the river basin system with incentive and pricing mechanisms that reflect the real costs and benefits involved in water consumption, pollution and rehabilitation. These mechanisms can drive industrial and agricultural development towards a sustainable level of exploitation balancing the needs of economic activity and wealth creation with the need to maintain environmental and ecological services, manage risks and protect against catastrophe.

Though China has a large water resource in absolute terms it is low on a per capita basis and unevenly distributed both geographically and seasonally. Overall China has about one quarter the world average water per capita, falling to one tenth in northern and western areas. Severe over utilisation of this small resource has resulted in scarcity and

36. Peasant farmers do not "own" land but lease it from the local authority for a 30-year period; most leases were let in the late 1990s following the introduction of such leases in the 1998 Land Law.

inequitable distribution leading to a major restriction to economic development and to the health and livelihoods of the population.

In almost all areas of China water courses, lakes and coastal waters are severely polluted as a result of agricultural, industrial and domestic discharges. This has resulted in severe degradation of aquatic ecosystems and is a major threat to human and animal health. It also restricts the utilisation of water without treatment and holding back economic growth, especially in the poorer, more disadvantaged regions.

With surface water scarce and polluted, demand for groundwater far exceeds the rate of replenishment in many areas and levels are falling rapidly in both rural and urban areas. As this stored resource becomes exhausted it becomes impossible to maintain the high and inefficient levels of agricultural and urban water consumption. The current massive investment in the development of distant sources and water transfer infrastructure may not be the most optimal approach if not associated with water extensive saving and conservation measures.

China's leadership and administration are aware of the serious nature of its water resource situation. A reasonably comprehensive legal framework for water resource and pollution management has been developed over the last two decades with clear standards to control discharges and abstractions and water quality objectives. The new Water Law of 2002 opens the way for integrated river basin resource and quality management, stakeholder participation and economic market mechanisms in water management.

Over the period of the last five year plan, total pollution loads discharged to watercourses through industrial and urban wastewater have been reduced in many areas, representing a decoupling of economic growth and pollution discharge. In the last three years the reported water quality of many rivers has improved slightly.

Also, enormous investment has been made in infrastructure to protect against flood damage. Flood risks have been reduced in many areas and communities are more informed about the risks they face. Planning laws are being strengthened to prevent further development on flood plains and there has been some return of reclaimed areas to flood storage functions.

However, China has very fragmented institutional arrangements for the management of water resources, water quality, pollution discharges and water supply and treatment infrastructure. The jurisdictions of institutions often overlap and conflict. Data sharing and coordinated planning is restricted. Though the planning, monitoring and enforcement of discharge standards and rates of abstraction has improved, far greater strengthening is required to collect, store and distribute reliable monitoring data, enforce standards and achieve objectives. In particular environmental inspectors have limited powers of access and may be subject to intervention by local interest groups which may compromise their independence in enforcing standards and ensuring that pollution control measures and equipment are fully utilised and the environment protected from accidents.

In some areas the criteria used to assess the performance of government departments and officials are incorporating water resource utilisation and pollution reduction targets. These are used in addition to the economic growth, primary production and population control targets commonly set but this is not common. Broader application of such measures could be very effective in motivating the administration to change to more environmentally considerate activities.

Chapter 3.

EFFECTS OF INTEGRATED ECOSYSTEM MANAGEMENT ON LAND DEGRADATION CONTROL AND POVERTY REDUCTION

Han Jun
Research Department of Rural Economy,
Development Research Centre of the State Council

Abstract

This paper focuses on a major concern that not only applies to China, but applies universally; namely how to achieve long term ecological improvement whilst addressing the problem of meeting short term needs of rural livelihoods. In the context of the People's Republic of China (PRC), Integrated Ecosystem Management requires: (i) cross sectoral planning; (ii) building human capital; (iii) protecting land users legal rights; (iv) improving policies for poverty reduction; and (v) promoting locally appropriate land management technologies.

Key points

The two most serious types of land degradation in China today are soil erosion and desertification. Soil erosion mostly occurs in the steep mountainous areas and on the Loess plateau. Desertification is mainly distributed in areas of deteriorating grasslands.

China's desert area has now reached a total area of 2.62 million km^2, expanding at a rate of 6 700 km^2 per year. The newly desertified areas in the west account for 9% of the nation's territory.

The volume of soil erosion in China totals 5 billion tonnes per year, two third of which comes from the Western Region.

The occurrence and severity of poverty are closely related to the ecological environment. Presently, over 90% of the population for the rural poor are living in a poor ecological environment, and the Western Region inhabitants are trapped in a vicious circle of ecological deterioration and aggravated poverty.

Many of China's most poverty-stricken areas are in regions where the ecological environment is under serious stress. Ecological reconstruction and environmental protection policies should simultaneously protect the Western Region's ecosystem and lift farmers in the region out of poverty. Therefore, the western development strategy should be adjusted to include ecological environmental protection as one of its primary goals.

There are a number of obstacles in the PRC's current systems and policies that deal with the land degradation and environmental deterioration. The predominant issues are a serious shortfall in investment necessary to deal with land degradation and implement environmental protection, and the absence of a stable mechanism for investment.

The application of integrated ecosystem management in ecosystem development

Integrated ecosystem management (IEM) is a cross-sectoral approach for integrated management. It is a new project framework created by the UN's Global Environmental Facility (GEF) under its Operational Program 12 (OP12) for integrated management to improve global ecosystems, particularly measures to address land degradation in dryland ecosystems. The first application of IEM in the PRC indicates that ecosystem development is a priority of the Great Western Development Strategy. Implementation of the strategy should follow international norms for sustainable development.

The overall objective of GEF OP12 project is to promote sustainable development of western China and protect the global environment by combating and mitigating land degradation in dryland region ecosystems. Under the national framework, the project will support the implementation of a 10-year plan to combat degrading ecosystems in Western China's dryland regions. The project emphasises integrated management and communication, and co-ordination in policies, laws, planning and activities among sectoral and administrative authorities by establishing partnerships to take unified actions to achieve integrated management of land degradation in the western region.

The project adopts middle and long term planning methodologies instead of individual project plans, providing sustained support for the development and ecosystem improvement of western China. The project also creates a great opportunity to help China to establish an integrated, cross-sectoral and cross-administrative regions management system for environmental resources. The implementation of the project provides a platform for all stakeholders and international organisations to participate in the development of the most optimum options for IEM.

GEF OP12 is not only a new concept but also a new challenge to adopt IEM in ecosystem development in China. To ensure IEM's success in China, the following points should be addressed:

- The Co-ordinating Group, comprised of representatives from different sectors and administrative regions, should develop a unified objective and plan to combat land degradation. They should also undertake to achieve closer co-ordination of policies, project and budget at all levels, and put forward recommendations for the plans of relevant agencies in land, agriculture, forestry, water conservancy, environment protection, poverty reduction, science and technology. They should also encourage administrative regions to improve the co-ordination among these agencies and administrative regions in order to improve the efficiency of such publicly funded projects. In the process of planning, attention should also be drawn to combining resources of agencies and administrative regions.
- A stable channel for government financial investment must be established, consolidating, combining and better co-ordinating existing projects and funds. Co-ordination of agricultural investment from different channels should be consolidated. The management of financial investment for agricultural infrastructure should be strengthened to avoid (i) the duplication of projects and investment and, (ii) the unnecessary scattering of funds, which increases overhead costs. The functions and responsibilities of all competent agencies should be well defined to ensure the effective allocation of agricultural investment.

- All relevant sectors and administrative regions should develop and improve policies, laws and regulations in IEM. This should include the provision of relevant policies and regulations to encourage the private sector, local communities and households to participate. Priorities should include:
    - Combating land degradation.
    - Developing strategies and policy systems to deal with a lack of knowledge regarding approaches to poverty reduction.
    - Developing an environmentally friendly poverty reduction strategy.
    - Developing an environment protection strategy conducive to poverty reduction.
    - Encouraging a poverty reduction strategy with broad participation of governmental agencies.
    - Developing a strategy which includes participation by civil society and poverty-stricken populations.
    - Developing the policies and regulations to compensate individuals who efficiently use resources and protect the environment to others' benefit.
- It is important to learn the experiences and lessons of IEM both domestically and abroad. It is extremely important to improve co-ordination among relevant sectors and administrative regions to avoid duplication of planning and investment.
- The advanced technologies and practices in combating desertification and land degradation at home and abroad shall be summarised and publicised.
- The success of IEM depends on the participation of all stakeholders. The local farmers, *i.e.* the beneficiaries, should be the main group involved in overseeing management restructuring. Thus, farmers should be encouraged to participate in the decision-making process, planning, management and supervision.

Constraints to the adoption of IEM in China: some observations

China faces severe ecological and environmental problems Past efforts have not always yielded good outcomes. The fact that governmental departments do not have clearly defined duties, that different policies and objectives are being pursued by different ministries and departments, different projects are managed by different departments, and funds and efforts are diverted from various approved purposes does not help. There is no unified system of monitoring, management and co-ordination. The efforts on the treatment of land degradation are often decided from a top-down approach and carried out in the mode of engineered construction, with little consideration of the environment or the human dimension. There is a lack of wide involvement of stakeholders, let alone the people – farmers and rural citizens – that are directly affected.

The enormous effort made by the Chinese government to solve land degradation has not reversed the trend of partial improvement and overall intensification

The Chinese government has focused its attention on combating land degradation and desertification, and has implemented a series of ecosystem protection projects

Since undertaking water and soil erosion conservation in 8 regions beginning in 1983, China has increased investment in soil conservation and erosion control. Several projects aimed at combating water erosion have been launched. Specifically, projects in the middle and upper reaches of the Yellow River and an ecological project for water and soil conservation along the Yellow River and on the Loess Plateau have been financed by World Bank loans. Since the 1980s, China has concentrated efforts and funds on combating water and soil erosion in Western China, with conservation, restoration and rehabilitation of natural ecosystems among the main tasks. All these efforts have greatly contributed to the improvement of local ecosystems.

To further contain the deterioration of ecosystems in the west, China has implemented other projects such as the natural forest protection project, afforestation project, and natural grassland ecosystems improvement project. The total investment on natural forest protection during the period from 2000-2010 will be RMB 96.2 billion, of which 78.4 billion will be sourced from the central government and 17.8 billion from local governments. During the period from 2000-2010, the afforestation investment is planned to be RMB 342.8 billion from the central government. By 2003, more than 6.6 million hectares of farmland have already been planted under afforestation project, and western China has taken the lead, planting more than 4.33 million hectares. According to the Regulations on Afforestation, the priority should be given to conversion of farmland on slopes greater than 25 degree, and farmland of slopes between 15-25 degree in regions of high ecological value, as well as severely degraded farmland. Since the implementation of the project, more than 12.4 million households (17% of rural households) of 52 million people (18% of rural population) have benefited from the project. By 2003, the central government has invested RMB 34.12 billion in the west region under this project, and increased total investment by 68.8%.

Table 1. Total tree planting under the project of conversion of farmland in forests from 1999 to 2003

(10 000 *mu*)

	Data from Ministry of Natural Resources					Data from SFA	
Regions	**Farmland in 1998**	**Total conversion from 1999 to 2003**	**Conversion for ecological purpose**	**Adjustment of structure**	**Proportion to farmland in 1998**	**Total conversion from 1999 to 2003**	**Proportion to total farmland**
National	194 463.6	10 377	8 116.5	2 260.3	5.34	10 631.4	5.47
Western	74 108.5	6 464.6	5 487.2	977.3	8.72	6 539.9	8.82
Inner Mongolia	12 004.1	1 638	1 626.9	11.0	13.64	982.6	8.19
Guangxi	6 616.7	162	66.4	95.9	2.45	270.3	4.09
Chongqing	3 802.5	266	205.7	59.9	6.99	436.9	11.49
Sichuan	9 899.9	739	565.9	173.5	7.47	1 175.3	11.87
Guizhou	7 340.7	445	401.0	44.2	6.07	543.4	7.40
Yunnan	9 638.0	379	245.6	133.6	3.93	412.8	4.28
Shanxi	7 651.4	1 277	1 026.6	250.9	16.70	1 134.3	14.82
Gansu	7 533.4	500	467.8	31.9	6.63	700.0	9.29
Qinghai	1 030.8	199	197.7	0.9	19.26	184.0	17.85
Ningxia	1 907.9	462	457.5	4.5	24.21	326.6	17.12
Xinjiang	6 136.1	388	217.5	170.9	6.33	373.7	6.09

The *mu* is a traditional unit of land measure equal to 1/15 ha (0.0667 ha).

China is a country with large areas of natural grassland, second in the world only to Australia. China's grasslands occupy 4 million km^2, taking up 41.7% of the national territory, 64% of the total national vegetation and 13% of the world's total grassland. Of the national total grassland, more than 3 million km^2 is of economic value, but livestock products arising from the grassland contribute relatively little to total livestock production. Because of the poor quality land and fragile ecology of natural grassland, the carrying capacity is quite low, with 2 hectares required to feed one sheep. It is estimated that the total stock volume, *i.e.* the population of sheep, is about 100 million across the natural grassland of 3 million km^2. This livestock population supports the livelihoods of fewer than 5 million herdsmen. In 2002, the State Council implemented further activities under its grassland restoration project by adding RMB 1.2 billion in the form of a bond. The trial project covers about 6.6 million hectares. The trial began in 2003 and was supported by the investment of RMB 2.2 billion from the central government.

China still faces a serious situation of land degradation and desertification, and the overall deterioration of the ecosystems in the west has not been reversed

The most deteriorated land in China can be grouped into two categories: water and soil erosion, and desertification. Water and soil erosion mainly takes place in steep mountainous areas and in the Yellow Plateau. Desertification mainly takes place in grasslands suffering ongoing degradation. Deserts cover 2.67 million km^2, accounting for 28% of China's national territory. They are expanding at an annual rate of 6 700 km^2. The expansion of desertification mainly occurs in the west, taking up more than 90% of the national total. As the most severe eco-disaster in China, desertification is increasing at an accelerating annual rate – 1 560, 2 100, 2 460 and 3 436 km^2 per annum in the 1970s,

1980s, mid-1990s and the period of 1994-1999 respectively. Desertification has also contributed to the frequency of disastrous sandstorms (8, 13, 14 and 23 times annually in the decade of the 1960s, 1970s, 1980s and 1990s respectively). More than 90% of natural grassland has been degraded to some extent, of which more than 50% suffers from degradation of medium severity. The degradation now increases at a rate of more than 2 million hectares annually.

According to the data of the second national remote sensing inventory released by Ministry of Water Resources in January 2002, the total area suffering from water and soil erosion in the west was about 2.937 million km^2, 82.5% of the national total. It is estimated that the annual erosion of soil is about 5 billion tonnes, of which two thirds occurs in the west.

At present, more than 90% rural poor live in poor ecological environments. The west is trapped in a vicious cycle of a deteriorating ecological environment that exacerbates poverty

By the end of 2002, the population of rural people living under the poverty line was 28.2 million, representing 3% of the total rural population and 58.25 million people, 6.2% of rural population, were either living in poverty or with low and unstable income. The net annual income under the poverty line was RMB 630 per capita, (around USD 80). The poverty line of low income was RMB 872 of net annual income (according to the standards set in 2001). Poor environmental conditions and poverty are intertwined. To some extent, poverty is an ecological problem and its occurrence and degree is closely correlated with the quality of the ecosystem.

Recognising that the most poverty-stricken areas in China are located in regions with the most degraded ecological conditions the Chinese government has adjusted its development strategy and polices to account for poverty reduction policies. Consequently, it has put forward concrete objectives of ecosystem protection in the Great Development of the West. This amounts to a task of balancing the relationship between rural livelihoods and ecological protection. For example, it is not wise to sustain the livelihoods of 5 million people by destroying the ecosystems on more than 3 million km^2 of land, one third of the territory. This approach has obtained an annual income of only RMB 10 billion for herdsmen in the North-West, while concurrently contributing to an annual net loss of RMB 50 billion from grassland ecosystems. Such measures as conducted under the programmes such as graze free seasons, rotational grazing, and year-round ban of grazing under the Grassland Ecosystem Improvement Project implemented in recent years have led to an effective but partial restoration of ecosystems. However, the objective to improve the livelihood of herdsmen has not been achieved. The task of trying to integrate poverty reduction in the west along with policies for protecting ecosystems has become an important task of policy making.

Overcoming the barriers to combating land degradation and ecosystem deterioration posed by administrative systems and policies

A serious shortage of investment in combating land degradation and ecosystem protection and lack of stable mechanism of investment

Since 1998, China has issued a large amount of state funds (as bonds), of which a large proportion has been invested in the forestry and hydrology sector. The funds raised from state bonds used in agriculture take up more than 70% of the central budget for agricultural infrastructure. It is unlikely that these funds will be available in the long term.

An overlap of functions of government agencies where there are too many sources of investment to concentrate on priorities

At present, there are many investment sources for agriculture, of which the sources for construction investment from the central government include agricultural infrastructure development investment, agriculture integrated development investment, poverty reduction fund, poverty reduction special fund and funds from finance agencies for small public infrastructures and etc. Investment for basic agriculture infrastructure construction is managed by the National Development and Reform Commission (NDRC) or jointly managed by agencies under the NDRC and the Ministry of Agriculture. The investment of agriculture research is managed by agencies responsible for finance and science or for science and agriculture. Expenditure for agriculture production, expenditure for agriculture enterprises, forestry, hydrology and gas supply and funds for integrated agriculture development are managed by agencies responsible for finance or jointly managed by line agencies responsible for finance and agriculture.

Funds for supporting agriculture from government revenue are usually allocated through financing agencies at all levels. The agencies responsible for agriculture, forestry and hydrology above the county level also allocate the same funds to agencies at the county level. These funds are thus allocated through different channels and in piecemeal fashion. At the county level, the agencies responsible for management of financial investment for agriculture may involve a myriad of bureaucracies. The Bureau of Development, the Bureau of Finance, the Office of Poverty Reduction, the Office of Integrated Agriculture Development, the Bureau of Agriculture, the Bureau of Forestry, the Bureau of Water Conservancy, the Bureau of Husbandry, the Bureau of Agricultural Machinery, the Bureau of Aquatic Products, the Bureau of Meteorology, the Bureau of State Land, the Bureau of transportation and other agencies may all be involved. It is quite common for a project (or components of it) to be managed by many agencies.

This can contribute to serious redundancies and contradictions in signals and incentives. Due to the lack of effective co-ordination amongst the agencies involved in managing the financial investment for agriculture, these agencies operate with their own understanding and control the implementations of policies and use of funds with no other agencies input. At present, there is no effective co-ordination mechanism for distributing financial investment for agriculture amongst the NDRC, the Ministry of Science and Technology, the Ministry of Finance, Ministry of Agriculture and other agencies or to their subordinate institutions. Each sector manages investment according to its own understanding, leading to investment being scattered and duplicated.

The pattern of agriculture investment management has the following weakness. First, it is hard to make wise overall allocation of limited financial investment to focus on priorities because financial investment in infrastructure, research, production and marketing is managed by different agencies. Second, investment for similar projects usually occur from different agencies. So, there will be redundancies in some instances and gaps in others. It is also problematic for localities to combine investment from different agencies and make a systematic plan for project implementation when they must comply with differing regulations on the use of the investment funds from the different sources. Third, a project with investment from different agencies will be plagued with supervision and inspection from all these agencies. In fact, it is difficult to conduct any proper monitoring as the agencies only care about the allocation and disbursement of funds, and care little about the wise management of funds.

Policy making in combating water and soil erosion and land degradation is top-down

Within the decision-making process, there is no broad participation of stakeholders, especially those stakeholders directly affected. Influenced heavily by the central planning system, the direction and quantity of agriculture investment and responsibility of governments at different levels are decided by upper levels of government. Large and middle scale projects are decided upon entirely by central or provincial authorities. This kind of decision making can help central and provincial government to concentrate investment on key projects, but always neglects the actual needs of local governments, communities and farmers. Thus, it is common that government support does not meet actual needs, especially the needs of farmers. A number of agriculture investment projects implemented by governments are not based on the actual needs of farmers, but on the will of governments. This kind of investment fails to solve the problems that farmers and herders care most about and fails to reflect the opinions and needs of the majority of land users.

Decision-making for projects of middle and large scale is highly concentrated at central and provincial level. In practice, it is not possible for agencies at the central and provincial to make correct and objective analysis of all project proposals as they are unlikely to have all of the relevant information for all projects. As a result, projects that should be implemented urgently are not approved, but those that should not be implemented or can be implemented later are approved. In the process of approving projects, the competent agencies at central and provincial level always proceed from previous precedents because of the limited information they have.

Priority is given to the application of funds, with less attention to the management of fund allocated and the implementation of projects. First, finance agencies and the line agencies responsible at all levels lack basic statistical data of agriculture projects and fail to conduct effective assessment of the use of funds and project implementation. Second, there are no strict regulations on the responsible management for implementing projects. The implementing agencies lack a sense of responsibility and pressure because the investment from government revenue comes in the form of a grant. Third, there is no incentive for the line agencies responsible to strengthen management of project implementation since application of new projects is not based on the implementation of existing projects. Rather, applicants and the line agencies responsible pay much more attention to packaging new projects, and less attention to supervising and managing existing projects.

Although the management system of agricultural fund has improve greatly, it is still hard to establish standardised supervision and management mechanism as funds are derived from, and managed by, different agencies that have their own regulations and requirements. It is, therefore, not uncommon to divert all kinds of investment for agriculture to other unintended uses. According to an audit conducted by the State Auditing Administration in over 50 counties in 2004, 50 counties diverted financial investment for agriculture in the amount of RMB 495 million. Instead, these funds were mainly used to balance the financial budget, lending, business, construction of building and purchase of cars, at a time when total financial investment in agriculture only amounts to 10% of the budgeted total. Our own investigation in two counties of a province showed that, during the period of 1993-2000, one county diverted RMB 34.58 million of financial investment allocated to agriculture to salary and expenditure of local agencies. Diverted funds originated from a special fund for hydrology (RMB 19.66 million), a special fund for forestry (RMB 3.56 million), a special agriculture fund (RMB 2.42 million), a special fund for developing vegetable production (RMB 2.67 million), a special fund for transportation (RMB 2.23 million) and a fund for agricultural education (RMB 0.5 million). All special funds from the central and provincial revenue run the risk of being diverted to salaries and other expenditure if allocated through normal financial channels. In another county we examined, diversion of financial investment was even more serious, with an annual diversion of RMB 20 million. Total diversion in this county alone was about RMB 56.2 million in recent years.

In addition, state owned institutions in China's agricultural sector are greatly overstaffed. A certain proportion of state investment funds for agriculture, rural area and farmers are misappropriated by these institutions and their staff, resulting in the funds available to invest in agriculture, rural development and farmer and herder assistance being reduced considerably.

The functions and projects of major government agencies in combating land degradation and reducing poverty are set out in the table below.

Ministry	Department	Function
Finance	Agriculture	Making financial expenditure budget
	Office of Integrated Agriculture Development	Organising and supervising implementation of policies, plans and projects of the state integrated agriculture development; managing and allocating funds from central government for integrated agriculture development. The projects include Bashang Ecological Agriculture, Shelterbelt in Middle and Upper Reaches of the Yangtze River, Combating Water and Soil Erosion in the Upper reaches of the Yangtze River and Taihang Mountain Greening.
Development and Reform	Regional Economy	Co-ordinating regional economic development, making regional economic development plan for the regions which was revolutionary bases, and is located in remote areas with minorities as main population and stricken by poverty
	Rural Economy	Putting forward proposals for the most important issues of rural economic development and strategy as well as reform of rural economic system
Water Conservancy	Relevant Departments	Investment for agriculture from water conservancy sector is mainly used for improving farmland, drought control, irrigation, restoration of ecosystem, and potable water for both humans and animals.
Agriculture	Relevant Departments	In 2000, investment in resource and ecosystem protection was about RMB 68 million, about 3.08% of the total investment from the ministry. In past two, investment in grassland was increased, but the function of the ministry in this regard is not strong enough
State Forestry Administration	Relevant Departments	In 2001, State Forestry Administration combined key forestry projects into six national projects: Natural Forest Protection, Shelterbelt, Farmland Conversion, Combating Desertification, Timber Base and Wildlife Protection.
Environment Protection	Relevant Departments	Planning, laws and regulations, monitoring and pollution control.
Science and Technology	Relevant Departments	Research on key technologies concerning ecosystem improvement, transfer of research results into practice
Office of Poverty Reduction, State Council		Poverty reduction funds are allocated by the central government for improving the basic livelihood of rural people and encouraging social and economic development in regions stricken by poverty. These funds include development funds for undeveloped regions, and increased financial investment and special agriculture funds for the west, northwest and southwest (state fund for poverty reduction).

Chapter 4.

WATER RESOURCES AND AGRICULTURAL PRODUCTION IN CHINA: THE PRESENT SITUATION

Ma Xiaohe, Department of Economic System Reform
and
Fang Songhai, School of Agricultural Economics and Rural Development, People's University of China

Abstract

China faces daunting, although not insurmountable, challenges with respect to its water resources. Moreover, its endowment of water resources per capita is low and badly distributed, with the North China Plain having one of the lowest per capita endowments in the world. Current farming practices, if they are to meet the needs of China's growing population, will also place increasing demand on irrigation capacity. A number of important river basins and watercourses also experience serious problems with pollution. Given these stark realities and emerging trends, current water management approaches are unsustainable. We make several suggestions on how to remedy the situation. Among these, we strongly suggest that correcting incentives and water management practices to raise the efficiency of its use should be a higher priority than reallocating water resources via the South to North water transfer project.

Relationship between agricultural production and water resources

Water is the source of life. Without water there can be no agriculture and no human life. Scarce water supplies not only affect production itself but also have a direct impact on national food security, which in turn depends on our agricultural productivity. Before addressing the main topic of this paper, let us first consider how a country like Israel, two-thirds occupied by desert and severely lacking in water supplies, developed its agricultural production.

Agricultural production and water use in Israel

Between 1949 and 1993, while Israel's agricultural production showed sustained growth, agricultural water consumption increased only until 1966, then remained stable from 1966 to 1986, and began to fall from 1987 (Figure 1). The rate of water consumption per unit output began to decline from 1958 both for food crops and all agricultural production (Figure 2). Indeed, it is this steady improvement in the efficiency of water use, with current per capita fresh water consumption of only 449 m^3, that has enabled Israel to stabilise and even reduce its total demand on water resources while maintaining sustained growth of agricultural output.

We began this paper with the example of Israel to illustrate a point: abundant water resources are good for agricultural production, and limited water resources constrain it; but increased technological input and more efficient water use can overcome this constraint. Moreover, the value of efficient water use is reflected in the relationship between irrigation capacity and agricultural output.

Figure 1. Water use and agricultural production in Israel, 1949-1993

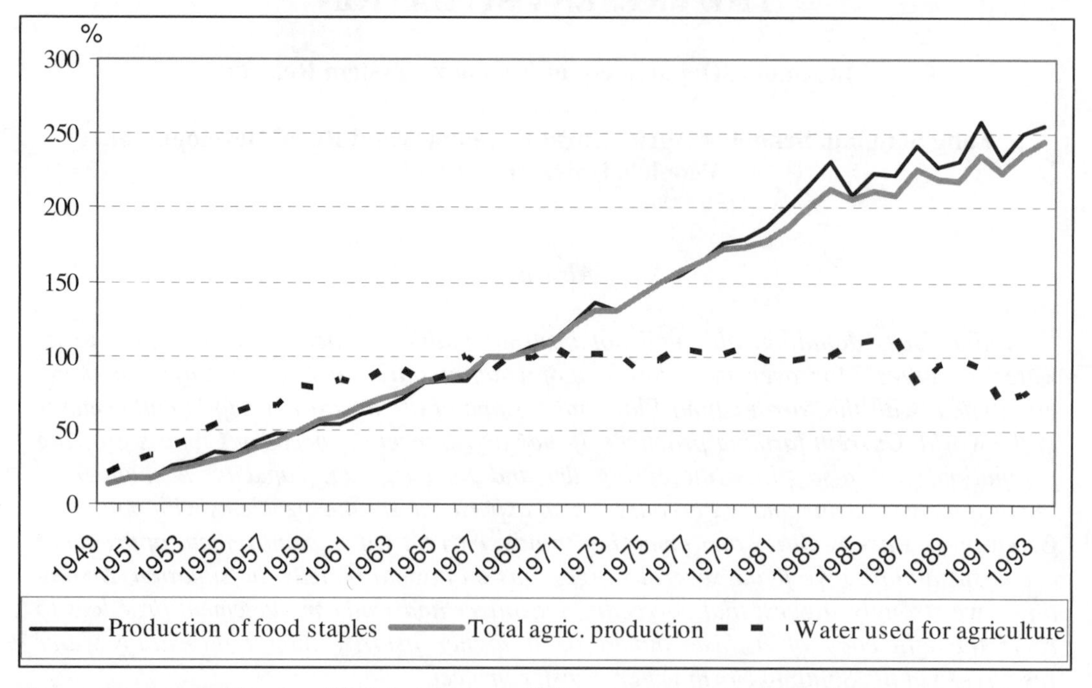

Source: Data compiled from Israel Statistical Abstracts 1993 as appended to Qin Zhihao (1996), *Agricultural Development in Israel.*

Figure 2. Changes in water consumption per unit of agricultural production in Israel, 1949-1993

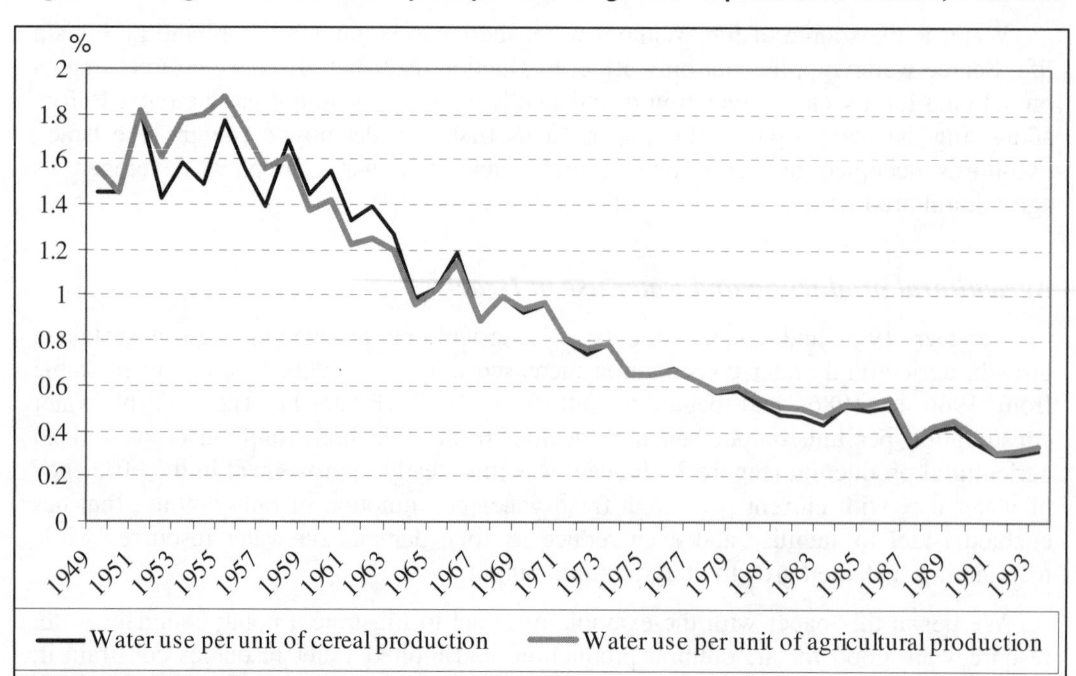

Source: Data compiled from Israel Statistical Abstracts 1993 as appended to Qin Zhihao (1996), *Agricultural Development in Israel.*

Water resources and agricultural production in China

Table 1 presents cross-sectional data on agricultural production and water use in China's regions in 2003. A simple correlation analysis and linear regression on these data showed that water use per unit of cultivated land was positively correlated with grain output per unit area to $p \geq 0.01$ in a two-tailed test, with a Pearson correlation coefficient of 0.502. Linear regression yielded the following relationship between grain output per unit of sown area[1] (y) and the amount of water used per unit of cultivated land[2] (x):

$$y = 3758.118 + 0.196x$$

However, this result does not necessarily mean that water use increases with the volume of production. If output increased with the volume of water used, then water use per unit area of sown land should be directly proportional to grain production per unit area of land. In fact, our empirical analysis showed no clear correlation between these two sets of data. The figures show that for Beijing, Fujian, Guangdong, Tibet, Qinghai, Ningxia, Xinjiang and other areas, water use per hectare sown is above 4 000 m^3, but these are not the provinces and regions where grain production is most efficient. We suggest that the figures for water use per unit of cultivated land reflect the level of access to water resources for irrigation in these regions[3], and that it is this factor – access to irrigation – which directly affects levels of grain production in those areas.

The above analysis highlights two key aspects of water use and agricultural productivity that warrant attention. First, while abundant water resources favour the development of agricultural production, they do not in themselves necessarily have a direct relation to output levels; they can only have a positive impact if the uptake of water is ensured through an efficient irrigation system and timely application. Second, while scarce water resources are a constraint on agricultural production, sustained output growth can still be achieved through constant and vigorous efforts to increase the efficiency of water use.

---

1. Food crops include grains, roots or tubers, and pulses, among which grains are dominant both in terms of dietary habits and production (Source: *China Statistical Yearbook* 2004). Between 1991 and 2003 grains, consisting mainly of paddy rice, wheat and maize, accounted for as much as 87-90% of all crop food production.

2. y = food output per unit of sown land (kilos per hectare), x = water use per unit area of cultivated land (m^3 per hectare).

3. Access to irrigation reflects factors such as the availability of water resources, presence of irrigation and hydrological infrastructure, etc, in each area.

Table 1. Water use in China by province, 2003

Province/ Region	Cultivated area	Grain yield	Water used for agricultural production	Average water use per sown hectare	Average water use per unit of cultivated land	Province/ Region	Cultivated area	Grain yield	Water used for agric. production	Average water use per sown hectare	Average water use per unit of cultivated land
	1 000 ha	kg/ha	million m^3	m^3/ha	m^3/ha		1 000 ha	kg/ha	million m^3	m^3/ha	m^3/ha
Beijing	333.0	4 455	1 292	4 182.08	3 878.16	Henan	7 854.0	4 279	11 335	828.30	1 443.19
Tianjin	470.3	5 001	1 117	2 227.50	2 375.31	Hubei	4 793.1	6 000	13 622	1 908.35	2 842.07
Hebei	6 665.8	4 256	14 956	1 731.28	2 243.64	Hunan	3 828.1	5 802	20 944	2 708.98	5 471.10
Shanxi	4 443.6	3 818	3 330	898.09	749.41	Guangdong	3 168.8	5 385	24 261	4 968.07	7 656.22
Neimenggu	7 941.8	4 489	14 611	2 539.86	1 839.78	Guangxi	4 268.6	4 687	20 544	3 271.84	4 812.84
Liaoning	4 042.9	6 042	8 354	2 246.33	2 066.45	Hainan	738.0	4 105	3 573	3 940.48	4 841.35
Jilin	5 402.1	6 074	6 752	1 431.49	1 249.88	Sichuan/ Chongqing	8 879.3	5 176	14 240	1 116.84	1 603.72
Heilongjiang	1 1401.0	4 634	17 137	1 748.20	1 503.12	Guizhou	4 748.5	4 427	5 216	1 125.55	1 098.46
Shanghai	305.1	6 895	1 632	3 893.22	5 348.32	Yunnan	6 218.7	4 043	10 961	1 904.20	1 762.53
Jiangsu	4 901.7	5 567	22 313	2 904.81	4 552.11	Tibet	351.1	5 298	2 262	9 678.59	6 440.43
Zhejiang	2 058.1	6 166	11 018	3 887.11	5 353.18	Shaanxi	4 978.1	3 540	5 068	1 249.58	1 018.07
Anhui	5 783.0	4 134	9 384	1 028.42	1 622.69	Gansu	4 865.9	3 387	9 644	2 663.53	1 982.04
Fujian	1 389.4	5 315	10 099	4 009.26	7 268.81	Qinghai	666.3	3 485	2 174	4 656.68	3 262.61
Jiangxi	2 898.8	5 024	10 410	2 083.10	3 591.14	Ningxia	1 228.7	3 935	5 845	5 174.68	4 756.80
Shandong	7 446.3	5 403	15 701	1 442.38	2 108.53	Xinjiang	3 859.7	5 872	45 493	12 869.29	11 786.58

Notes: The most recent *China Statistical Yearbook* does not provide the latest figures for all regions. The data in this table are based on 2003 figures for the total land area of China published by the Ministry of Land and Resources in 2005. The figures for each region/province being calculated on the basis of their share of total national land area according to the 2004 *Statistics Yearbook*. To allow comparability with historical data, data for Sichuan and Chongqing [now an autonomous urban area] have been merged in this table.
Source: Compiled from *China Statistical Yearbook* 2004 and 2005 bulletins of the Ministry of Land and Resources.

Table 2. Analysis and projections for China's water resources

Chinese Academy of Engineering Project Group Report

Region	River Basins	% of national total				Per capita water volume (m^3/person)			Average. volume of water per *mu* (m^3 / *mu*)
		Water Resource	Population	Cultivated land	GDP	1997	2010	2050	
North	North East	6.9	9.6	20.2	10.4	1 646	1 501	1 287	660
	Hai and Luan rivers	1.5	10	11.3	11.6	343	311	273	259
	Yellow River	2.7	8.5	12.9	6.7	707; 517	621; 454	526; 385	400; 293
	Huai and Shandong	3.4	16.2	15.2	14.1	487	440	383	437
	subtotal	14.5	44.3	59.6	42.8	747; 732	674; 620	582; 540	471; 447
	Yellow-Huai-Hai river basin subtotal	7.6	34.7	39.4	32.4	500; 453	449; 407	389; 352	373; 338
South	Yangtze	34.2	34.3	23.7	33.2	2 289	2 042	1 748	2 783
	South-East coastal region	9.2	5.6	2.5	8.1	2 885	2 613	2 231	5 344
	Pearl River and South China	16.7	12.1	6.7	13.5	3 228	2 813	2 377	4 501
	South-West	20.8	1.6	1.8	0.7	29 427	25 056	20 726	23 090
	subtotal	80.9	53.6	34.7	55.5	3 481	2 592	2 634	4 317
Inland rivers		4.6	2.1	5.7	1.7	4 876	4 140	3 331	1 589
All China		——	——	——	——	2 220	2050	1 760	1 888

Notes: 1. *Mu* = 1/15 ha; 2. Inland waterways include Eerqisi river; 3. The South-East coastal region does not include rivers in Taiwan Province; 4. China's total population in 1997 was 1.236 billion (1.267 billion including Hong Kong, Macao and Taiwan); 5. Cultivated area is based on 1993 data; 6. In the data for the Yellow river, the second figure for water per capita and per *mu* is calculated after subtraction of the 20 billion m^3 of water needed for sediment flushing.
Source: Liu Changming, Chen Zhikai (2001), *Evaluation of the Present Situation of China's Water Resources and Analysis of Trends in Supply and Demand*, p. 4.

Supply and demand of water resources for agriculture in China

The current water situation: tight per capita supply and interregional structural imbalances

According to a report by China's Academy of Macroeconomic Research, China has the world's largest water resources at 2.81 billion m^3. Water resources per capita are approximately 2 000 m^3, only one quarter of the world average, making China one of the poorest countries in terms of per capita water resources, at 121st of 153 countries worldwide. Twenty per cent of China's water resources are utilised, but this ratio rises to 50% for several regions in the north, well above the internationally recognised danger threshold of 40%. The ratios for the main river basins are 90% in the Hai River basin, 67% for the Yellow River basin, 59% in the Huai river basin, and 40% for all inland waterways. In an average year, drought affects more than 400 million *mu* of cultivated land, or about one fifth of all cultivated land in China. A typical year sees a shortfall of about 30 billion m^3 in water requirements for irrigation, and a shortfall of about

6 billion m^3 for urban and industrial water requirements. Foregone food production due to water shortages induces economic costs of about RMB 50 billion and RMB 200 billion in lost industrial production. A study of nationwide river basin water resources by a project group of the Chinese Academy of Engineering Project Group Report (Liu and Chen, 2001) has shown vast differences between North and South. In the North, water is very scarce, with only 14.5% of the country's water resources for 59.6% of its cultivated land area, of which 39% is concentrated in the Yellow-Huai-Hai river basin region which has access to only 7.6% of the country's water resources. In the South, water is relatively abundant, with 80.9% of the country's available water for only 34% of its total cultivated land area. Overall, the total volume of available water per capita in China will continue to decline for a considerable time into the future, as shown in Table 2.

Supply and demand of water for agriculture: no grounds for optimism

China's water resources will not support sustained expansion of the current model of water use. Because of the special characteristics of China's economic and demographic structure, agriculture and daily survival needs have for a long time accounted for the lion's share of water use. But with increasing industrialisation and urbanisation, this share has steadily declined, from 97.1% in 1949 to 88.2% in 1980, 75.3% in 1997 and 67.6% in 2004. Both the relative shares and absolute volume of industrial and urban residential water use have risen steadily: in 2004 industry used 22.2% of the nation's water (up from 10.3% in 1980) as the share of urban residential consumption reached 10.3% (of which 1.48% is 'ecosystem' water use). Since 2000, the total consumption of water has been falling yearly. However, this has been due entirely to the drop in rural and agricultural water use, from 419.8 billion m^3 in 1997 to 373.7 billion m^3 in 2003, while over the same period industrial and urban residential water use has been climbing steadily (Table 3 and Figures 3 and 4).

Table 3. Growth in water consumption in China since 1949

Year	Agriculture and rural		Industrial		Urban residential		Total billion m^3	Water use per capita (m^3)
	water use billion m^3	% of total	water use billion m^3	% of total	water use billion m^3	% of total		
1949	100.1	97.1	2.4	2.3	0.6	0.6	103.1	187
1959	193.8	94.6	9.6	4.7	1.4	0.7	204.8	316
1965	254.5	92.7	18.1	6.6	1.8	0.7	274.4	378
1980	391.2	88.2	45.7	10.3	6.8	1.5	443.7	450
1993	405.5	78	90.6	17.4	23.7	4.6	519.8	445
1997	419.8	75.3	112.1	20.2	24.7	4.5	556.6	458
2000	406.1	73.9	113.9	20.7	29.8	5.4	549.9	435
2001	411.5	73.9	114.2	20.5	31.1	5.6	556.7	438
2002	403.4	73.4	114.2	20.8	32.1	5.8	549.7	429
2003	373.7	70.2	117.7	22.1	40.6	7.6	532.0	413
2004	374.9	67.6	122.9	22.2	57.0	10.3	554.7	427

Notes: Values for 1949, 1959 and 1965 are estimated; data for urban residential water use in 2002 and 2003 are extrapolated from their share of total household use in 2002 (51.8%); 2003 data contains an additional ecosystem component, amounting to 7.95 billion or 1.5% of total water use in that year, which has been attributed to the urban residential category for the purposes of the calculation.

Sources: Data for 1949-1997 are compiled from p. 10 of Liu and Chen (2001); and for 2000 and succeeding years, from the *China Statistical Yearbooks* for 2004 and 2005 and government bulletins on hydrological statistics and water resources.

Figure 3. Changes in water resource use in China since 1949

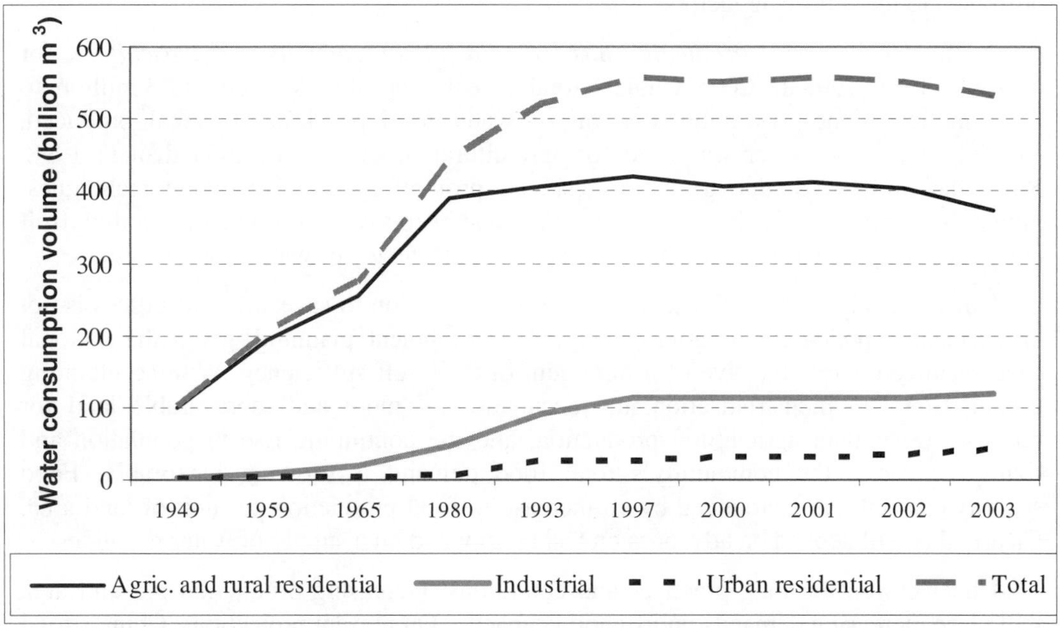

Source: Same as Table 3 above.

Figure 4. Structural changes in water resource use in China since 1949

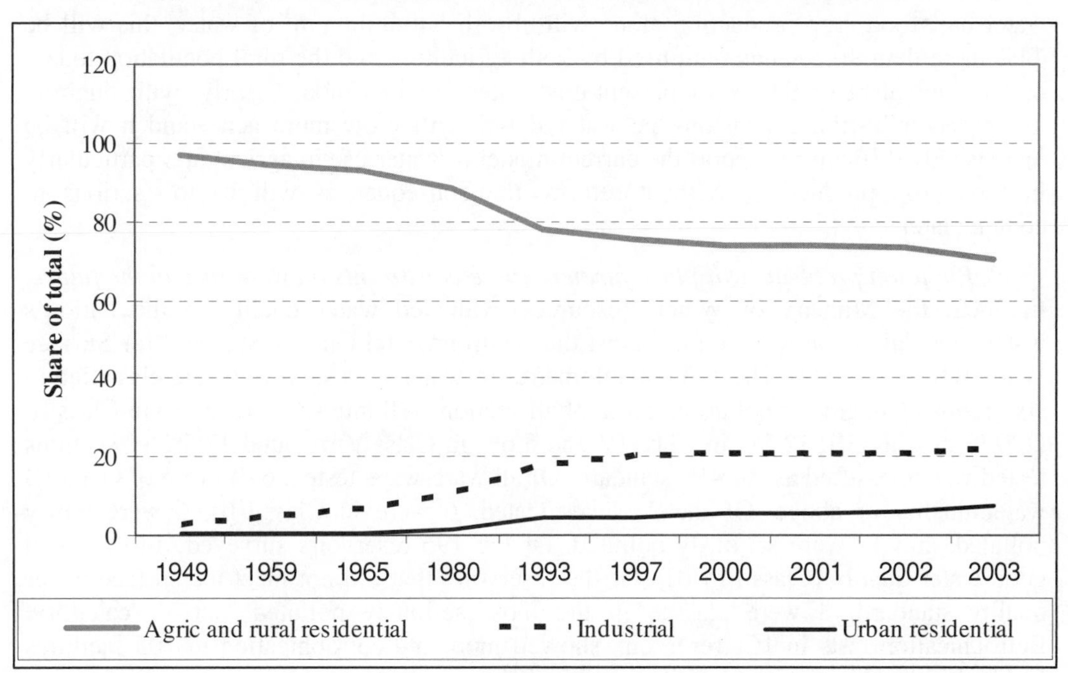

Source: Same as Table 3 above.

Overall, the fall in agricultural water use in China since the mid-1990s can be attributed to the following factors:

1. *The fall in cereal production has led to a fall in water use for agriculture*; for example, from 1998 to 2003 China's total cereal crop shrank from 512.3 million to 430.7 million tonnes; assuming one m^3 of water used per kilo of cereal produced, 81.6 billion m^3 less water was used for agricultural production in 2003 than in 1998. Since the mid-1990s, due to the widespread promotion of water-saving techniques, hundreds of millions of cubic metres of water have been saved in food crops, cotton, fruit and vegetable production, markedly increasing the efficiency of water resource use.

However, relying on cutbacks in total food production to save water resources is not an acceptable policy for long-term national development planning, given the national food security policy objective of a minimum of 95% self sufficiency. With accelerating urbanisation and industrialisation, the occupation of more and more arable land for purposes other than agricultural production, and the continuing rise in population and living standards, the community's total food demand is growing inexorably. Food security can only be guaranteed by an increase in food production per unit of land area. Clearly this will depend on advances on technology and on a supply of water resources.

Under current farming practices and conditions, increasing production per unit area will place increasing demands on irrigation capacity. On present projections, China's total food staple demand will reach 599.61 million tonnes by 2020, requiring domestic production to reach 544 million tonnes to achieve even 90% self sufficiency. Assuming 100 million hectares of land are still under cereal crops at that date, productivity will have to rise to 5 400 kg per hectare. In addition, without a change in our model of agricultural water use, food crop production alone will absorb 540 billion m^3 of water; this will be 44% more than the total amount used by both agriculture and the rural population today, or the equivalent of 97.3% of present-day water use in China. Clearly, with ongoing urban and industrial expansion, the water deficit will grow more acute and it will be increasingly difficult to support the current model of water use in agriculture, particularly in food crop production. Without reform, the consequences will be too serious to contemplate.

2. *Pollution problems will place further severe constraints on water use in the future.* In 2002, the Ministry of Water Resources evaluated water quality in the nation's waterways, lakes and reservoirs, against the Environmental Quality Standard for Surface Water (GB3838-2002). The 123 000 kilometres of waterways surveyed were classified in six groups by degree of pollution: 5.6% of all sections fell into Class I, 33.1% in Class II, 26.0 % in Class III, 12.2% in Class IV and 5.6% in Class V. Around 17.5% of sections tested were classified as 'sub-V' standard. Of all waterways tested, 64% were of Class III water quality or above. Of the 24 lakes tested, 6 were in Class I-III, 6 were partly polluted, and 12 were seriously polluted. Of the 196 reservoirs surveyed, 146 showed good water quality (Class I to III). Of the reservoirs that did not meet the surface water quality standard, 8 were classed in the most seriously polluted 'sub-V' category. Eutrophication tests in 161 reservoirs showed moderate eutrophication in 105 facilities and high eutrophication in 56 (65.1% and 34.8% respectively).

The situation in the various river segments was as follows: in the South-West, all inland waterways, and the South-East Yangtze and Pearl River basins, water quality was good or acceptable, with 92.6%, 90.3%, 76.9%, 76.8% and 75.8% of sections respectively in Classes I-III. In the Song and Liao, Hai River, Yellow River and Huai River basins, water quality was relatively poor, with only 43.1%, 40.6%, 38.6% and 30.8% of sections respectively in Classes I-III. Results for the three lakes enjoying national priority treatment were as follows: for the Taihu, 16.5% of the lake's area was classed II or III, 75.3% in Class IV and 8.2% in Class V; medium eutrophication was found in 16.5% of the lake's total area, high eutrophication in 83.5%. Dianchi Lake in Yunnan was classed V or sub-V, with high eutrophication. The eastern half of Chaohu Lake was classed IV, the western half V or sub-V, with mid- and high-level eutrophication in the eastern and western halves respectively (Ministry of Water Resources, 2002).

Table 4. Pollution intensity in China's main watersheds, 2000

	Songhua	Liao	Hai	Huai	N-West Yellow	Yangtze	Pearl
COD as % of national level (A)	3.7	5.2	11.5	8.2	5.5	33.1	15.8
% of total national water resources (B)	4.8	2.1	1.5	3.4	2	34.2	16.8
Intensity of watershed pollution discharge, % (A/B)	77	248	767	241	275	97	94

Note: COD is the pollution load carried by a body of water; pollution intensity is the pollution load by river unit.
Source: COD values are taken from SEPA's Environmental Protection Plan within the 10th Five-Year Plan. Water resource data from Environmental Engineering Manual, taken from Qian Yi (ed.), Research on a Strategy for Sustainable Development and Ecological and Environmental Infrastructure for Water Resource Allocation in North-West China: Water Pollution Prevention and Treatment, p. 7.

As Qian Yi's (2004) study shows, pollution discharges in four major watersheds, the Hai, Liao, Huai and North-West Yellow River basins, are all well above the national average. In the Hai and North-West Yellow River basins, pollution intensities are, respectively, 7.67 and 2.75 times the average national level-and this heavily polluted Huai-Hai-Yellow River Region comprises 39.4% of all China's cultivated land area. Water pollution severely reduces the total volume of the resource available for use, aggravating the serious supply-demand imbalance in this area. Even in the Pearl River Delta and other regions where water is relatively plentiful, pollution problems have led to shortages in available freshwater resources (Academy of Macroeconomic Research Project Group Report, 2003). The very recent incident in which fish died in Baiyangdian Lake (reported on the Xinhua News Network on April 10th) should be seen as yet another wake-up call.

General conclusion and recommended measures

General conclusion

Under present irrigation conditions, the elasticity of irrigation water use with respect to food crop production is greater than 1, *i.e.* a unit increase in food crop production requires more than a unit increase in water use. By 2020, assuming productivity of 360 kg per *mu* of cultivated land, water demand for cereal production alone will reach 540 billion m^3. This is equivalent to 1.44 times the present combined total of agricultural and rural residential water demand, or almost the entire national consumption of water. This does not even take into account the needs of non-cereal agricultural production. If we add in steadily growing industrial and urban residential use, plus interregional imbalances, water pollution and other factors, then achieving this level of irrigation in the agricultural food sector is clearly a pipedream. The only possible course of action is the widespread dissemination of water-saving practices in agriculture.

Recommended measures

Based on the foregoing analysis and comparing the Israel's strategy with China's present concerns, it is our view that long-term planning, and the policy priority placed on agricultural food production and national food security, mandate urgent measures to restructure agricultural production. The highest priority must be placed on reducing as far as possible the cultivation of high-water-consuming crops. Secondly, we must accelerate the promotion of water conservation and pollution prevention and management methods. This is the only way to ensure sustainable agricultural development. In this context, we propose the following methods for promoting water conservation and pollution prevention and management:

Support the installation of agricultural water conservation facilities

The focus of such efforts should be on North and North-West China. A comparison between national rainfall distribution and regional agricultural water demand shows huge water supply gaps in north-western provinces like Gansu, Xinjiang, Neimengu and Ningxia. In these areas, rainfall meets less than 50% of agricultural water needs, while the North China region is also very close to failing to make ends meet (Table 5). These are the two regions of China where water needs are most critical. Yet annual quotas for field irrigation in the North-West can be as high as 16 537 m^3 per hectare, or 1.4 times the national average. In Ningxia, the figure is as high as 32 550 m^3, or 2.8 times the national average. In the North the quota for areas under irrigation ranges from 7 500 to 12 000 m^3 per hectare, 2 to 5 times the real need for agricultural crop production (Wang Jiashu 2002). In such regions, where unscientific cultivation methods aggravate the water resource imbalance, promoting widespread use of water conservation facilities, restricting extensive and uncontrolled water use, and promoting intensive and water-saving irrigation practices should all take a higher priority than reallocating water resources from South to North.

Table 5. Annual rainfall and agricultural water demand by region

Region	Rainfall (mm)	Agricultural water demand (mm)
South China	1646	1 426
Lower and Mid Yangtze	1394.1	997
Yunnan Plateau	1 236	965
Sichuan Basin	1 038	935
South Region 4	1 337.2	1066
North China	747.3	711
North-East Region	581.2	451
Shanxi & Shaanxi Region	616.2	483
Northern Region 3	637.9	577
Neimenggu & Liaoning Region	290.9	523
Gansu & Xinjiang Region	235.4	563
North West Region 2	258.7	543
Qinghai & Tibet	509.3	439
All China	648.4	757

Source: Shi and Lu (eds.), *Agricultural Water Needs and High-Efficiency Water Saving Infrastructure* (September 2001), p. 247.

Simultaneously increase irrigation utilisation coefficients, control paddy-field irrigation and promote the re-use and recycling of irrigation water

In gravity irrigation networks, the water utilisation coefficient is usually only around 0.5, and the rate of water use in the field is only about 0.7, typically resulting in an overall water use coefficient of only 0.3 to 0.4. In areas using lift irrigation, the coefficient is usually around 0.5 to 0.6 while, in well irrigation systems, it typically falls between 0.60 and 0.65. A series of water-saving irrigation projects launched in the last ten years has successfully raised water utilisation rates to around 0.7-0.9 (Shi & Lu 2001). Even if no other action is taken, simply raising the water utilisation rate of an irrigation system from 0.5 to 0.7 results in a 28.6% saving in agricultural water use.

As for the vast area under paddy fields, if controlled paddy-field irrigation can be combined with water recirculation methods (cyclical irrigation), a clear improvement in efficiency and water saving can be achieved. Taking the distribution of irrigation in the central and lower Yangtze region as an example, shallow-water flood irrigation in all paddy fields without recirculation yields a water utilisation coefficient of 0.5 and a net water deficit of 55.35 billion m^3.[1] If the utilisation coefficient is raised to 0.7, the water shortfall drops to 4.1 billion m^3. Recirculation and re-use of 30% of the water used would result in a water surplus of 36.8 billion m^3 Even controlled paddy-field irrigation without recirculation, at a utilisation coefficient of only 0.5, would ensure a surplus of 1.8 billion m^3 for the region. With recirculation of 30% of the water volume, the surplus would reach 51.84 billion m^3. If, the water utilisation rate could be raised further to 0.7, the irrigation surplus would reach 86.9 billion m^3 (Table 6). This is equivalent to 16.3% of the total volume of water used for irrigation throughout China in 2003 (532.04 billion m^3), or almost a quarter of the total consumption for agriculture and rural household use. So, such efficiencies are definitely within reach with proper planning and incentives.

Table 6. Irrigation balance in the middle and lower Yangtze basin

Unit: billion m^3

Irrigation technique	% of water recirculated	Water utilisation coefficients					
		0.5		0.6		0.7	
		Surplus	Deficit	Surplus	Deficit	Surplus	Deficit
Paddy fields shallow water flood irrigation	No recirculation	5.8	61.2	8.5	35.4	12.9	17.2
	20%	10.3	42.5	16.5	17.0	24.5	2.4
	30%	14.1	33.2	20.7	8.0	36.8	
50% of paddy fields under controlled irrigation	No recirculation	8.2	27.3	13.6	7.6	240.	
	20%	16.4	9.1	32.4		50.3	
	30%	21.9	1.5	45.6		63.5	
100% of paddy fields under controlled irrigation	No recirculation	15.7	4.9	32.0		46.6	
	20%	38.4		58.9		73.4	
	30%	51.8		63.0		86.9	

Source: Shi Yulin & Lu Liangshu (eds.), *Agricultural Water Needs and High-Efficiency Water-Saving Infrastructure* (September 2001), p. 42.

1. Including Hunan, Hubei, Jiangxi, Jiangsu, Anhui, Zhejiang and Shanghai.

Start with water conservation projects in non-food staple agriculture to drive water savings throughout the entire agricultural sector

Comparison of the relative shares of water use per unit of cultivated area for different food crops shows a strong negative correlation (-0.574, 0.001) between water use per unit area and food crop type. A simple linear regression gives the following result:[2]

$$x = 10918.662 - 11716.1\alpha$$
$$\Leftrightarrow x = -797.438 + 11716.1 \times (1-\alpha)$$

The above analysis provides a reminder that a significant part of the increase in agricultural water use is due to the expansion of non-food crops. Investment in water conservation projects in non-food agriculture would drive savings throughout agriculture and contribute to raising productivity in agricultural food production, as well as being a relatively efficient approach to the problem. This proactive approach would result in more efficient water conservation and is also in line with the policy-driven appeal for investment.

Support research and development of water-saving farming techniques

Water conservation must be addressed system-wide, cutting across the whole agricultural sector and calling for the implementation of water saving techniques in both cash and food crop cultivation in order to guarantee the future availability of water resources. The advanced use of water sources for irrigation of the kind achieved in Israel will not be possible without a clear commitment to R & D and widespread dissemination of the technologies. If breakthroughs can be achieved both in saving water on crops and keeping down the cost of irrigation facilities, it will be much easier to solve the water resource problem.

Raise the price of water for agricultural use to an appropriate level

Water prices in China are currently too low, which is a major factor contributing to wasteful and inefficient water use. The scarcity of water resources and wastage due to low prices are both factors at odds with China's present needs and situation. The price of agricultural water must be adjusted by a sufficient margin to encourage the rural population to adopt water saving techniques. It is sometimes objected that this will result in higher costs and lower income for farmers. The experience of the last few years in China has shown that although water-saving measures have in part raised the cost of farming, they have also clearly improved efficiency to a degree that easily offsets the additional costs incurred. However, at the same time as water prices are raised, the administration may wish to implement a policy of financial subsidies to help farmers to adopt water conservation technology.

Combat and control water pollution

This issue must unavoidably be addressed in any agenda for survival and sustainable development. In China the management and control of water pollution is a long-term task. To combat industrial, residential and agricultural pollution requires comprehensive and system-wide action with a solid and comprehensive legal basis, supported by administrative measures.

2. Analysis based on *Chinese Statistical Yearbook* 2004.

REFERENCES

Department of Economic System Reform Project Group: Research on Some Important Issues Concerning the Agricultural Use of Water Resources in the Context of the Goal of Quadrupling GDP between 2010 and 2020, December 2003, pp. 1-2.

Liu Changming, Chen Zhikai, (2001), *Evaluation of China's Water Resources with an Analysis of Trends in Supply and Demand*, China Water Power Press (Zhongguo Shuili Shuidian Chubanshe), December 2001, no. 1-1, p. 4.

Ministry of Water Resources (2000, 2001, 2002, 2003, 2004), *Bulletin on China's Water Resources.*

Ministry of Water Resources (2002, 2003), *China Water Resources Statistical Bulletin* (Zhongguo Shuili Tongji Gongbao), www.cws.net.cn/cwsnet/gazette-new.asp.

National Bureau of Statistics (2004), *China Statistical Yearbook* 2004, China Statistical Publishing House (Zhongguo Tongji Chubanshe).

National Bureau of Statistics (2005), *China Statistical Yearbook* 2005, China Statistical Publishing House (Zhongguo Tongji Chubanshe).

Project Group Report: Research on Measures to Address China's Food Security and the Role of Development Banks.

Qian Yi (ed.), *Research on a Strategy for Sustainable Development and Ecological and Environmental Infrastructure for Water Resource Allocation in North-West China: Water Pollution Prevention and Treatment,* Science Press (Kexue Chubanshe), April 2004, vol. 1-1, p. 7.

Qin Zhihao (1996), *Israel's Agricultural Development*, China Agricultural Scientific Press (Zhongguo Nongye Kexue Chubanshe), 1-1, pp 180-181

Shi Yulin & Lu Liangshu (eds.) (2001), *Agricultural Water Needs and High-Efficiency Water Saving Infrastructure*, September, p. 247. Special Reports on Research on Sustainable Development of China's Water Resources, vol. 4. China Water Power Press (Zhongguo Shuili Shuidian Chubanshe), August, Vol. 1-1, pp. 3, 42, 172, 288.

Wang Jiashu (2002), Water Resources and National Food Security, Geological Publishing House (Dizhen Chubanshe), August, Vol. 1-1, p. 108.

Xinhua News Network (2006), Who Polluted Baiyangdian Lake, Jewel of the North? The Full Story of the Investigation into the Dead Fish Incident, 11 April 2006.

http://news3.xinhuanet.com/society/2006-04/11/content_4408639.htm

PART II.

EXPERIENCES IN AGRICULTURAL RESOURCE MANAGEMENT AND ENVIRONMENTAL PROTECTION IN OECD COUNTRIES

Chapter 5.

AGRI-ENVIRONMENTAL POLICIES IN OECD COUNTRIES AND NATURAL RESOURCE MANAGEMENT

Wilfrid Legg
Agriculture Directorate, OECD

Abstract

Agriculture is the major user of renewable land and water resources in OECD countries and contributes to the provision of ecosystem services, including biodiversity. Governments give a high priority to the sustainability of agriculture such that food and fibre be efficiently produced to can meet present and future needs while maintaining the health of the environment. A wide range of economic instruments, regulations and voluntary approaches have been employed across OECD countries to address agri-environmental issues, and there has been some converge in policy approaches recognising that improvements involve both reducing the environmental harm and enhancing the benefits from agriculture, as well as changing agricultural practices and avoiding agriculture on land with high ecological value. But only a small share of agricultural producer support is directly targeted at agri-environmental concerns. Moreover, the long history of production-linked support to agriculture in many countries, which has increased pressure on natural resources, has often led to offsetting policy measures and regulations to improve the environment. While overall there have been improvements in the environment, these have cost more than would have been the case with lower levels of production-linked support.

Introduction and background

Agriculture has a major impact on natural resources in OECD countries. It accounts for around 40% of total OECD land use and nearly 45% of water use, is closely linked with the provision of ecosystem services such as biodiversity and, in many countries, dominates and shapes the landscape. A major challenge is to ensure that natural resources contribute to the sustainable and efficient production of enough food and fibre while providing ecosystem services.

Agriculture is the sector that produces most of the world's food, beverage crops, and leather; much of its fibres, and a number of its chemical feedstocks. Its relative economic importance in OECD countries has been steadily declining over the past century, but it still contributes around 3% to member countries' GDP on average, and three to five times that percentage in Greece, Iceland and Turkey. A much larger proportion of economic activity is directly dependent on agriculture, however, as suggested by the share of food in total consumer expenditure - which ranges from around 7% in the United States to over 20% in the Czech Republic, Hungary, Mexico, Poland, Slovakia and Turkey. Despite its modest size as a sector, total support to agriculture (support to individual producers and support to the sector)was equivalent to 1.1% of GDP (USD 384 billion) for the OECD as a whole in 2005 (OECD, 2006).

Natural resources underpin sustainable agricultural development. They provide the raw materials for meeting human needs for food and fibre, as well as playing an important role in providing the basis for reproduction, food and habitat for living resources, and meeting ecosystem functions like carbon and nitrogen fixation, water catchments and temperature buffers, and in meeting societal demands for cultural landscapes. They also provide recreational services for increasing numbers of people. Owing to increasing demands for natural resource use, many of them are being degraded, together with pollution and waste from natural resource exploitation. Some resources are being depleted, often at local level, although this is being offset by resource saving technologies and substitution.

Agriculture's environmental significance owes much to the land area it occupies and the biomass that grows upon it, the amount of water that it uses, and the minerals and gases that cycle through it. Agriculture is the leading consumptive use of water in the OECD; an important source of nitrate, phosphate and pesticide pollution; and a source and sink of greenhouse gases. In much of the OECD, farmland dominates, and to a large extent shapes, the landscape. It serves at some point in their life cycles as a home or feeding place for a significant percentage of wild flora and fauna, particularly vascular plants, insects and birds. In short, agriculture and the ecosystems of which it forms a part, controls or interacts with various stocks of natural capital - stocks that are of value not only to the sector but to the rest of society as well.

The agricultural industries of OECD countries as a whole have succeeded in meeting not only the growing demands of the OECD countries for food and fibre, but also demand growth in other parts of the world. Overall, output has continued to expand on average in the OECD area by 3% since 1990s despite a 3% decline in total agricultural area, 10% in agricultural employment, 2% in fertiliser use and 8% in pesticide use. On the other hand, energy use has increased by 6% and water use by 3% over the same period. Thus, because of rising productivity, this growth has occurred without corresponding real increases in commodity prices. As a consequence, much less of the value-added in the chain between farm and final food consumer is contributed by farmers.

In order to supply ever-greater quantities of food and fibre, farmers have in the past expanded along the extensive margin (bringing new land under cultivation) and the intensive margin (increasing yields). In many OECD countries, most of the land that could be farmed *is* being farmed, so future growth will mainly have to be accomplished through productivity gains. If current trends continue, the capacity of OECD countries to meet the demands placed on agriculture will be increasingly determined by the informational and technological component of manufactured capital - the stock of embodied knowledge - rather than simply its mechanical or chemical force. Some producers in OECD countries, particularly those in areas deemed to have high nature-conservation value, may continue to use extensive farming practices or reduce the amount of capital and intermediate inputs they use on their farms, as they respond to incentives to practice "low-intensity" or similar types of farming systems. But the majority of producers will continue to adopt new technologies and techniques, many of which have the potential to conserve resources more effectively while maximising returns from farming.

Given the wide diversity of production systems employed in OECD agriculture, the possible developments in technologies and farming practices that are likely to have a bearing on the future demand for land and other resources are vast. Those technologies include those that improve biological potential, make crops and livestock more resistant to pests and diseases, more precisely target pests and diseases, administer nutrients and water more efficiently, and reduce wastage following harvesting.

In all OECD countries the sustainable management of natural resources is a policy goal. This paper first outlines the characteristics of natural resources, particularly those associated with agriculture and then describes and assesses the policies in OECD countries that attempt to achieve the goal of the sustainable management of natural resources in agriculture. In particular, the efficiency and coherence of policies is highlighted.

Characteristics of natural resources in agriculture

The natural resources associated with agriculture - land, water and ecosystems - are largely (renewable) flow resources that are, or can be, naturally replenished within a sufficiently short time-span to provide an indefinite stream of benefits. But they can be reduced or removed through human activity if not managed sustainably. Some of these resources are mobile, and that can complicate their management.

Resources are also distinguished by the ownership regime. At one extreme are open access ownership regimes, in which there is no means of excluding others from resource exploitation. At the other extreme is private ownership, in which rights to exploit the resource are clearly defined. Common property regimes are in an intermediate position, and can take a variety of forms. Access is restricted, but by means of conventions, norms and rules, not through ownership *per se*. Water resources often fall within this category. Public ownership is a fourth form of regime. In theory the same resources can be subject to different ownership regimes. In practice the characteristics of the resource itself (location, mobility, and degree of linkage among different resources), information shortfalls, transaction costs and historical precedent and practice often constrain the practicable choice of ownership regime.

Governments have always played a key role in determining access to natural resources. The nature and degree of government control over commercially valuable natural resources varies among resources, nations, and sub-national units. In most OECD countries, the majority of agricultural land, apart from semi-arid range-land, is vested in private hands. Water rights range from purely public to purely private, typically being more elaborately defined in areas with low rainfall. Governments have also become owners of large areas of land and water managed or set aside as nature reserves, or for recreation and other uses where the mining or harvesting of natural resources is usually limited or forbidden. Access to these parks and nature reserves varies widely, ranging from free and unlimited entry to fee-for-access.

Government involvement in the harvesting of natural resources is less common and declining. Today in OECD member countries, private individuals and companies carry out most farming activities. The main exceptions are in energy and water. Because of the large scale of investments required, or the strategic nature of the resource, many governments at the beginning of the 20th century formed state-owned enterprises to develop large-scale irrigation. In recent years, many of these enterprises have been sold to private investors. However, most governments continue to regulate access to resources under their control, through licences and permits.

A central condition for sustainable development is the preservation and enhancement of the capital base of societies for current and future generations. Natural resources in agriculture are a key part of the capital base. Three features that distinguish them from other sorts of capital are:

- If depleted or degraded natural resources cannot always be replaced or restored.
- Natural resources are an integral part of the larger ecosystem. Natural resource depletion and degradation can lead to environmental harm and reduced ecosystem services but the impacts of this environmental harm are uncertain especially in the long run.
- While any natural resources are valued directly as inputs in the production of economic goods and services (use values), other natural resource values are often not (or only poorly) reflected in markets, including biodiversity and ecosystem services like carbon and nitrogen fixation, and water catchments, which have public good characteristics.

Impacts of natural resource use in agriculture

The fundamental objective of sustainable natural resource management policy is to contribute to enhancing welfare through optimising economic efficiency and net environmental benefits within the context of relevant societal values in each country. This means taking into account commercial and non-commercial use, on-site and off-site effects, and the balance between use by current and future generations. The challenges for natural resource management policy imply minimising resource degradation and addressing discrepancies between public and private returns from resource use.

Concerning the *net environmental* impacts of natural resource use and resource degradation the challenges are to:

- Minimise the impact of agricultural pollution and waste.
- Minimise the threats posed by unsustainable farming practices on the capacity for resource renewal, the productivity of resource use, and the environmental services that the resources provide.

Concerning the *policy approaches,* the challenges are to:

- Provide the right structure of incentives and disincentives to optimise economic efficiency and net environmental benefits, including defining clear, enforceable property rights.
- Ensure that the policies implemented are coherent, work with market signals and are delivered at the appropriate level of administration.

Despite the fact that there has been little evidence of overall physical or economic scarcity in the overall supply of natural resources, this is not the case in all places at all times. Moreover, optimising the rate of use and minimising resource degradation are important issues. For renewable resources, a central policy question is the choice between using them up now, leaving them in a natural state, or pursuing intermediate strategies.

In concrete terms, policies in OECD countries play a role in relation to the sustainable development of natural resources associated with agriculture in the following areas:

- Developing property rights and markets.

- Reforming policies that provide production-linked support.
- Reducing resource degradation and enhancing ecological service provision.
- Managing publicly owned natural resource based assets.
- Dealing with information shortfalls.
- Addressing distributional implications of natural resource management policies.

Developing property rights and markets

Comprehensively assigned, clearly defined, secure and transferable property rights are a necessary condition for efficient allocation of resources among competing uses. Clearly defined property rights are also necessary to indicate who bears responsibility for the environmental consequences of resource use: who should be held to account for resource degradation and pollution, and who should be rewarded for resource enhancement and the provision of environmental services. The establishment of property rights for natural resources gives the owners of rights an incentive to optimise the exploitation of resources for different uses through time. In principle tradable property rights will be allocated to the most efficient resource users, leading to optimum productivity of the resource. In the absence of property rights for resource outputs, there may be no incentive to develop a particular use of a resource because returns cannot be secured, such as for some environmental services, and allocation of the resource may not be optimal. Alternatively there may be an incentive for farmers to exploit resources as much as possible in the short term, before they are exploited by others. This often leads to greater than optimal rates of exploitation, such as for water and land resources.

Reforming policies that provide production-linked support

Production-linked support (subsiding inputs and commodity outputs) can lead to sub-optimal use of natural resources and environmental damage in three respects:

- Excess productive capacity and effort are encouraged as producers face lower production costs, reduced risks, or increased output prices. Higher levels of output in the short-term often come at the expense of production in the future, distorting activity in favour of current production.
- The efficiency of resource allocation within the economy is reduced.
- Users receive distorted signals regarding scarcity, retarding the development of resource-conserving technologies, products and practices.

There is a long history of government intervention in agricultural resource use. Incentives for natural resource use may take form of direct budgetary monetary transfers, preferential taxation structures, subsidised provision (under pricing) of resources, and preferential rights to resource use. Examples are grants and tax incentives for land clearing. Under priced water will favour excessive rates of water abstraction increasing pressures in fragile water ecosystems. While water is often described as socially important public goods, wealthy people often benefit more than the poor, who also suffer more from adverse environmental impacts. The reduction or elimination of perverse incentives will increase economic efficiency and can improve environmental and social outcomes.

Production (commodity and input) linked support measures still remain dominant in OECD countries, on average currently accounting for over 70% of overall support of EUR 280 billion to OECD producers in 2005 as measured by the Producer Support Estimate (PSE), although this has decreased from a share of over 90% in the mid-1980s. However, only a small share of this support can be identified as directly targeted at environmental improvement (on average currently in the OECD area around 4-5% as a share of the PSE or 15-20% as a share of that part that is accounted for by budgetary support. Expenditures on agri-environmental programmes have increased rapidly in most OECD countries – from around 1% of the PSE in the mid-1980s.). Nevertheless, other payment measures are often conditional on farmers adopting environmentally friendly practices (cross compliance); services available to farmers, such as research, education, training, and information often have a high environmental content; and agriculture is subject to environmental regulations (polluter pays principle) on, for example nutrient loading in water courses or pesticide residues in food. Although agri-environmental policies can reduce environmental degradation and conserve natural resources, they can also alter relative prices and thus affect production and trade patterns.

Historically water has often been provided free or at less than full cost in OECD countries so that prices have not covered the costs or operation, maintenance and investment. This does not provide appropriate signals to bring about the efficient and sustainable water use. Studies indicate that increasing water prices provide incentives for water conservation. Farmers react to water price levels, water application costs, and water shortages. OECD work has also shown that the adoption of water-conserving technologies depends crucially on what users are charged for the resource.

Reducing resource degradation and enhancing environmental service provision

Degradation of water and other renewable resources has arisen from the impacts of pollution and waste, and inappropriate resource management practices. Examples of the former include the effects of effluents on water. The under-provision of environmental services such as habitat for biodiversity and environmental sinks is largely due to the public good characteristics and imperfect development of markets and pricing mechanism to provide an incentive to supply enough of these services.

The main strategies adopted to deal with degradation have been regulation and instruments to enforce the polluter pay principle. Management practices have gradually improved linked to the experience of the costs of sub optimal management. Support for the provision of environmental services is more recent and experience is more limited, but is important in some countries with respect to land based activities, including agriculture. A range of regulatory, market based and voluntary instruments have been used (Box 1). While regulations remain important, emphasis is shifting more to market based mechanisms and voluntary partnership based arrangements.

Largely as a result of public pressures, stricter controls have been imposed in most OECD countries on many of the larger sources of freshwater pollution, and many of the worst polluted water bodies are now being cleaned up (albeit often at fairly high cost). Non point source pollution such as run off of nutrients from agricultural land remains a problem. Few OECD countries meet baseline quality standards for inland waters (*i.e.* suitable for fishing and swimming). Most countries have also had difficulty to protecting groundwater quality, especially from non-point sources. Nitrate concentrations in excess of WHO drinking water guidelines are widespread in some European and North American aquifers. Available evidence suggests that there will be a continued worsening of aquifer water quality in the future. Once groundwater sources are contaminated, they can be very difficult to clean up because the rate of flow is usually very slow, and purification measures are often costly.

Approaches to biodiversity conservation follow the general trend with a gradual switch from regulation to market based instruments (MBIs). Much environmental-related legislation has been focused on command-and-control measures (CAC) such as different types of zoning and banning agricultural activities in specified areas, such as national parks. CAC measures are being effectively complemented and enhanced by MBIs. Examples of MBIs affecting prices are visitor's fees in national parks, taxes (Netherlands) on groundwater abstractions which can be associated with biodiversity loss, and hunting rights on agricultural land.

Box 1. Instruments for natural resource management

Regulations (command-and-control) measures: impose requirements on producers to achieve specific levels of environmental performance, and primarily consist of mandatory regulatory requirements, such as environmental restrictions, permit requirements, and maximum rights or minimum obligations, which are enforced through the legal system. Cross-compliance mechanisms require farmers to meet specific environmental conditions in order to be eligible for agricultural support programmes. In cases where support payments are relatively high, cross-compliance measures effectively serve as de facto regulations for most farms that are eligible for payments, as the potential sanction of losing that support provides strong incentives for farmers to meet the specified environmental conditions. Regulations play a key role in managing access to resources and enforcing the polluter pays principle, especially to prevent damage to human or ecosystem health. Examples include ambient standards for air and water quality, and waste disposal regulations. Regulation is also used to implement the precautionary approach for managing resources without established markets or values such as biodiversity, and for managing phenomena with uncertain impacts, such as climate change. A particular form of regulation – zoning – is used in many countries to forbid agricultural activities in specified areas of high nature value. However, regulations are often costly to implement and enforce. They often impose rigid solutions that do not reflect the diversity of situations and can deter innovation or progress beyond regulated limits.

Market-based economic instruments: affect costs and benefits of alternative actions open to farmers, with the intended effect of influencing behaviour in a way that improves environmental outcomes. They include monetary transfers – payments and charges/taxes – and the creation of new markets – tradable rights relating to the use of natural resources or pollution. There is an increasing emphasis on fees, charges and taxes for resource use or pollution. These are preferred to regulations because they maximise incentives and economic efficiency. Two types of market based instruments are price and quantity rationing. In the case of pollution control price rationing sets a charge for specific emissions without controlling total emissions. Comprehensive information on emissions is needed for implementation, and charges work best if there are relatively few polluters or pollution sites – which is often not the case in agriculture. Charges have been frequently used to handle point source air and water pollution. Quantity rationing involves establishing an overall quantity of emissions that are then allocated to producers, for example in a tradable rights scheme. Implementation requires establishment of the permit market and its rules, and monitoring trades to ensure the rules are maintained. Administration of the permit market can be expensive if there are a large number of traders, but the market may be inefficient and inequitable with a small number. Tradable permits have been used less often than charges, but they may prove more effective for dealing with non point source pollution. Information requirements are the major constraints on both types of instrument.

Voluntary agreements between farming and government can provide a flexible framework for cost effective country specific approaches to natural resource management. Examples include land-care type schemes in watersheds, whereby the government provides set up funds to the farmers concerned. Although voluntary agreements rarely produce major environmental improvements, and may be weakened by free riders they can be a valuable complement to regulation and economic instruments. In addition, Information and advisory measures include measures to improve information flows to promote environmental objectives, from the creation of knowledge to its application – for example research, extension services and product information.

One of the biggest natural resource policy challenges is incorporating non-commercial considerations into resource management decisions. Traditional water management policies often ignored the ecosystem effects of regulating water flows. Agricultural management did not take account of the role of agriculture in providing habitat and maintaining biodiversity. Such water management practices, especially in arid and semi-arid areas, have often ignored the effects of regulating flows of previously untamed rivers on aquatic and associated wetland ecosystems, leading to the retreat or disappearance of unique ecosystems once dependent on larger and irregular water flows. That situation is changing (Box 2). The Australian Water Reform Framework acknowledged "the environment as a legitimate user of water." In the United States, steps are being taken to restore some of the water flow to the Everglades, a unique wetland ecosystem threatened by agricultural pollution and urban development.

Box 2. Water resources and agriculture

The doubling of the global population, combined with generally increasing per capita demands, has resulted in approximately a four-fold increase in global water abstractions over the past fifty years (much less in OECD countries). However, global water withdrawals still represent only 8% of annually renewed water resources. None of the major regions of the world are currently suffering from high water stress (greater than 40% usage) at an aggregate level, although this is increasing in some areas. Several OECD countries have stabilised or reduced per capita (and total) water abstractions since the late 1980s while GNP rose, suggesting that per capita water use can be decoupled from growth.

Technology improvements such as more efficient irrigation systems and reduced pipe leakage are contributing to increased efficiency in water use, and reduced environmental pressures, but there is scope for much improvement. For example, only about one-third of the water withdrawn for irrigation purposes actually reaches the crop. Although drip irrigation has increased 28-fold since the mid-1970s it is still employed in less than 1% of the world's irrigated areas, primarily in OECD countries.

Continuing water quality is not assured. Many areas are already subjected to environmental and health related problems related to water pollution – contaminated drinking water (*e.g.* excess levels of nitrates, persistent organic pollutants, and heavy metals), eutrophication, and acidification. While the situation is most serious in the developing countries, especially in areas undergoing rapid urbanisation, there are problems in OECD countries too. Inappropriate water management practices exacerbate soil erosion in many regions, and lead to waterlogging or salinisation of the soils through the excessive application of irrigation water. Overdrawing groundwater resources can result in various undesirable (and usually long-term) environmental consequences - subsidence, desertification, salinisation (including sea water intrusion), wetland destruction and build-up of heavy metal concentrations.

Tradable permits for water abstraction and water pollution provide a promising approach to these problems. A number of OECD countries have successfully introduced water abstraction rights trading. The most active trading has taken place in the agricultural sector, where it has been facilitated by longstanding institutional stability, and water supply emergencies. Tradable pollution rights are more complex, and examples are confined to specific pollutants such as salt, organic and nutrient pollution. Water trading affects many people other than buyers and sellers, and ecosystems. This means that the overall implications of trading schemes are complex have to be assessed on a case by case basis. More generally, the costs of introducing tradable rights are uncertain, and schemes require substantial information, strong management institutions and consideration of equity issues.

The OECD and the Australian authorities hosted a ***Workshop on Water and Agriculture: Sustainability, Markets and Policies*** in Adelaide on 14-18 November 2005 Workshop, Adelaide, Australia, 14-18 November 2005. The major challenge identified for the sustainable use of water resources in agriculture is to manage community expectations to meet social and environmental aspirations, while ensuring that food and fibre is produced competitively and profitability.

Irrigated farming accounts for a major and increasing share of agricultural production, but overuse of water resources is an increasing concern. In addition, the growing incidence and severity of droughts linked to climate variability and climate change is placing pressure on farming and water resources. Overexploitation of water resources by agriculture, within some regions across certain OECD countries, is leading to:

- Reduced environmental flows in rivers and lakes.
- Natural recharge rates of aquifers being exceeded.

- Increased competition for water resources between farmers and other demands for water, including the maintenance of aquatic ecosystems.
- Higher agricultural energy intensity, as the expansion of irrigated farming usually leads to an increase in the energy requirements to support this system of farming.

Over recent years, there has been a shift by farmers and policy makers in most OECD countries from water resource exploitation to water resource and environmental management. This is associated with changing societal demands, as farmers seek to both improve their efficiency in the use of water resources and also address the growing societal interest in the conservation of aquatic ecosystems.

There is also a greater public awareness that water used by agriculture is not a "free" good for personal benefit, but one that imposes costs and generates benefits. Although water application rates per hectare irrigated have been improving in many cases, wastage and inefficiency in water use remain high, associated with poor maintenance of irrigation infrastructure and a low rate of adoption of efficient irrigation technologies, such as drip emitters. Under some farm management practices and farming systems agriculture maintains and enhances certain ecosystem services related to water, such as maintaining water meadows and facilitating groundwater recharge.

Production and input subsidies continue to misalign farmer incentives and aggravate overuse and pollution of water across many OECD countries. Market price support provides incentives to intensify agricultural production, while support for irrigation infrastructure capital (construction and depreciation costs), operation and maintenance costs (including institutional costs) together with support to lower water supply charges, discourages the more efficient use of water resources. Energy subsidies to agriculture in some countries, by lowering pumping costs, are aggravating the depletion of aquifers and increasing the energy intensity of irrigated agriculture.

Policies and actions are beginning to shift toward more sustainable agricultural water management in OECD countries as policy makers are giving higher priority to water issues in agriculture and are using a mix of market-based, voluntary and regulatory approaches to address these issues. There is a widespread recognition of the need for greater use of market based instruments, such as better pricing structures and tradable permits, accompanied by government regulations, as well as cooperative efforts among water users. But the adoption of these measures should take into account the frequent regional imbalances of water resources within countries and the negative and positive environmental externalities arising from agriculture's use of water. A growing concern is the impact of agricultural policy on opportunities to mitigate or adapt to climate change and climate variability as they affect the water sector.

Countries are at different stages in reforming their water policies, partly reflecting the varying importance of water related issues in agriculture across OECD countries and current systems of property rights and management structures. But all countries need to reinforce the monitoring and evaluation of current water policy reform initiatives to ensure that these reforms are moving toward sustainable agricultural water management.

The Workshop recommended a number of issues that could be addressed by policy decision makers ranging from decision makers at the watershed through to national levels, including:

- Using an appropriate mix of instruments and tools aimed at addressing agriculture resource management issues to ensure the achievement of coherent agricultural, environmental and water policy goals as well as cost effective implementation (*e.g.* integrated policy treatment of water and energy input use by agriculture), including coordinated policy responsibilities and structures at different levels from the watershed to national level.
- Integrating and expanding current scientific research and data collection capacity to underpin improved policy making, including better water accounts.
- Identifying property rights attached to water withdrawals, water discharges and ecosystem provision.
- Establishing clear lines of responsibility in the institutional framework to manage water – who does what, who pays for what, who monitors and evaluates – underpinned by a long term commitment from governments to resource the necessary actions, especially with the growing concerns related to climate change and climate variability.
- Strengthening water policy reforms to provide a robust regulatory framework to allow, for example, for water pricing and trading, and water service competition or benchmarking performance where competition is limited, and nutrient trading for pollution abatement.
- Raising the capacity for stakeholders (farmers, industry and community groups) to participate in the design and delivery of policy responses for integrated water management.

Managing publicly owned natural resource assets

Governments continue to be heavily involved in the management of some natural resource related infrastructure, especially for water, for which large investments are likely to be required in coming decades. Greater involvement of water users, including private sector firms and communities, is a critical element of successful strategies for the sustainable management of water resources. Social instruments designed to modify water user behaviour to increase the efficiency of use and to conserve water: education, information, and partnership strategies are needed which foster both user "ownership" and responsibility in water resources management. In a number of cases, governments have gone beyond simply a "public consultation" process in making water management decisions, and developed a broader multi-stakeholder process involving water users in the development and implementation of water management policies. In some cases, the full management of irrigation water systems has been transferred to Water User Associations, placing the responsibility for the sustainable use of water resources and the recovery of the costs of supply in the hands of the irrigation users.

Dealing with information shortfalls

Good and effective natural resource management policy requires sufficient and relevant information. These include information and data on:

- Commercial resource stocks, flows and depletion.
- Technologies, recycling and substitution.
- Physical properties and values of environmental services provided by natural resources (including the effects on services of reductions in resources), and links between resource exploitation and these services.

Information on commercial resources and their values is readily available but there are shortfalls in other elements such as new technologies, substitution and recycling possibilities, links between natural resource exploitation and ecosystems and the values of non commercial outputs. In particular there is a great deal of uncertainty about the long term environmental and economic impacts of natural resource degradation. Such shortfalls could favour more (or less) rapid depletion, or incorrect assessments of the environmental implications of policies.

In the 1990s, OECD countries made considerable progress in providing information to the public on new technologies and the environment. For example, provision of environmental data, indicators, state of the environment reports and other types of reports as well as electronic access to environmental information is now often done on a routine basis. However, little research has been done on the links between natural resource exploitation, environmental services provided by natural resources and ecosystem productivity.

Concerning monitoring and enforcement, the pressure to improve environmental quality came primarily from government regulators. Continuing practical difficulties have been experienced in providing information in the public domain, for example because of scattered sources of information, inadequate electronic communication infrastructure and confidentiality constraints.

Addressing distributional questions implications of natural resource management policies

There are tradeoffs between conserving resources to preserve the capital base for future generations, and consuming natural resources to benefit current generations – whether these current benefits are used to reduce poverty and address distributional issues is a further question. A central issue in the context of sustainable development is whether the earth's ecosystem can sustain the high pressure on natural resources that would result if all countries were to adopt food consumption patterns similar to those prevailing in most OECD countries. This raises difficult questions, but the solutions depend in part on the resource management decisions in OECD countries through markets and policy approaches over the coming decades and whether it proves possible to enjoy similar levels of well-being as in the past at lower environmental cost.

When resources are privately owned and traded, rents tend to become explicitly incorporated into the purchase price of farm land. When the resources are vested in the state, rents tend to become implicitly incorporated in the value of resources and government managers (or political governing bodies) must decide on how to collect and spend (or invest) the rent.

Practices vary among countries and across different natural resources. In the case of privately owned agricultural land, it is common for many governments to tax the market value of the property. Access fees are typically charged for publicly owned range-land. Examples of government attempts to capture resource rent from water users (especially farmers) are rare. While a few countries collect special fees to pay for government-provided services (such as monitoring and research costs), these are not directly related to the economic rent derived from the resource. Understanding the establishment, valuation and expenditure of resource rents is an area in which further analysis is needed.

Much progress has been made in developing policy frameworks for the sustainable management of natural resources. Yet the actual transition to such policies is likely to give rise to adjustment and distributional problems, creating resistance to policy change. Ways of "managing the transition" is an area where the trade-offs between natural resource sustainability and social implications are most evident.

Managing the transition to the sustainable use of natural resources is complicated by the complex web of laws, traditions and property rights that have evolved along with them. Inappropriate incentives over time create expectations that are costly and painful to change. Many government financial transfers, including those provided through sector-specific policies, may have contributed to structural imbalances. Yet adjusting use levels may create short-term redundancies in capital and labour.

Concluding comments

Collectively, the agricultural sectors of OECD countries have managed to achieve unprecedented growth in the output of food and non-food commodities since the middle of the 20th century. They have done this with fewer workers and on slightly less land, but using more water, chemicals and machinery. Increased levels of pollution, however, have often accompanied agricultural growth, as well as greater homogenisation of the landscape, and destruction of wildlife habitat. Problems of soil degradation, loss of genetic resources and water depletion persist in some areas. But agricultural activities have also maintained ecosystem services and traditional landscapes in some areas. Many agricultural and associated trade policies have depressed world prices and market access, reducing agricultural sustainability, especially in the developing countries. However, a combination of policy and trade reform, agri-environmental policy measures, international commitments, market and public pressure have led to some improvements in the overall environmental performance of agriculture, albeit with areas of concern with regard to water resources and localised environmental problems.

There is a consensus that an increase of up to 50% in the projected demand for food and fibre could be expected in the next two decades, which could be met with no increase in real prices should yield growth from the application and dissemination of existing and new technologies continues. However, projections by the FAO suggest that growth in both population and per-capita consumption mean that global food output will need to nearly double by 2050. This will place enormous pressures on resources, especially land and water, and will present significant challenges for research and technology. Key requirements for achieving increased output in a sustainable way are implementing of appropriate agricultural and other policies, in tandem with markets and the adoption of best practices by farmers.

Policies in OECD countries are giving high priority to sustainable agriculture, although only a small share of support to the sector can be directly attributed to achieving this objective. In many countries there is a lack of coherence between the signals given to resource owners and users from production-linked policies and those more closely targeted to sustainable resource management. There appears to be some convergence in policy objectives and instruments used across OECD countries but there is often a dampening of incentives to explore market-based solutions to agri-environmental concerns in countries with high levels of production–linked support. Thus the cost of achieving environmental improvements through specific agri-environmental measures is higher than would be the case in the absence of production-linked support, which still predominates in many OECD countries. The emphasis in natural resource management policy is shifting towards preventing maintaining resource degradation and non commercial outputs such as environmental services. Some OECD countries have taken steps to recognise environmental uses of water. A range of market and partnership based mechanisms to conserve biodiversity are being developed to complement regulation.

How the agricultural sector manages land and water is crucial to sustainable development. With proper market signals, policies and management, agriculture can make more positive and fewer negative contributions to the overall welfare of society. To do so, however, many of the policy measures currently influencing agriculture need to be changed. The risks to agriculture's resource base of delaying reforms will grow even larger in future years, as pressures on agriculture increase in line with rising populations and competition for other uses of the resources on which agriculture depends.

The over-riding policy objective should be to maximise agriculture's net benefit to society. These benefits derive from contributing efficiently (including through trade) to satisfying current and future demands for adequate, safe and reliable supplies of food; non-food commodities; and environmental services. In order to meet these demands, the sector has to be able to respond efficiently and innovatively to market signals; to take into account external costs or benefits it imposes on or provides to other segments of society; and to maintain sufficient flexibility to cope with change. These requirements in turn are most likely to be fulfilled if:

- Markets for natural resource exploitation, inputs and commodities are not distorted by inappropriate government policies or excessive market power.

- Market failures and public goods are addressed.
- Sustainable technologies for resource use and waste reduction are encouraged.
- The resource base on which agriculture depends is maintained in a state commensurate with the demands that it is likely to fulfil.
- Effective property rights – ownership and access rights – are a pre-condition for the development of markets and efficient resource allocation.
- Clear and consistent signals from markets and policies are provided to producers and consumers of agricultural products and services as to the environmental consequences of their actions.
- Farmers can benefit from gaining access to research and knowledge on how to make more efficient use of the natural resources used in agriculture, while at the same time avoiding degradation of both those and other resources with which agriculture comes into contact.
- Opportunities are available for resources (including farmers) to move in and out of the sector.

Agricultural reforms would permit resources to be used more efficiently as market signals, not government allocation rules, would guide producer decisions and stimulate income and employment growth in the rest of the economy. Equally, fewer trade restrictions could widen consumer choice at competitive prices. Such reforms would also allow more-specific targeting of individual policy measures to clearly defined outcomes and intended beneficiaries. Developing country exporters stand to gain from greater access to OECD country markets and from reduced competition with subsidised exports. As importers, some of them may be faced with higher prices. But this potential cost could be outweighed by improvements in the allocation of domestic resources and the fact that some developing countries are net importers because of import barriers preventing them from exporting commodities in which they have a comparative advantage.

In many cases, agricultural trade liberalisation can also contribute to overall improvements in environmental performance. The direction and magnitude of the overall impact will depend on several factors, including the trade liberalisation-induced changes in agricultural production patterns; the state of the environment at the time of the trade policy reform; and the environmental regulations and policies in place to preserve and improve environmental quality. Also, because of the considerable diversity of agricultural production systems and natural conditions, the differing regulatory approaches within and between OECD countries, and the fact that many environmental effects are site-specific, the environmental impacts will vary from one location to another. A reduction of trade barriers could therefore end up increasing environmental quality in some countries, but reducing it in others, in particular if the trade policy reforms in the latter lead to changes in patterns or levels of production that, because of insufficient safeguards, lead to major increases in negative environmental externalities or reductions in positive externalities.

In conclusion, the following are the key policy recommendations to enhance the sustainability of natural resources:

- *Ensure that market signals guide production decisions*. A central requirement is to further reform agricultural and trade policies so that market signals play a greater role in allocative decisions, encourage the more efficient use of scarce resources, and thus reduce pressure to farm in environmentally damaging ways. Markets provide key information for the agricultural sector to adjust and adapt to changing conditions of supply and evolving demands, which is essential to achieve sustainable agriculture in a dynamic context.

- *Facilitate the creation of markets to take account of externalities*. Externalities and public goods arising from agriculture need to be taken into account in farmers' decisions. That means the clear designation of property rights so that the polluter-pays principle should be applied to deal with negative externalities, and developing innovative ways for markets to return to farmers the extra costs that they incur in providing the public goods and services that people want but currently do not fully pay for through market transactions. Where markets cannot be created to deal with these problems, the challenge will be to identify who causes pollution and who generates public goods, and to find targeted, more cost-efficient policy measures to provide them. Policies will be more effective in achieving sustainable agriculture if they are designed to work *with* and not *against* market signals.

- *Strengthen the agricultural knowledge system*. Research and development, education, training and advisory services in agriculture continue to evolve to meet the increasing opportunities from emerging technological developments, and the demands and concerns of a well-informed and discriminating public. A key aim should be to encourage farmers to take initiatives to adopt sustainable farming methods, and to develop new market opportunities. Particular consideration needs to be given to the needs and perspectives of developing countries, and for targeted capacity building.

- *Improve policy coherence*. Moving towards sustainable agriculture has to be viewed in the wider context of sustainable development. In several countries there are often conflicting policies - some result in environmental damage that lead to other policies to redress the damage, while others lead to unintended distributive effects within agriculture and on society. Improving the coherence of policies within agriculture and with other areas is crucial. Providing better data on, and scientific understanding of, the processes that are shaping the long-run potential of agriculture and the natural resources with which it interacts, and the effects of changing patterns of production and technologies on the environment and social conditions could aid policy makers. Achieving greater policy coherence requires entering into dialogue with stakeholders, with the aim of elaborating better policy measures, and identifying where institutions might need improvement. A corollary is the need for better indicators on trends in the sustainability of agriculture, and associated analytical tools. Establishing regular monitoring and evaluation procedures and building in greater flexibility to adapt and change policies is also essential.

Annex

Agri-Environmental Policies in Selected OECD Countries

AUSTRALIA

A range of policies have been implemented to address agri-environmental concerns. The *National Landcare* programme, in which about 40% of farmers participate, provides funding for community based approaches, with nearly AUD 160 (USD 117) million over 2004-08, through promoting the exchange and transfer of information. Landholders can claim a tax deduction for expenditure relating to *Landcare* operations and water storage. Funding of AUD 18 (USD 13) million is available under the *Environmental Management Systems* programme to improve farm management.

Agriculture is implicated by a range of environmental policies. The *National Strategy for Ecologically Sustainable Development* provides the framework for most environmental and natural resource policies, and funding to states/territories to enact legislation supporting national strategies. The *National Action Plan for Salinity and Water Quality* aims to reverse salinity and water quality problems, with funding of AUD 1.4 (USD 1.0) billion over 2000-08, while *The Natural Heritage Trust,* focuses on biodiversity and sustainable natural resource management, with funds of AUD 1.3 (AUD 0.9) billion over 2004-08. Under the *National Water Initiative* funding of AUD 2 (USD 2.5) billion is provided for programmes, which include irrigators, to move toward full cost recovery for water, expand trade in water, improve access entitlements, plan for environmental needs and enhance water management. Farmers are provided rebates for on-farm diesel use, equal in 2005 to AUD 660 (USD 500) million of budget revenue foregone, while biofuels are subject to lower taxes than fossil based fuels. The *Greenhouse Challenge Agricultural Strategy* is a voluntary based farmer aiming to reduce greenhouse gas emissions. Farming is affected by Australia's commitment under the *Montreal Protocol* to eliminate by 2005 methyl bromide use, an ozone depleting substance.

CANADA

The key agricultural policy legislation in Canada is the federal-provincial-territorial Agricultural Policy Framework (APF). This five-year policy (2003-08) has five pillars: food safety and quality, science and research, sector renewal and skills development, risk management and environment. For environment the objective is to accelerate efforts to reduce agricultural risks and provide environmental benefits, supported by on-farm action and measured against established clear and measurable agri-environmental targets, using agri-environmental indicators.

Measures aimed specifically at agri-environmental concerns vary considerably, and include, regulations, financial incentives, voluntary measures and research. Through the APF, there are various programmes such as the Environmental Farm Planning Program, the National Farm Stewardship Program, the Greencover Canada Program, and the National Water Supply Expansion Program. In addition, tools are being developed to help farmers make better land-use management decisions through, for example, the National Land and Water Information Service.

The main environmental policies affecting agriculture include legislative and regulatory measures through such federal acts as the Canadian Environmental Protection Act; the Canadian Environmental Assessment Act; the Species at Risk Act; and the Pest Control Products Act. In general, the objectives of these laws are to conserve Canada's environmental resources and minimise public health risks caused by environmental degradation and pollution. Agricultural practice is also regulated at the provincial and municipal level.

FRANCE

Expenditure on agri-environmental programmes increased over the 1990s, and now accounts for 15% of total national agricultural expenditure. Since 2000 agri-environmental programmes have covered the whole country and most farmers, aiming to promote diversified cropping patterns, crop rotation, and sustainable farming practices. The programmes provide payments over 2000-06, such as for hedge maintenance and converting arable to grassland. Also, support is provided for integrated farm management; conversion payments for organic farming, which occupies less than 1% of agricultural land; and eco-food labels are promoted.

Subsidies and taxes address agricultural pollution. A programme to control agricultural water pollution covers 60% of the costs of constructing manure and slurry storage facilities for 50 000 farmers and amounts to EUR 1.28 billion over 2000-06, a nine fold increase since the early 1990s. Pollution taxes are levied on phosphates for all farmers, and on nitrates for large livestock producers based on emission estimates. Some herbicides are banned and a pesticide tax, introduced in 2000, levied relative to toxicity, involves pesticide producers paying EUR 40 million annually. Voluntary initiatives, such as *Ferti-Mieux,* encourage improved farm nutrient management.

Farming is subject to economy-wide environmental measures. A biofuel production scheme from 2005 aims to raise the share of biofuels in transport fuels to nearly 6% by 2010 through production support and fuel tax reductions, while biomass and animal waste used for energy generation benefit from higher tariffs into the national grid. A diesel tax concession (about one seventh of the normal rate) is provided to farmers involving about EUR 950 million annually of budget revenue foregone. Irrigation is supported, both infrastructure costs (up to 65%) and water charges (about one-fifteenth of household charges). Commitments under international environmental agreements, such as lowering nutrient loadings (into Lake Geneva, the Rhine, and North Sea), ammonia emissions (*Gothenburg Protocol*), and methyl bromide use (*Montreal Protocol*) affect farming.

GERMANY

Expenditure on agri-environmental programmes in Germany has risen substantially and is largely administered at the Länder level. Currently spending on agri-environmental measures is about 13% of total farm expenditure. This expenditure is largely aimed at providing payments to farmers for environmentally friendly farming practices (*e.g.* to reduce water pollution and enhance biodiversity conservation, and organic farming). Also there are regulatory measures that enforce certain environmental farming practices, such as on fertiliser application and livestock densities; while the 1998 *Federal Soil Protection Act* requires farmers to adopt soil conservation practices. Organic farming is also encouraged through the *Federal Organic Farming Scheme*. Organic farming accounts for over 4% of farmland in 2003, and the aim is to increase this to a share of 20% by 2010.

Agriculture is affected by a number of economy-wide environmental measures. Farmland in nature conservation areas is exempt from property tax. Farmers are also provided an 80%

exemption on the standard rate of tax on fuels, although this exemption will be reduced to 40% in 2005. Under the Renewable Energy Sources Act, grid operators are obliged to buy electricity using a differentiated feed-in tariff. Biofuels have tax exemptions and biomass installations for heat production are supported by the Government. Farming is also affected by commitments under international environmental agreements, in particular, the reduction of nitrate pollution into the Northeast Atlantic (*OSPAR Convention*) and the Baltic Sea (*HELCOM Convention*), and ammonia emissions under the *Gothenburg Protocol*.

ITALY

Expenditure on agri-environmental programmes has risen substantially, accounting for 10% of total agricultural payments in 2002, of which over 80% was EU co-financed, with 90% provided to farmers in central and northern Italy and 10% for the south. About 90% of payments were provided for conversion to organic farming, adoption of integrated farming, and grassland management. Other measures aim to reduce erosion, limit water use, and enhance biodiversity conservation, such as payments of EUR 202/head for endangered cattle species.

Agriculture is affected by a number of economy-wide environmental measures. The 1992 *Hunting Act* requires that 20-30% of agricultural and forestry land should be devoted to fauna protection. Water abstraction charges were introduced in 1994 under the *Galli Act* at very low rates for farmers of EUR 36/100 litres/second compared to EUR 1 550 for households and EUR 11 362 for industry in 2001, while subsidies are also provided for irrigation capital and operational costs. A pesticide tax, introduced in 1999, is 2% of the retail price; and a reduction of 22% of the full fuel tax is provided for agriculture, estimated to cut variable costs by about 14%. Incentives for biofuels are provided, mainly for biodiesel, through exemptions on excise duties amounting to EUR 300 million over the period 2002-05. Farming is also affected by commitments under international environmental agreements, such as lowering ammonia emissions (*Gothenburg Protocol*) and methyl bromide use (*Montreal Protocol*), and addressing desertification (*UN Convention to Combat Desertification*).

JAPAN

Japan relies mainly on budgetary payments to address agri-environmental issues. Expenditure on agri-environmental programmes more than doubled over the 1990s, but represents only 10% of total payments to farmers. Adoption of sustainable agricultural practices is encouraged by concessionary loans, tax relief and payments to farmers' groups to help reduce fertiliser and pesticide use, and also a mandatory code of practices for pesticide application. Payments to mountain farmers aim to prevent conversion of farmland to other uses and maintain a range of ecosystem services associated with farming in these areas. Irrigation and drainage infrastructure is also financed from the national budget.

Tax exemptions, low-interest loans, regulatory standards and other policy instruments are also used to address agri-environmental issues. Farmers, and some other users, are exempt from water charges, as well as energy and fuel taxes. In 1999 measures where introduced to enhance livestock manure management, including the use of subsidies, low-interest loans and credits, and regulatory standards for manure storage. Some Prefectures (local governments) finance facilities that recycle farm waste, such as manure, and others set targets to reduce farm nutrient pollution of water. Regulations under the 1970 *Water Pollution Control Law* set upper limits for agricultural pollution, such as from pig and cattle units, and the 1972 *Offensive Odor Control Law* covers livestock.

KOREA

"The Agro-Environmental Policy towards the 21st Century" was launched in 1996 to address environmental issues in agriculture. The initiative seeks to limit harmful impacts of agriculture on the environment and encourage wider use of practices which can reduce environmental pressure, such as Integrated Pest and Nutrient Management, and organic farming. While fertiliser and pesticide inputs are subsidised, Government plans are to reduce their use by 30% by 2005 from 1999 levels. Irrigation water charges, investment, operation and maintenance costs are also subsidised. The Government is seeking to address biodiversity concerns related to agriculture, in particular, by halting a number of projects that would have reclaimed wetland and tidal habitats for farm use, and by introducing wetland preservation schemes in cooperation with the Global Environmental Facility.

Cross compliance conditions have been linked to budgetary payments to reinforce existing agri-environmental measures. The Direct Income Support for Paddy Field programme imposes environmental cross-compliance conditions to the payments granted to paddy fields on a per hectare basis, which are conditional on farmers reducing the use of fertilisers and pesticides. Since 1999 Direct Payments for Environmentally-friendly Farming were introduced to restrict the use of fertilisers and pesticides in drinking water conservation areas and also for soil conservation practices. The measure was broadened in 2002 by making payments available nationally, with eligibility based on the amount of chemicals used and in the case of soil conservation practices according to local soil fertility and climatic conditions.

MEXICO

Policies addressing agri-environmental are limited. Agri-environmental payments are possible under PROCAMPO, for soil and water conservation, although farmer uptake of these payments has been limited to date. A number of programmes support forestry but only one aims at the reforestation of farmland, and eco-certification of shade-grown coffee plantations is being developed. Budget transfers to the National Water Commission reduce farmers' irrigation costs, while an energy subsidy also reduces farmers' power costs for pumping water.

Economy-wide environmental policies also affect agriculture. Under the *Federal Law on Water Taxes* (1982), a system of water abstraction charges was established, but farmers were exempt from these charges up to 2003, although they are liable for water pollution charges introduced in 1992 under the same law. The *International Boundary and Water Commission* resolves water issues at the Mexican – United States border, including sharing water resources for irrigation, while the *North American Commission for Environmental Cooperation*, established under NAFTA in 1994, addresses regional environmental issues, for example, transgenic maize. The *National Environment Programme* also provides a framework for biodiversity and natural resource conservation.

NEW ZEALAND

Regulation is the main policy measure used to achieve agri-environmental objectives. Two nationwide overarching policies address environmental concerns: the *Resource Management Act* (RMA, 1991) and the *Hazardous Substances and New Organisms Act* (HSNO, 1996). The RMA integrates measures governing resource management, and its key themes are: sustaining the potential of natural resources; safeguarding the quality of soil, water, air, and ecosystems; and avoiding, remedying or mitigating adverse effects on the environment. The HSNO aims to protect the environment by preventing and managing the adverse effects of hazardous substances, including pesticides and new organisms - any animal, plant or microbe - not currently present in New Zealand.

The environmental policy framework affecting agriculture is characterised by decentralisation of decision-making and devolution of responsibility. The RMA and HSNO are implemented by 74 territorial local authorities, which are empowered to determine their own priorities for managing the environment. The local authorities charge farmers in order to recover costs associated with programmes and applications, while responsibility for resource management is with farmers.

UNITED KINGDOM

Expenditure on agri-environmental programmes increased 5 fold over the period 1993-2004, rising to GBP 245 (EUR 360) million. Following the government's 2002 *Strategy for Sustainable Farming and Food*, together with the *Rural White Paper* and CAP reforms, agri-environmental programmes are being developed to encourage sustainable practices across all farms and to continue with conservation of high priority habitats and landscapes. Support is also provided for conversion to organic farming, while voluntary *Codes of Good Agricultural Practice* (soil, water, air) encourage farmers to minimise water pollution.

Agriculture needs to respect national and international environmental policies. The *Bioenergy Infrastructure Scheme* provides farmers grants to expand biomass and bioenergy production, linked to consumer tax reductions for biodiesel and bioethanol. Diesel fuel tax is reduced by nearly 90% for farmers, involving around GBP 180 (EUR 265) million annually of budget revenue foregone. National agricultural targets are included under the *Biodiversity Action Plan*, such as reversing the decline in farmland birds by 2020. Farming is affected by commitments under international environmental agreements, including lowering: nutrient loadings into the North Sea (*OSPAR Convention*), ammonia emissions (*Gothenburg Protocol*), and methyl bromide use (*Montreal Protocol*).

UNITED STATES

Agri-environmental programmes form a growing dimension of agricultural policy. The *Conservation Reserve Program* (CRP) aims to remove from production highly erodible cropland (HEL) and the *Wetlands Reserve Program* (WRP) seek to re-convert farmland to wetlands. Both the CRP and WRP retire land for periods varying from 10 to 30 years in return for annual payments. Under the *Environmental Quality Incentives Program* (EQIP) and the *Wildlife Habitat Incentives Program* (WHIP) payments defray costs for respectively adopting sustainable farming practices, such as for soil and water quality conservation, and providing wildlife habitat. The *Farmland Protection Program* (FPP) aims to avoid farmland being converted into urban use by purchasing development rights to farm properties. *Cross-compliance* provisions also require that to receive payments under commodity programmes farmers must not cultivate HEL (*sodbuster*) or drain wetlands (*swampbuster*).

The 2002 Farm Act substantially increases funding for agri-environmental policies. For the period 2002-07 funding will be USD 3.5 billion annually, a 75% increase over the annual spending for 2000-02 of USD 2 billion annually which was 8% of budgetary payments. The *Farm Act* expands the CRP and WRP but its emphasis shifts to supporting conservation practices, especially under EQIP. Two new measures, the *Conservation Security Program*, pays farmers to adopt or maintain practices to improve soil and water quality or wildlife habitat; and the *Grassland Reserve Program* aims to preserve and improve native grass species. The *Farm Act* also supports technical advice and research to promote sustainable farming. Of the nearly USD 17 billion irrigation construction expenditure considered reimbursable by the Federal government, irrigators have been allocated USD 3.5 billion to repay over 50 years at zero interest. Water charges are considerably lower than those paid by industrial and urban users. The *International Boundary and Water Commission* resolves water resource allocation issues, including for irrigation, at the US – Mexican border.

Economy-wide environmental policies also impact on agriculture. Between 1994 and 1998, seven agencies provided USD 3 billion annually to address nonpoint source pollution. The *Clean Water Act* (CWA) has responsibility for reducing water pollution, but nonpoint sources of pollution such as agriculture are not directly covered by the CWA, although large confined animal feeding operations require pollution permits and implement comprehensive nutrient management plans. Policies affecting agricultural water pollution are mainly implemented at the State level, using a mix of measures which vary across States, such as restrictions and taxes on fertiliser and pesticide use and payments for adoption of best management practices. Some Federal agri-environmental measures directly (*e.g.* EQIP) and indirectly (*e.g.* CRP, WRP) affect water quality, as adoption of soil and water conservation practices can help to reduce off-farm flows of soils, nutrients and pesticides into water bodies. Also the *Great Lakes Water Quality Agreement* between the US and Canada addresses concerns related to agricultural water pollution. To reduce urban vehicle carbon monoxide pollution the *Clean Air Act* requires the use of oxygenates in about a third of national petrol use, which has increased the demand for ethanol, mainly produced from maize, already used as a fuel extender and octane enhancer. A tax exemption is provided for the use of ethanol and assistance granted to develop ethanol production facilities, while there are exemptions on Federal fuel taxes for on-farm machines and vehicles. The US is a signatory to the *Montreal Protocol,* which provides a phase out period for the ozone depleting methyl bromide pesticide, and the *Gothenburg Protocol* on long-range transboundary air pollution, which includes ammonia.

REFERENCES

This paper has drawn on work in the OECD in recent years, in particular, *Agriculture and the Environment: Lessons Learned from a Decade of OECD Work* (2004); *Analytical Report on Sustainable Development* (2001); *Environmental Indicators for Agriculture Volume 4* (forthcoming 2006); *Inventory of Agri-Environmental Policy Measures* (OECD website); and *Monitoring and Evaluation of Agricultural Policies in OECD Countries* (annually).

Chapter 6.

MARKET MECHANISMS IN WATER ALLOCATION IN AUSTRALIA

Seamus Parker[1]
Department of Natural Resources, Mines and Water, Queensland, Australia

Abstract

Market mechanisms are now a central element of the framework for the allocation and management of water resources in Australia. Extensive reforms have given rise to a new system of water management aimed at ensuring the environmental sustainability of river systems while allowing greater flexibility for water users.

In many catchments, water entitlements are now detached from land and able to be sold to other water users on either a temporary or permanent basis. New enterprises in need of water in fully allocated catchments can purchase water from the market, and there are now significant economic drivers for existing water users to make the most efficient use of their current supplies.

This paper considers how water trading has been introduced to the water allocation process in Australia, which is a federation of State Governments and a Federal Government with responsibilities defined by its Constitution. Significantly, water resource management in Australia falls within the Constitutional responsibilities of the State and Territory governments. The reform process has been characterised by high level intergovernmental agreements to policy outcomes implemented by the States, supported with financial incentives from the national government. The paper outlines the many steps taken along the way to create the institutions and the legal and planning frameworks necessary to allow for a functional water market.

Historical perspectives and the need for reform

The history of water resources law, allocation and development in Australia is in many ways similar to that in other western democracies – State control of water through vesting provisions in laws around the start of the 20th century, licences granted by the State to take water, and the establishment of government departments with a focus on engineering and resource development through the construction of State-subsidised infrastructure, mostly for irrigated agriculture.

Water was typically allocated through the grant of licences to water users, with licences assessed on an individual basis without the use of any broader planning mechanism. In the case of supplemented water – that is, water supplied from a (typically

1. The author gratefully acknowledges the comments and contributions from Simon Smalley, Fiona Bartlett and Robert Speed, Department of Agriculture, Fisheries and Forestry, Australian Government, and Beatrix Brice, Department of Natural Resources, Mines and Water, Queensland Government. Further information on the reforms as they have been implemented in the State of Queensland can be found at www.nrm.qld.gov.au.

State owned and operated) dam – the licence described both the right of the holder to water as well as the terms under which it would be supplied.

Licences for unsupplemented water – that is, to take water from natural river flows – detailed the works that could be used to take the water (for example, a 12-inch pump) and the location of the works, described by reference to the related land title. The size limit on the works provided a *de facto* limit (often the only limit) on water extraction.

Licences to take water were normally granted for set periods, usually between two and ten years, with no statutory right of renewal, and licences could be cancelled, amended or changed at any time with no rights to compensation. In addition, the terms of the licences gave the chief executive of the administering department wide discretion in terms of the management of licences from dams (supplemented supply), and little certainty in terms of the entitlements from non-supplemented supplies.

In the 1990s, a growing demand for water resources, heightened environmental pressures, and a changing economic environment highlighted the inadequacies of this management framework. In particular:

- "Incremental licensing", where licences were dealt with on a case-by-case basis rather than assessing the cumulative effects of the grant of many licences over time, became unsustainable as demands increased. While the State continued to grant licences, in some places this may have affected the reliability of those licences previously issued.
- In fully allocated catchments, because licences were attached to land, the only way for someone without a water licence to obtain one was to buy land. The alternative was to allow for trading of entitlements but the existing framework did not allow for this.
- In making decisions about granting licences, there was no specific requirement to consider how much water was required for environmental purposes.
- Uncertainty associated with licences threatened investment in water infrastructure and water-dependent developments.
- Growing conflicts over water resources highlighted the limited rights of third parties to be involved in decisions that could ultimately affect a whole catchment.
- Where there were disputes, the limitations of an adversarial approach to addressing what were really planning issues.
- Pricing reforms – to address externality issues and to recover some of the cost of subsidised water schemes – required that the product (*i.e.* the water entitlement) be better defined.

The water reform process undertaken across Australia over the past decade has sought to address these issues through a combination of legislative, institutional and pricing reforms.

The water reform process

The reform process began in earnest with the agreement of the Council of Australian Governments (COAG) on water reform,[2] part of a major economic reform agenda under

2. National Competition Council, *Compendium of National Competition Policy Agreements – Second Edition,* June 1998. Available at www.ncc.gov.au/publication.asp?publicationID=99&activityID=39.

the title of the National Competition Policy aimed at increasing the efficiency of the Australian economy. Under the agreement, the State and Territory governments – which have constitutional responsibility for managing water resources – committed to implementing a range of measures aimed at achieving an efficient and sustainable water industry, in return for payments from the national government.

The States and Territories were required, amongst other things, to:

- Establish and clearly define the roles and responsibilities of regulatory and service provider institutions.
- Implement a comprehensive system of water entitlements backed by separation of water property rights from land title.
- Implement trading arrangements for water entitlements.
- Make allocations of water for the environment as a legitimate user of water, including ensuring allocations are made on the best scientific information available.
- Implement water pricing based on cost recovery and the volume of water used.

More recently, the water reform agenda has been given further energy by an updated agreement, the National Water Initiative (NWI)[3] and the establishment of the National Water Commission to oversee its implementation.

Implementation of the reforms has required changes to institutional, pricing and entitlement arrangements. These three key areas of reform are interrelated, and one of the strengths of the approach adopted in Australia is how they have been dealt with as a "package". For users in particular, while the pricing reforms have been unpopular, the long term benefits of better defined, tradable water entitlements has lead to a degree of support for the reform package. The connections between these reform areas are shown diagrammatically in Figure 1 below.

Figure 1. An integrated approach to water reform

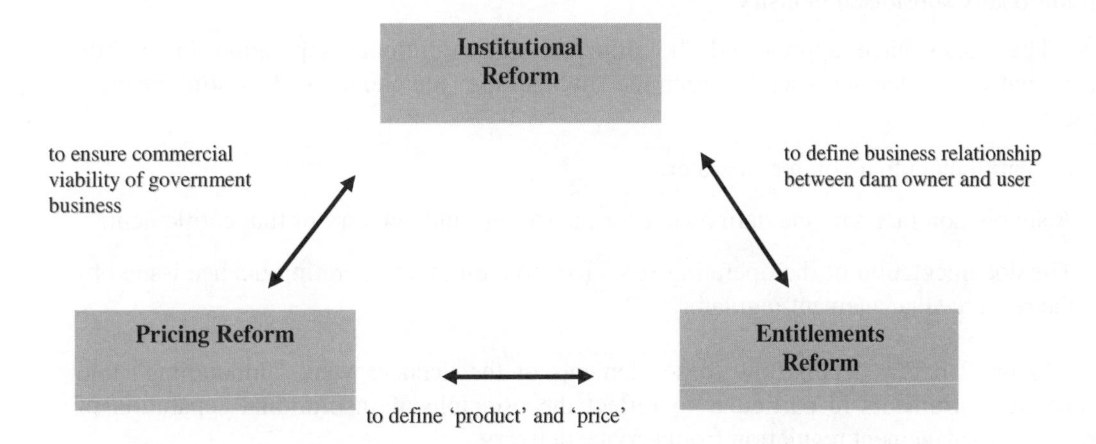

3. The National Water Initiative is an agreement between the State and Commonwealth governments. A copy of the National Water Initiative is available at www.nwc.gov.au/NWI/index.cfm.

Historically, the State water resources departments performed the conflicting roles of:

- Resource manager (*i.e.* protecting the resource and determining what is available for extraction).
- Regulator (setting and enforcing standards for the operation and maintenance of the water infrastructure).
- Infrastructure operator/service provider (operating infrastructure and charging users for a water storage and delivery service).

To address these conflicts, commercial functions have generally been separated and either sold to end users (as has occurred with some of the larger irrigation channel schemes in the State of New South Wales) or transferred to commercial business units (such the government-owned corporation SunWater which now owns and operates the majority of dams and irrigation channels in the State of Queensland).

These types of arrangements allow the commercial side of the water businesses to be operated independently from regulatory aspects, but require changes to how the water entitlements are defined, and the prices for storage and delivery services.

Under the previous arrangements, where the state department managed the resource and the delivery infrastructure, there was little need to define with any certainty the total amount of water available or who owned it. The original dam designs may have included estimates of yields, but the operations were never formalised in a regulatory instrument. Similarly, the way in which the dams were operated and water supplied was not clearly specified. Regional officers and local engineers who operated the schemes did so with wide discretionary powers.

With the separation of existing water supply operations from regulatory arrangements, these local administrative arrangements and discretions were not sustainable. There was a need to specify total volumes available and rules for operations. These operating rules needed to be documented because they determine the product being sold to end users through specifying things such as reliability of supply. At the same time, allowing these new institutions to operate commercially required pricing reforms within a traditionally subsidised industry.

The States have approached the principle of institutional separation in slightly different ways. One approach has seen the "unbundling" the elements of existing licences into:

- An entitlement to a volume of water.
- A supply contract with the dam owner for the storage and delivery of that entitlement.
- The documentation of the operating rules for the dam in an operating licence issued by the resource management regulator.

Figure 2 below details how these elements of the licences were "unbundling" into separate instruments in particular to reflect the principle of institutional separation of resource management regulation from service delivery.

Figure 2. The "unbundling" of water licences

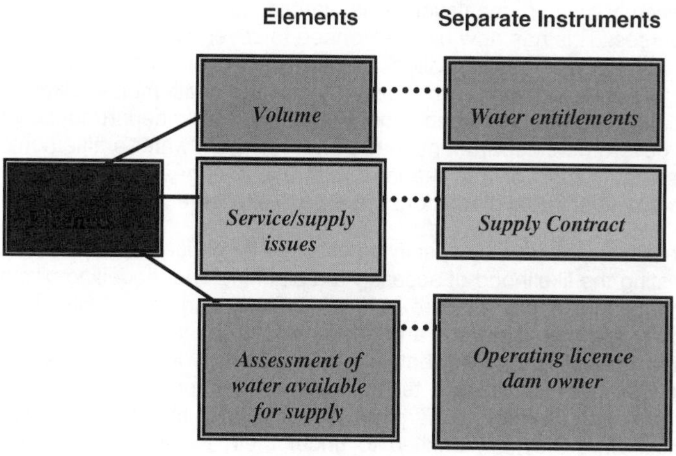

Elements	Separate Instruments
Volume	*Water entitlements*
Service/supply issues	*Supply Contract*
Assessment of water available for supply	*Operating licence dam owner*

In addition to the "unbundling" of the elements of the water licences, other approvals have also been separated. In the past, a regulatory approval on how the water may be used – for example, restrictions on application rates of water to prevent rising water tables – may have been a condition of the extraction licences. Now it is the subject of a separate regulatory approval.

While this "unbundling" and "separating" may at first seem unnecessarily complex, it is integral to a system of water trading separate from land.

Defining water entitlements

The cornerstone of any water trading system is the water entitlement. Listed below are the critical elements required of a water entitlement for it to form part of a viable water trading market.[4] Also listed for each element is a policy approach to achieving these goals.

4. See for example: National Competition Council (2001), *Assessment of Governments' Progress in Implementing National Competition Policy and Related Reforms*, National Competition Council, June. Also: Productivity Commission (2003), *Water Rights Arrangements in Australia and Overseas.* Commission Research Paper. Available at: www.pc.gov.au/research/crp/waterrights/.

Element	Policy considerations
Universality – all available water resources covered by the system of rights.	Until recently, water regulation was limited to underground water and water in watercourses. This has now been extended to cover overland flow water (that is, water flowing across land having fallen as rain) to prevent the taking of this water from affecting planning outcomes and reliability of water entitlements. Likewise, responsibility for allocation of water has been consolidated into single departments, where in the past various government departments had powers to grant water entitlements. The NWI also requires States monitor and manage land use changes that intercept water (such as large commercial forest plantations) to ensure entitlements are secure.
Predictability of volume — users have a reasonable expectation of the volume of water that they can extract from a source.	Predictability is provided by specifying volumes for entitlements (rather than just a pump size), testing the likelihood of access via a planning and modelling processes (details of the water planning process are outlined below), and then protecting access to that water by capping entitlements and ensuring future water allocation decisions (such as approving trades) are consistent with the specific volume. These objectives specify levels of reliability, which are tested against hydrology models, and which must be maintained during the life of the plan. Through these mechanisms, water users should both know what they are entitled to under their allocation, and will have a level of certainty, within the limits of the plan, that the water will be available.
Enforceability — the right can be protected from encroachment by others.	The inclusion of what were previously administrative arrangements within the legal framework has lead to a greater transparency and hence accountability. Contracts for supply, operating rules of dam owners, metering and a greater focus on compliance activities by the administering departments provide a package of measures to ensure entitlements are protected.
Certainty of title — legal recognition and protection of entitlements.	Allocations in some States are recorded on the same database, and in the same manner, as land titles, providing the high level of quality assurance and certainty of ownership that accompanies land title dealings.
Duration — entitlements are granted for a specified period.	Water allocations are granted in perpetuity. They are guaranteed for the life of a water plan may involve compensation payable for losses due to any changes (compensation issues are discussed further below).
Exclusivity — the benefits and costs of a water entitlement accrue to the owner.	Pricing reforms are designed to ensure that where appropriate the costs associated with the storage and delivery of the water entitlement are paid by the owner, and not the broader community. Only the owner of the entitlement is allowed to take water (unless they choose to assign that right to others).
Detached from land title and use restrictions	Allocations can be held independent of land ownership. Water use is also regulated independently of the right to take. A separate regulatory regime may exist for water use depending on the risks. This should ensure water resource planning can focus on watercourse issues, and not (for example) on suitability of land for irrigation, which will be determined via land-use planning. Equally, approvals to construct works (*e.g.* a pump on a river bank) are independent of the right to take water. The key factor in deciding an application is the impact of the actual installation of the works rather than the impact of taking water, which is assessed in the planning and allocation process.
Divisibility and transferability	Water entitlements may be subdivided – and sold to others in whole or part.

Water planning

The NWI[5] includes detailed provisions outlining the range of matters that need to be addressed in water plans – both in terms of content and process, The detail of the way in which these requirements have been implemented vary between jurisdictions, but the outcomes are consistent.

To give a specific example of the how this has been implemented, the *Water Act 2000* in the State of Queensland has detailed provisions on water resource planning. Catchment based water resource plans are made at a strategic level and determine the consumptive/non-consumptive balance and specify outcomes (including environmental outcomes) sought for the plan area.

Extensive technical assessment of the affected area is undertaken as part of the planning process, including the use of hydrological models to model historic rainfall, flow patterns and water extraction. These assessments generate a set of long-term flow statistics (usually around a 100-year period) at key points throughout the basin. The flow statistics are used to describe the extent to which particular allocation and management arrangements (current and future) would vary flow patterns from assessed "pre-development" conditions. Through a benchmarking approach, such flow statistics can be used to predict likely ecological outcomes associated with particular flow patterns. The statistics become the basis for defining environmental flow objectives that, based on best scientific information, would provide the flow patterns consistent with achieving ecological outcomes.

Similar modelling is used to determine relevant statistical benchmarks in respect of reliability of supply and these are used to set water allocation security objectives for the plan.

Water resource plans remain valid for up to 10 years, at which time they must be reviewed. Compensation is payable to a water entitlement holder if an amendment to a plan during the life of the plan decreases the value of the entitlement. Plans are reviewed and replaced at the time of their expiry, and no compensation is payable at that time for any change to a holder's entitlement.

The water planning/trading framework is underpinned by extensive monitoring and public reporting to ensure that:

- Information is available to inform water planning and to determine whether a plan's outcomes and objectives are being achieved.
- Water infrastructure operators comply with their operational rules.
- Water users comply with the limits of their entitlement.

Increases in the value of water have increased the incentives for non-compliance. At the same time, the shift away from works-controlled take to volumetric limits increases the challenges of policing water extractions. A key component of the NWI, and a practical implementation issue currently being managed is how to meter all users. As its simplest an entitlement and trading framework amounts too little in the absence of an effective measuring and water accounting systems.

5. The NWI provides that water plans should have duration of at least ten years.

Adaptive management versus entitlement security

The two competing objectives for water resource managers are:

- To retain State control over water, and on the basis of emerging science, allow for adaptive management of a resource which is still regarded in all jurisdictions as a public good – while
- Providing commercial certainty to water users and headwork operators so that they know that entitlements granted are sufficiently certain: (a) to secure finance (in terms of its "security") for investment decisions, and (b) in terms of its reliability (which influences the suitability of the water entitlement for particular uses).

As described above, certainty for users is provided by the clear definition of their entitlement and the protection of their rights through robust water plans. Fundamentally though, entitlements depend on the certainty of water plans. Thus assignment of risk for changes to water plans is critical to the ultimate security of entitlements.

In most Australian states, water allocations are guaranteed for the 10-15 year life of water plans. However, beyond that period no compensation is payable for changes that reduce the amount of water available under the allocation. Regular reporting on the progress of plans during their 10-15 years is designed to inform water users of the progress of the plan in achieving its outcomes and should provide an indication of the likely need for reductions in allocations at the end of the plan.

These compensation provisions are being remeasured, as under the National Water Initiative all States have agreed to a standard framework for assigning risk for changes. Under this model, which will apply from 2014, within the 10-15 year planning cycle government will bear the risk (*i.e.* pay compensation) for changes as a result of bona fide improvements in the knowledge of water systems' capacity to sustain particular extraction levels. At the end of the 10-15 year period, water users will bear the risk of the first 3% of any reduction in allocation, with government paying compensation for any reduction beyond that mark.

Water trading

The water planning process culminates with the creation of tradable water entitlements and the establishment of trading rules under jurisdictional planning frameworks. Water trading can be either:

- Temporary, which involves the sale of a person's right to take water under the entitlement for that particular season.
- Permanent, where ownership of the entitlement itself is sold; or
- Via lease, where the right to take water under the entitlement is leased for a set period of time, often a number of years.

These types of dealings (leasing or selling) generally do not require the consent of the resource manager – they can simply be registered – as there is no change to the conditions under which water can be taken, and hence no resource issue to be considered. In the case of temporary trades of water within a water supply scheme, it is usually the scheme operator who administers the dealing. Water brokers and Exchanges are progressively becoming more prevalent in the Australian water market both for temporary and permanent trades.

A person buying or leasing an allocation may need to apply to "change" the entitlement to be able to use it in the way, or take it from the place, they require. "Change" refers to a change to one of the elements of the entitlement, such as the location from which water may be taken under the entitlement.

Locations are typically specified with reference to "zones", or stretches of a watercourse, within the catchment. No change is necessary if someone sells to another party wishing to take the water within the same zone. If water needs to be taken from a different zone, then the owner will need to apply to the resource manager for a change to the allocation. Applications for changes are assessed for compliance with the requirements of the water plans.

Extensive modelling during water planning can provide for a faster approvals process of trades, by pre-testing compliance of certain changes with environmental flow and water allocation security objectives. In this way, both the regulator and water users know that certain volumes of water can be traded from one zone to another while meeting the plans objectives. Where a change is outside the pre-tested limits it may still be permitted, provided that on testing it is shown to comply with the requirements of the applicable plans.

The extent to which pre-tests are done is determined on a cost-benefit basis. Thus, in less developed catchments where there are fewer entitlements or where trading is less likely, most assessments will be done at the time of the application, rather than during the planning phase.

Water users may also apply for changes to other aspects of their entitlement. Within supplemented schemes, water entitlements may be assigned a priority group (*e.g.* high priority, medium priority, etc) which is used in allocating water between groups, and is especially important at times of scarcity. Thus an entitlement priority group will determine its reliability. Again, these sorts of changes can be pre-tested during the planning process, which can result in a conversion factor for changing water between priority groups.

An entitlement holder can apply to change their priority group, usually in exchange for a reduction in available volume. In this way, a water user requiring a smaller but highly reliable supply – such as a power station – can obtain it.

Water trading in action

Across all jurisdictions, the number and volume of temporary trades has exceeded the number and volume of permanent trades. Table 1 provides an indication of the magnitude of permanent and temporary trades in some Australian States.

Table 1. Permanent and temporary trades in Queensland, Victoria and New South Wales

	Permanent		Temporary		Permanent trades as % of total volume traded
	Number of trades	Total volume ML	Number of trades	Total volume ML	
Queensland[1]	86	9 250	1 772	190 598	4.6
Victoria[2]	n.a.	59 279	n.a.	380 497	13.5
New South Wales[3]	52	31 927	1 880	434 934	6.8

1. Queensland Government, Department of Natural Resources and Mines, *Annual Water Statistics 2004-05*.
2. Victorian Government, Department of Sustainability and Environment, *State Water Report 2003-04*.
3. New South Wales Government, Department of Natural Resources, cited at www.wma.dipnr.nsw.gov.au/wma for 2005/06.

Prices paid for permanent transfers of water entitlements vary significantly between catchments due to local conditions. For example, in New South Wales, prices for permanent intra-scheme transfers ranged from AUD 450/ML to nearly AUD 2 800/ML[6]. These price differentials are likely to reflect the availability of water in the respective catchments and the differing fortunes of the industry and the entitlements priority group. For example, in Queensland's sugar growing areas prices are around AUD 1 300/ML compared to over AUD 2 000 in areas where water is used primarily for cotton.[7] Furthermore, where water has become a limiting factor to coal mining operations, prices paid per megalitre are significantly higher at AUD 7 500/ML.

While trading provides the opportunity for water users to benefit from water savings initiatives, this is equally true for water service providers. Water infrastructure operators are typically allocated water to account for the losses associated with water delivery (for example, the water required to fill supply channels or the water lost through evaporation or leakage from a channel). Where the operator can reduce those losses – say through more efficient operation of the scheme or through infrastructure upgrades such as channel lining – the operator may be able to sell the saved water, either on the temporary or permanent market.

Release of unallocated water

The release of unallocated water is a further part of the water allocation process that can be subject to market mechanisms. Unallocated water is water identified during a planning process as being available for allocation without threatening environmental outcomes or the security of existing water entitlements.

The National Water Initiative requires that, to the extent practicable, releases of unallocated water should be via market mechanisms.[8] Queensland has developed its own principles on the release of unallocated water, which also requires market based releases

6. New South Wales Government, Department of Natural Resources, cited at www.wma.dipnr.nsw.gov.au/wma for 2005/06.
7. Queensland Government, Department of Natural Resources, Mines and Water.
8. National Water Initiative, section 72. Available at: www.nwc.gov.au/NWI/docs/iga_national_water_initiative.pdf.

(auction, tender, etc) where possible.[9] To date there have been limited releases via such processes.

Challenges from water trading

The implementation of water trading can have significant, incidental consequences. Water entitlements traded away from water supply infrastructure, such as channels, can leave stranded assets, and infrastructure owners with a reduced customer base to meet operational costs. This can ultimately threaten the commercial viability of a scheme, but also gives a market signal that the infrastructure is losing economic value. Generally these risks are addressed by requiring the payment of exit fees to the infrastructure owner prior to trading away from the asset. Again, the arrangement between the entitlement owner and dam owner is important. Preferably, the terms and conditions of these exit fees would appear in the supply contracts - the issues being ones concerning infrastructure pricing – not resource management conditions.

In the past, the most common strategy to address the equity and long term investment issues associated with stranded assets has been to prohibit or limit water trade out of irrigation assets. In the spirit of opening up the Australian water market, States and Territories have agreed that, subject to their design, exit fees are a suitable mechanism to address the impacts of stranded assets provided they do not block trade or completely dampen market signals for structural change. Work is currently underway on developing national principles to ensure that exit fees do not become institutional barriers to trade and that trade is open and competitively neutral.

Similarly, there is the possibility of water trading away from a local community or a particular sector impacting on that community or sector. Such possibilities are considered through socio-economic assessments at the planning stage and where necessary limits on trade can be set to protect such interests. However, the imposition of restrictions is generally seen as running counter to the high-level objectives of water trading - as one means of structural adjustment through water moving to high value uses. While some user groups have objected to trading for fears of water shifting from agriculture to industry, the potential gains that may accrue to individuals from trading has generally seen a softening of opposition. There are few indications of adverse affects from inter-sectoral trades, and few of the trading rules in place have placed limitations on these types of trades.

There are other less obvious but equally important consequences from water trading to consider. Changes to the nature of water entitlements have implications for wills and the laws of intestacy. While water licences were attached to land, the licences would pass with the land under a will. As a separate property right, water entitlements now need to be expressly dealt with under a will in Australia.

Separation of land and water titles can alter land valuations, which are used in most states as the basis for levying local rates. Likewise, separation can have capital gains tax implications. All are issues that must be considered carefully in implementing a new water trading framework.

9. The Queensland policy position on the release of unallocated water is available at www.nrm.qld.gov.au/wrp/pdf/general/unalloc_water_info_sheet.pdf.

The Australian water market is characterised by a multitude of products which offers market participants the opportunity to diversify their water entitlement portfolio. While the mix of water products across state borders may complicate cross border trading, the Australian States and Territories are working towards opening up interstate trading under the National Water Initiative. It is proposed that permanent trades across state borders will be modelled on a "tagged" trading system. This means that each State's water entitlements retain their characteristics when traded into another jurisdiction and remain registered in the State of Origin. Tagged interstate trading therefore has the capacity to deal with different product specifications in each State.

Conclusions

Market mechanisms, and water trading in particular, provide significant opportunities for improved water management outcomes. They also require major structural adjustments to be accommodated. In Australia, more than 10 years into the water reform process, there is still significant work to be done. While the legislative and administrative frameworks to allow trading are now in place in most jurisdictions, trading is only becoming available in individual catchments as water planning activities are completed.

Nonetheless, the benefits of this long process are slowly but surely being reaped as we see improved governance arrangements, a greater focus on environmental requirements for water, the development of water markets and water entitlements shifting to where they are most needed.

Application of Australian experiences to China

The Australian Government is presently undertaking an activity with the Chinese Ministry of Water Resources which is looking to apply some of the Australian experiences in the water reform process to a Chinese context. The activity is aiming to develop a whole-of-country framework for implementing water entitlements and water trading.

For further information on the activity, please contact Robert Speed, Australian Team Leader, Water Entitlements and Trading Activity, at Robert_a_speed@yahoo.com.au.

Chapter 7.

THE DUTCH APPROACH TO WATER QUALITY PROBLEMS RELATED TO FERTILISATION AND CROP PROTECTION

Peter J.M. van Boheemen
Ministry of Agriculture, Nature and Food Quality

Abstract

Intensive development of Dutch agriculture between 1950-1985 was coupled with an increased loading of the environment with nutrients and crop protection products. Since 1985, the soil surplus of phosphate and nitrogen has been reduced from 103 to 40 kg P_2O_5/ha/yr and from 265 to 145 kg N/ha/yr respectively. The use of crop protection products declined from 21 to 10 kg/ha/yr of active substances. The current environmental standards on water quality will be attained by around 2015, thirty years after the reduction policy was begun.

The set of measures taken by the government sought to encourage farmers and supply businesses to use their entrepreneurship and innovation capacity to improve mineral and crop protection management. Those who initially participated in these endeavours were also encouraged to share their experiences. Research, extension and education programmes were also promoted. In general, farmers received sufficient time to adapt to new regulations and to implement the necessary investments into the normal investment schedule. Those who invested in new techniques received a subsidy.

Farmers increasingly certify their production process in so far as the environmental quality of these products is concerned. This "license to deliver" offers them a competitive advantage.

Characteristics of the Netherlands

Geologically, the Netherlands is built on deposits from rivers and estuaries. Most of the country has a shallow ground water table (<2 m below surface) and a dense network of mainly artificial watercourses. Without dunes and dikes, 65% of the land would be flooded by high seas and river water levels. The lowest place is about 7 m below sea level and the highest about 330 m above seal level.

The Netherlands is densely populated (456 inhabitants per km^2) and has a high degree of urbanisation; 89% of the Dutch population lives in urban areas.

The country covers an area of 34 000 km^2, of which 70% is used for agriculture (including small roads and waterways), 13% for urban areas and traffic, and 14% for woodlands and nature reserves. These figures demonstrate that agriculture plays an important role in rural areas. However, the countryside is also used in a variety of other ways. It accommodates numerous economic activities that are not always related to agriculture given the increasing social needs, for example, for nature and recreational areas, as well as living and working spaces.

Development of Dutch agriculture

The lower parts of the Netherlands consist mainly of clay and peat soils; the higher parts have sandy soils, in addition to some loess soils. The climate is characterised by a regular rainfall pattern (700-900 mm/yr), a mild temperature pattern, and a long growing season (April-October). These factors explain why the Netherlands are highly suitable for agricultural production.

Before 1950 livestock was almost exclusively fed with farm grown crops. All the manure was stored properly and applied to own land. There was a cycle of nutrients and an almost complete maintenance of soil fertility. The rather small losses of nutrients were balanced by some inputs of mineral fertilisers and compound feed.

In the period 1950-1985 livestock numbers increased rapidly (Table 1) due to the following developments:

- Increased forage production, mainly due to a larger use of mineral fertilisers and improved water management.
- A large increase in the use of compound feed, produced of raw materials from abroad.
- Better disease control for large groups of animals.
- Large increase in labour productivity.
- Founding of the European Common Market.

As a result, there was much more manure available than was needed for fertilisation. Today, nearly all poultry manure is produced in solid form, and all cattle and pig manure in liquid form (slurry).

At the same time, yields of arable and horticultural crops increased sharply, coupled with a larger use of chemical fertilisers and crop protection products.

All these developments were supported by an adequate knowledge system from existing research, extension and educational institutes.

The expansion also brought specialisation. This meant a choice between arable farming, horticulture or animal husbandry. In the case of the latter, one type of livestock was usually selected. The pig and poultry production became largely independent of the area of own land.

Today, dairy farming is the sector with the largest number of farms which are evenly spread over the whole country. Grass and silage maize are the main crops. On most dairy farms the cattle graze in summer (May-October), sometimes only by day. There is also a small but growing group of farms housing the cattle all year round; this is related to unfavourable parcelling in relation to the numbers of cows and to automatic milking.

Intensive livestock farming (pig and poultry farming) occurs mainly in the sandy regions in the east and the south. Arable farming is important in the coastal clay areas; horticulture occurs in many regions.

Total agricultural area is decreasing slowly (Table 1), but the number of farms is declining substantially faster due to scaling up. In 1990-2003, the total production value decreased by 3%, but the mean production value of the farms increased by 70%. Scaling up was combined with the implementation of new technology, which otherwise would not be feasible economically.

Table 1. Development of Dutch agriculture in 1950-2003

	Units	1950	1960	1970	1980	1985	1990	1995	2000	2003
Number of farms	10^3	339	291	185	145	136	125	113	97	86
Area of grassland	10^3 ha	1 317	1 326	1 330	1 198	1 164	1 096	1 048	1 012	985
Area of arable and horticultural land	10^3 ha	1 020	990	812	823	855	909	917	944	948
Cattle stock	10^6	2.7	3.5	4.3	5.2	5.2	4.9	4.7	4.1	3.6
Pig stock	10^6	1.9	3.0	5.6	10.1	12.4	13.9	14.4	13.1	11.2
Poultry stock	10^6	23.7	43.9	57.2	82.6	91.1	94.9	91.6	106.4	*81.1
Production of compound feed	10^9kg	...	4.7	8.6	14.6	16.5	16.2	16.5	4.6	12.1
Manure production										
Nitrogen (N)**	10^6kg	...								
Phosphate (P_2O_5)	10^6kg	...	...	356	483	542	539	571	415	373
			...	170	230	259	226	209	183	162
Use of mineral fertilisers										
Nitrogen (N)	10^6kg	156								
Phosphate (P_2O_5)	10^6kg	120	212	405	483	500	400	406	339	300
Use of crop protection products; active substances	10^6kg	...	113	110	83	81	74	66	52	40
			...	...	...	21.3	20.3	12.6	11.4	9.6

Notes: * Smaller stock due to Influenza. ** Exclusive gaseous losses from stall and manure storage.
Source: Land-en tuinbouwcijfers, LEI en CBS, 2005 and other years.

In addition to strong primary production, there has also been the development of a strong business in processing and distribution of agricultural products as well as in supplying of all kinds of products and services to farmers. Together, these amount to about 10% of both national production value and employment. In 2003, total primary production on farms amounted to EUR 20.0 billion (USD 15.4 billion); the gross added value was EUR 8.1 billion (USD 6.2 billion), corresponding to 2.0% of the national gross value added. Further, it accounted for 71% of the surplus on the balance of trade, but for only 2.6% of national employment (LEI, 2005).

In the Netherlands and in Europe as a whole, agriculture, especially livestock farming, is undergoing a process of momentous change. A new balance is being sought in the areas of environment, animal health, animal welfare and food safety. The transition is currently taking place without undermining the prospects of agriculture's economic viability.

Emission of nutrients

The amounts of nutrients applied in the form of manure and other fertilisers are examined first followed by an examination of the changes observed in the quality of ground and surface water. The Dutch approach to reaching a new balance is then presented. Finally, the regulatory measures on fertilisation are specified.

Extend and effects

Table 2 shows the loading of agricultural land with phosphate and nitrogen, calculated as the difference between the amounts applied in the form of manure (exclusive gaseous losses), mineral fertilisers and other inputs such as deposition on the one hand, and discharge in the form of harvests on the other hand.

The surpluses reached their maximum level around 1985. In 2003, the surpluses were already lower than in 1970.

Table 2. Phosphate and nitrogen loading of agricultural land in 1950-2003 (kg/ha/yr)

	1950	1970	1980	1986	1990	1995	2000	2003
				Phosphate (P_2O_5)				
Manure	...	80	115	128	108	101	94	72
Mineral fertilisers	...	50	39	41	37	32	32	27
Others	...	5	6	7	8	4	5	5
Total input	...	135	160	176	153	137	131	104
Harvests	...	50	66	73	71	65	68	64
Surplus	...	85	94	103	82	73	63	40
				Nitrogen (N)				
Manure	...	133	190	241	239	252	205	177
Mineral fertilisers	...	185	240	249	201	201	169	147
Others	...	14	17	19	19	19	20	19
Total input	...	332	447	508	459	472	394	343
Harvests	...	167	210	243	248	228	212	198
Surplus	...	165	237	265	211	244	182	145

Source: Milieucompendium, CBS/RIVM, 2004 and other years.

The surpluses lead to emissions to ground and surface waters by leaching, drainage and run-off. The emissions to surface waters disturbed the ecology (eutrophication), often resulting in less biodiversity. This also concerned species for which the Netherlands has accepted an international maintenance responsibility.

Eutrophication has also damaged recreational functions of surface waters and has obliged companies providing drinking water to engage in high extra costs for water purification.

In surface freshwater, the phosphate concentration is the critical factor for eutrophication to occur. The nitrogen concentration is only of importance at a low phosphate concentration.

In 1950, there was hardly any emission of phosphate from agriculture to surface water. This situation has changed dramatically due to manure surpluses which were applied under unfavourable weather circumstances (run-off).

Leaching of phosphate did not take place at that time; the soil laying above ground water level continued to have a large phosphate absorption capacity. In the last four decades the remaining phosphate absorbing capacity of much of the agricultural land has decreased rapidly; however, phosphate leaching to ground water is now starting to become more serious in that phosphate leaching occurs on 50% of agricultural land (Schoumans, 2004).

Due to surplus nitrogen loading in numerous places, groundwater today has nitrate concentrations exceeding the standard for drinking water purposes.

The intensification of agriculture has also been accompanied by increasingly larger emissions of ammonia (Table 3). This is directly relevant to water quality. Most of the ammonia emitted is deposited on short distances and contributes to eutrophication and acidification.

Table 3. Ammonia emission from agriculture (million kg NH_3/yr)

	1950	1970	1980	1986	1990	1995	2000	2003
Stall and manure storage	...	...	77	86	89	86	73	57
Manure application	...	...	114	125	105	63	45	41
Grazing	...	...	14	16	16	14	10	8
Mineral fertilisers	...	...	15	14	13	13	11	9
Total	...	...	220	239	223	179	139	115

Source: Milieucompendium, CBS/RIVM, 2004 and other years.

Water quality trends

Ground water

The Netherlands adheres to the European Nitrates Directive 91/676/EEC, which came into force in 1991. This Directive prescribes that nitrogen leaching to groundwater may not lead to a nitrate concentration of more than 50 mg NO_3/l. For the Netherlands, the nitrate concentration in the upper ground water (1 to 2 m below ground water level) is critical. Figure 1 shows how this concentration evolved from 1992-2002.

Figure 1. Development of nitrate concentration at 1-2 m below ground water level in 1992-2002

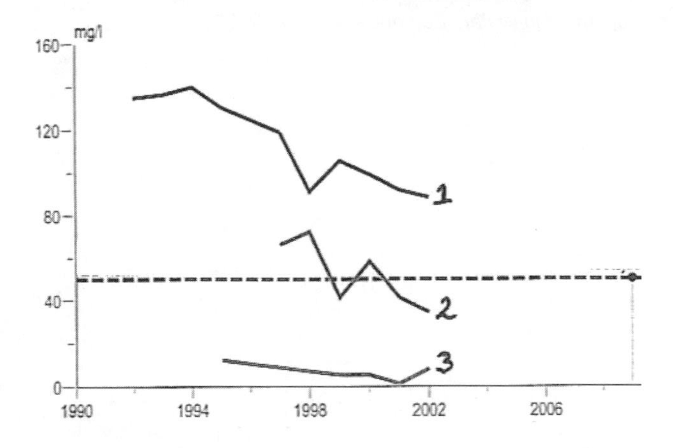

1. Sandy soils; 2. Clay soils; 3. Peat soils.
Source: RIVM, 2004a.

In the clay and peat areas forming the lowlands, the nitrate concentration of the upper ground water complies today to the European standard of 50 mg/l. In the sandy regions, the concentration of the upper ground water declined between 1992 and 2002 from 135 to 90 mg NO_3/l, but that is still not enough (RIVM, 2004a).

Surface freshwater

There are many freshwater lakes which are sensitive to eutrophication. For these lakes, the following standards have been defined: the phosphate concentration can not exceed 0.15 mg total-P/l and the nitrogen concentration can not exceed 2.2 mg total-N/l (V&W, 1998). These environmental quality standards (EQS) are indicative of the situation for all the other surface freshwaters in the Netherlands because nearly all of them eventually drain off to a lake.

The standards above do not fully do justice to the differences in conditions and types of Dutch freshwaters (V&W, 1998). The degree of meandering, the velocity pattern, the temperature, and the degree of shadowing can have more influence on ecology than the actual nutrient concentrations. Consequently, flowing waters are less sensitive to eutrophication than still waters or stagnant waters.

The European Water Framework Directive 2000/60/EC came into force in 2000 and will lead to a new differential set of standards in 2009. According to the Guidance on Eutrophication (EC, 2006), the following groups of species will be the quality elements: phytoplankton, phytobenthos, macrophyten, macrofauna and fishes. The new standards must be reached by 2015; respite can be obtained until at the latest 2027. Table 4 shows the indicative values for the new phosphate standard (MNP, 2006b).

Table 4. For some surface freshwater types indicative values of the new phosphate standard

Type of water	Phosphate standard (mg total-P/l)	
Brooks in sandy region	0.15	(range 0.06-0.75)
Ditches and small canals in peat and clay regions	0.23	(range 0.19-0.42)
Shallow lakes	0.12	(range 0.04-0.12)

Source: Welke ruimte biedt de Kaderrichtlijn Water? Een quick scan, MNP, 2006b.

Water courses in agricultural areas

RIZA (2004) have analysed data on water quality for 378 measuring points scattered over all Dutch agricultural areas (Figure 2). The data concern the period 1985-2002.

Figure 2. Mean phosphate (left) and nitrogen concentrations (right), measured in surface waters of agricultural areas in 1985-2002

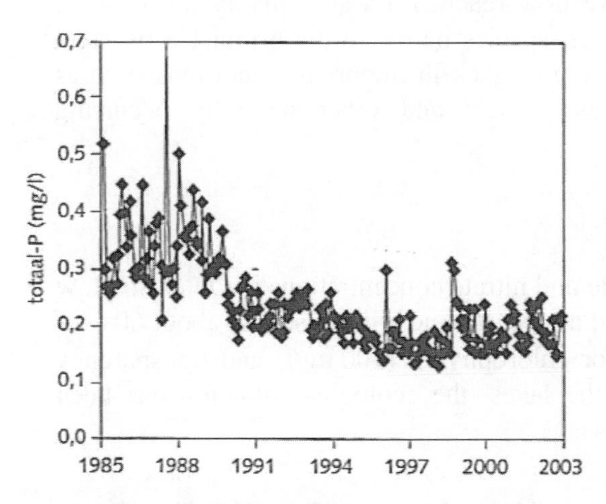

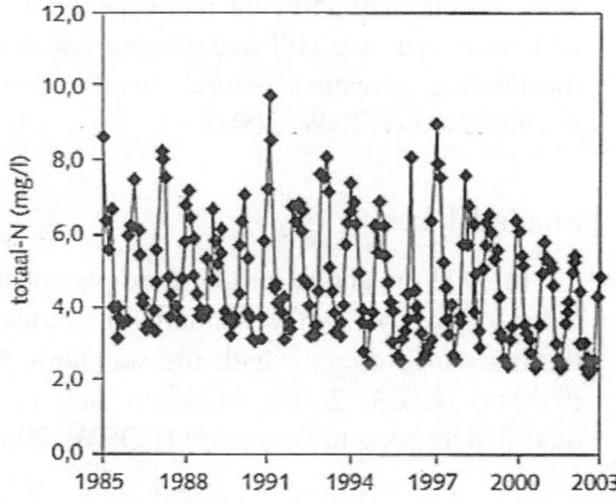

Source: RIZA, 2004.

The seasonal fluctuation of phosphate concentration has declined dramatically. Rainy autumns like 1998 appear to lead to a peak due to higher run off levels.

Between 1992 and 2003 the annual mean of the phosphate concentration has been halved from 0.4 mg/l to 0.2 mg total-P/l. Most of the decline took place in the period 1988-1990, mainly by a significant drop-off during the winter period. This could be related to measures that came into effect during that period and which were intended to reduce run off.

In 1991-2002, the phosphate concentrations were stable. This means that the phosphate saturation of agricultural top soils as described earlier has not yet led to a phosphate leaching to surface waters. This corresponds with the long response time of the Dutch soil water system for phosphate (at least 30 years).

The data sets have been split up into parts corresponding with the same landscape type. The analyses for the different landscape types delivered no additional insights or conclusions, except for the magnitude of the phosphate concentration due to differences in seepage with a high phosphate content. In the clay areas near the coast, the annual mean of the phosphate concentration amounts to 0.5 mg total-P/l. However, in 1997-2002, many points in the sandy areas recorded concentrations that were lower than the standard of 0.15 mg P/l.

The analysis of the nitrogen concentrations delivered significantly different results. The lowering over the period 1985-2002 was rather small and seasonal fluctuation remained large.

In the first years of the period 1985-2002 there was hardly any decrease, but in the 1990s the annual mean concentration decreased steadily, both during winter and summer and in all landscape types. This reflects the reduction in nitrogen application rates since 1987. The response time of the Dutch soil water system for nitrogen (about 10 years) is shorter than for phosphate.

The number of locations with an annual mean of nitrogen concentration lower than the standard of 2.2 mg total-N/l remains small. In many places in the sandy regions the concentration amounts to as much as three times the acceptable standard.

Water courses in agricultural areas have now reached a water quality level where some species are showing recovery. However, species native to the original ecosystem and water type are still under-represented due to a lack in important conditions such as meandering streams, natural banks, leaves packets and other naturally occurring phenomenon (LBOW, 2004).

Shallow lakes

In 2002 the annual mean of the phosphate and nitrate concentrations in Dutch shallow lakes fell just below the standards mentioned above. It is now the case that about 60% of the lakes also comply with the standards for chlorophyll-a (100 µg/l) and transparency (0.40 m) (RIZA, 2004). In nearly half of the lakes, the ecological situation has been qualified as good to very good (LOBW, 2004).

Figure 3. Summer mean concentrations of phosphate in mg P-total/l (left) and nitrogen in mg total-N/l (right), measured in Dutch shallow lakes in 1980-2002

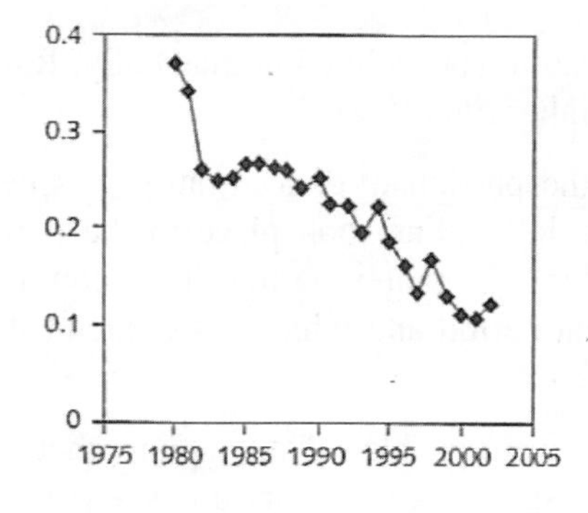

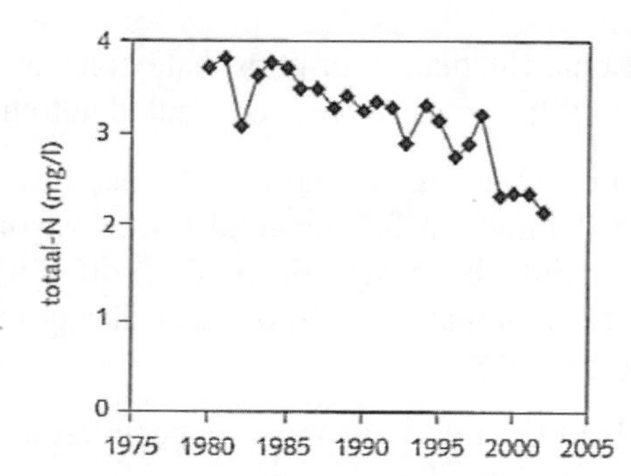

Source: RIZA, 2004.

Large rivers

The nutrient concentrations of the large rivers, measured at the boundary, are very important for the water quality in the Netherlands. Figure 4 shows the development for the Rhine, the largest river in the Netherlands. The phosphate and nitrogen concentrations have been lowered substantially since 1985, but are still above current standards. The decrease is due mostly to reductions by industry and waste water treatments. In recent years, nutrient loading and concentrations in the Rhine have been relatively stable.

Figure 4. Phosphate (left) and nitrogen (right) concentration of some large Dutch freshwaters and the coastal zone, compared to the current environmental quality standards (0.15 mg total-P and 2.2 mg total-N/l)

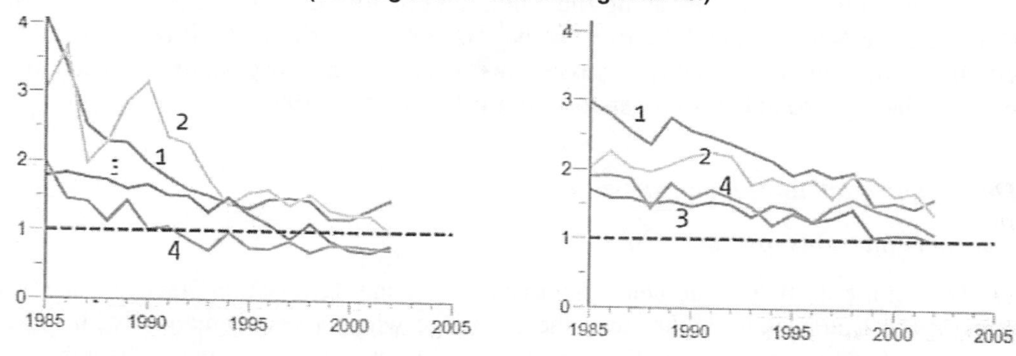

1. River Rhine; 2. Coastal zone; 3. Lake IJssel; 4. Other lakes.
Source: MNP-RIVM, 2004a.

Marine waters

Within the scope of the Oslo and Paris Convention (OSPAR), Dutch coastal waters also have a eutrophication problem. The main nutrient sources are the discharges coming from the large rivers (Rhine, Meuse and Scheld) and atmospheric deposition.

In saline waters, the nitrogen component is the most critical factor for eutrophication (MNP, 2006b). Environmental quality objectives have not yet been set. Presently proposed standards concern 1.5 times the natural background concentrations (OSPAR, 2005). The European Commission is in the process of preparing a Marine Strategy Directive (EU, 2005).

The nitrogen concentration (DIN-value) has declined very little; it amounts today to 1.5 times the standard proposed (Figure 4). The phosphate concentration (DIP-value) has already reached the standard proposed (MNP-RIVM, 2004a).

Emission to total surface freshwater system

Table 5 examines the nutrient emission to the total surface freshwater system. The loading imported by the rivers is much higher than the inland loading, but the total catchment area abroad is seven times larger than the inland catchment area.

Table 5. Nutrient loading of the Dutch surface freshwater system in 2002 (million kg/yr)

	Industry	Waste water treatments	Rain water sewers	Atmospheric deposition	Rural areas	Total inland loading	Loading from abroad
P-total	1.0	3.0	0.1	0	3.8	8.0	26.8
N-total	10	28	2	34	70	144	380

Source: Milieucompendium, CBS/RIVM, 2004 and other years.

In 2002, the loading by rural areas, including natural loading and emissions from agriculture land, has contributed the most. This has been the case since the mid-1990s.

In 1987, all the countries along the Rhine agreed to reduce nutrient loading in the Rhine and the North Sea by 50% over the period 1985-1995 (by OSPAR postponed until 2010). In 2002, the Netherlands surpassed this objective for phosphate (69% reduction), but only managed to reduce nitrogen loading by 30% (V&W, 1998).

Dutch approach to reduce emissions of nutrients

Nutrient emissions to water system

The main objective of the Dutch approach is to attain fertilisation practices whereby nutrient application is no higher than the discharge with harvests, enlarged with losses acceptable for the environment. Thereby, soil fertility must be maintained. The approach consists of two strategic initiatives:

- Controlling the extent of manure production.
- Improving mineral management at the farm level.

As a first step, a ban on enlarging pig and poultry livestock was proclaimed in 1984, leading to a quota system for pigs and poultry husbandry. At almost the same time, a system of milk quotas was introduced in the European Community which placed a ceiling on the growth of the dairy industry.

Standards on manure application were introduced in 1987. Farmers with a large surplus of manure were forced to ship manure off their farms and transport it to others without a manure surplus. Farmers could also choose to buy additional land or reduce their livestock numbers (Boheemen, 1987).

In later years, the amount of manure that can be applied on a parcel of land was gradually lowered. This encouraged more and better spreading of surplus manure. Farmers with less land in relation to own manure production were confronted with costs for storing and shipping manure off. This situation stimulated affected farmers, as well as supported supply businesses, to search for possibilities to reduce the nutrient excretion of the livestock through changes in nutrition and production methods. It also encouraged them to look for possibilities to sell their livestock manure to other farmers (Boheemen, 1990, 1992).

Farmers with relatively less manure production were stimulated to look for possibilities to substitute manure for commercial mineral fertilisers without affecting crop yields and soil fertility. This not only saved expenses on chemical fertilisers, but also obtained some earnings in circumstances where the manure market was stretched.

All these measures were implemented in such a way that farmers and supply businesses received sufficient time to adapt to new regulations and to integrate the necessary investments into their normal investment and write off procedures.

In fact, a market-based mechanism was introduced under conditions imposed by the government, often in concert with farmers' organisations. This was possible due to the collective starting-point of maintaining a vital and competitive agriculture. The measures appealed to farmers and supply businesses and encouraged their entrepreneurship and innovation capacity. Those who were among the first to invest in new techniques received subsidies.

The government supported the adaptation process through an extensive research programme designed to improve mineral management at the farm level. At the same time, the government started an extension programme, with a focus on promoting higher awareness of environmental problems and to communicate new insights with respect to mineral management. Farmers who were dubious and did not embrace this approach were supported in their choices and given assistance to leave farming, and finding income elsewhere.

To accelerate adaptation and maintain a balanced manure market, the government enforced production quotas on pig and poultry farmers for the period 1999-2001. This was done under the condition that stalls and other production facilities were dismantled. Quota sales from one farmer to another were also reduced somewhat. In 1995, the Netherlands started the so-called mineral accounting system (MINAS) based on the difference between inputs and outputs of minerals on a farm (MNP-RIVM, 2004c). This difference (*i.e.* the surplus) should be lower than the loss standards defined by the government. In the case of a surplus, a levy was imposed. The MINAS-system focused directly on losses and gave farmers much freedom to reduce them. However, the European Court of Justice decided in 2003 that the European Nitrates Directive required application standards, and not loss standards. It is now the case that regulations and practices are in accordance with the mineral management system as stipulated by this directive.

In 2002, a manure contract system for ensuring a balance of the manure market was implemented. At the beginning of each year a farmer should have sufficient contracts with other farmers for delivering his manure surplus. If not, animal numbers must be reduced.

The manure contract system was terminated in 2005 and the animal quota system will soon be abolished. This will be done when the nutrient levels and loading come into balance with the environment which should probably be attained by around 2015. This means that it will take about 30 years to rid the environment of the problems caused in the previous 35 years.

Environmental awareness by both farmers and citizens is now highly developed and knowledge-based; violation of regulations and standards on fertiliser applications no longer meets with any understanding. Many farmers, often in cooperation with the buyers of their products (traders and retailers) go about it in a commercial way by placing their practice under a certificate programme with independent inspection by a third party. Due to heavy competition they do not get higher prices, but they expect to put themselves in a preferred position with a "license to deliver". An example of this is the EUREPGAP certification developed by the Euro Retailer Produce Working Group.

Compliance with the application standards will be inspected both administratively - on the basis of the data sent to government agencies by the farmer and the databases available to government agencies - and physically, on the farms themselves. If, in a specific case, it is determined that the application standard has been violated, sanctions will follow.

The way in which farms are inspected depends on the risk profile for keeping the administrative burden on farmers and government as low as possible and to allow inspection services to function efficiently. Shipping manure off the farm leads to high costs. Therefore farmers with a manure surplus have to provide the most detailed reports and are subject to the closest inspection. The latter also concerns intermediary businesses involved in transporting, storing, processing or exporting manure. They have to record manure flows and are therefore required to sample, analyse and weigh manure brought on or shipped off to farms and document each of these activities. Other considerations, such as being situated on sandy and loess soils and the extent to which a farm has to fertilise below the optimum production level, are also taken into consideration.

The primary responsibility for ensuring compliance with the regulations lies with the police, the municipalities and the General Inspectorate of the Ministry of Agriculture and is part of their basic inspection tasks in rural areas. The idea is that the police can easily catch people in the act of contravening a regulation without an additional administrative investigation.

The European Court of Justice made it clear that the application standards must be enforced with deterrent, punitive sanctions. Pollution must be prevented and it must not be possible to "buy it off" by paying a levy having a compensatory or reparatory nature.

Contravening the regulations constitutes an economic offence. Sanctions have the form of a combination of administrative fines and criminal justice, with the emphasis on administrative fines. Criminal prosecutions are, in principle, reserved for very serious infringements and for fraud.

Emission of ammonia to the atmosphere

Most Dutch natural areas suffer from nitrogen deposition. Many of them may accept at most 1 400 mol/ha/yr; the most sensible ones at least 400-700 mol/ha/yr. In the sandy regions the nitrogen deposition is more than 2 400 mol/ha/yr; in areas with many pigs and poultry farms, this increases to more than 3 400 mol/ha/yr (MNP, 2005).

For reasons of public health, the Netherlands have undertaken to reduce ammonia emission to 128 million kg/yr in 2010 (EU, 2001). The latter can be reached by measures lowering ammonia emission from livestock barns, manure storage and manure application. The reduction to 128 million kg/yr protects sufficiently only 20-30% of the natural areas (MNP-RIVM, 2004a). Therefore, efforts will be made to reduce emissions to 100 million kg/yr in 2010 (VROM, 2001). To protect the most sensitive natural areas, a reduction to 30-55 million kg/yr is necessary (VROM, 2002).

Regulatory measures on fertilisation

In 2006, the regulatory measures were adjusted to bring them into accordance with the European Nitrates Directive and other European Directives related to nutrients. The regulations now include:

- Standards for manure application.
- Standards for total nitrogen application.
- Standards for total phosphate application.
- Additional measures for reducing run off, leaching and ammonia emission.

Standards for manure application

No more than 170 kg N/ha/yr of manure may be applied. An exception has been made for dairy farms with more than 70% of the farm area on grasslands, where 250 kg N/ha/yr is allowed at most. For these categories, it has been proved that the higher standard does not lead to more than 50 mg NO_3/l in the upper groundwater and drainage water. The standard for manure application is related to the total quantity of nitrogen applied with manure, including manure and urine excreted during grazing and covering both organically- and inorganically-fixed nitrogen.

There are no lower standards for particular types of manure because the standards for total phosphate application are also determining the amount allowed. For example, use of pig manure will generally not exceed 140 kg N/ha/yr because of its high phosphate content and the phosphate application standard mentioned later.

Overall, the sum of manure production plus manure brought in, minus manure shipped off, corrected to stock differences, may not be greater than the product of the application standard and the farm acreage.

The amount of nitrogen in the manure produced by cattle and other grazing livestock has been linked to milk production (Table 6). The intention is to further refine the forfeits for dairy cows on the basis of the urea content of the milk.

Table 6. Net nitrogen production of dairy cows minus gaseous losses from stall and manure storage

Milk production (kg/yr)	Net nitrogen excretion (kg N/yr)
5 500	99
6 500	107
7 500	115
8 500	123
9 500	131

Source: Third Dutch Action Programme (2004-2009) concerning the Nitrates Directive, LNV, 2004.

For pigs and poultry housing, the manure production must be determined on basis of a stall balance. Hence, nutrient production amounts to the difference between mineral input (feed and animals) and mineral output (animals and animal products), allowing for gaseous losses from stall and manure storage. Table 7 clarifies the method for the net nitrogen production in the form of manure. In the same way, a calculation can be made for phosphate. Gaseous losses have been defined per type of stall. Below certain numbers of pigs and poultry, farmers have to use forfeits instead of a stall balance.

Table 7. Balance for calculating net nitrogen production in a stall with 2 000 pigs (kg N/yr)

Input		Output	
Piglets	3 970	Pigs	18 090
Feed	37 400	Gross nitrogen excretion (unknown balance item)	23 280
Sum	41 370	Sum	41 370
		Gaseous losses (29% of gross excretion)	-6 750
		Net nitrogen production	16 530

Source: Third Dutch Action Programme (2004-2009) concerning the Nitrates Directive, LNV, 2004.

In the Netherlands there are central recording systems for data on farm inputs and outputs. Feed inputs are recorded not only by farmers, but also by the feed suppliers who deliver the records (per individual farm) to the government.

Standard for total nitrogen application

The sum of total nitrogen applications in the form of manure, mineral or other fertilisers must be equal to or less than the standard for nitrogen application. The application standard is calculated as the aggregated product of the application standard per hectare per crop, multiplied by the acreages of the crops.

Where the sum threatens to exceed the standard, the farm must reduce its fertiliser usage. In special cases this may compel the farm to reduce its usage of animal manure further than allowed by the application standard for animal manure.

The standard for total nitrogen application is set equal to the level leading to the economic optimal crop yield, until this level leads to more than 50 mg/l NO_3 in the upper ground water or drainage water. If the latter occurs, the standard is lowered in the period 2006-2009 to the level fitting on the European Nitrates Directive. Table 8 gives the standards for some important crops.

Table 8. Standards for total nitrogen application for some crops in 2006-2009 (kg N/ha/yr)

Crop	Soil type	Application rate for economic crop yield	Standard for total nitrogen application			
			2006	2007	2008	2009
Grassland, grazed	- Clay	345	345	345	325	310
	- Peat	265	290	290	265	265
	- Sand and loess	315	300	290	275	260
Grassland, only mowed	- Clay	385	385	385	365	350
	- Peat	300	330	330	300	300
	- Sand and loess	355	355	350	345	340
Maize	- Clay	160	160	160	160	160
	- Sand and loess	160	155	155	155	150
Consumer potatoes	- Clay	250	275	275	250	250
	- Sand and loess	265	265	250	…	…
Winter wheat	- Clay	220	240	240	220	220
	- Sand and loess	160	160	160	…	…
Sugar beets	- Clay	150	165	165	150	150
	- Sand and loess	150	150	145	…	…

Source: *Third Dutch Action Programme (2004-2009) concerning the Nitrates Directive,* LNV, 2004.

Mineral nitrogen available in spring, net mineralisation and nitrogen deposition have been incorporated into the calculations of the standards. Furthermore nitrogen fixation by leguminous crops has been taken into account. Table 9 gives the percentages to be used in calculating the total nitrogen application.

Table 9. Fraction of nitrogen to be used in calculating the total nitrogen application (%)

Type of manure	2006	2007	2008	2009
Own cattle manure on grassland with grazing	35	35	45	45
Own cattle manure on grassland without grazing	60	60	60	60
Cattle manure from other farms on grassland	60	60	60	60
Pig manure	60	60	60	60
Manure applied in autumn on arable clay soil	30	40	50	Banned

Source: Third Dutch Action Programme (2004-2009) Concerning the Nitrates Directive, LNV, 2004.

Standards for total phosphate application

To avoid phosphate saturation of the soil and phosphate emission to surface water there exist defined standards for total phosphate application. The objective is to achieve equilibrium between application and harvest by 2015 (Table 10).

Table 10. Phosphate application standards in kg P2O5/ha/yr

Land use	2005	2006	2007	2008	2009	2010	2011	2012	2013	2014	2015
Grassland	130	110	105	100	95	95	95	95	95	95	90
Arable land	115	95	90	85	80	75	70	70	65	65	60

Source: Third Dutch Action Programme (2004-2009) concerning the Nitrates Directive, LNV, 2004.

Additional measures for reducing run off, leaching and ammonia emission

- Ban on application of animal manure on sandy and loess soils in autumn and winter (September-January). For clay and peat soils, the ban is restricted to grassland; in 2009 this ban will be extended to arable land on clay and peat soils.
- Ban on application of chemical fertilisers in autumn and winter (September-January). For some crops an exception has been made (field vegetables, fruit, and flower bulbs).
- Manure must be applied with specified techniques, by which it is placed directly on or into the soil. On fallow land the manure must be covered immediately by ploughing or any similar operation.
- Restrictions on fertilising of sloped land and soils (> 7%). Strict rules governing erosion must be respected. It concerns only 0.4% of the agricultural area in the Netherlands.
- Restrictions on application of fertilisers near water courses. Many European countries specify a fertiliser-free zone of at least 5 m to reduce leaching and run off. The hydrological situation in a flat country like the Netherlands justifies the question whether wide fertiliser-free zones are useful, especially because no fertilisation will take place in periods with most leaching and run-off (autumn and winter). Furthermore, due to the high density of surface waters, the establishment of wide fertiliser-free zones would reduce substantially the area that can be cultivated, with huge economic consequences. Based on both arguments the fertiliser-free zones have been limited provisionally to the cultivation-free zone prescribed by crop protection regulations (see next chapter). Recently, research on the effectiveness of buffer zones under Dutch conditions has been set up, both in highland and in lowland regions. Depending on the results the restrictions will be readjusted.
- Fertilisers have to be spread as uniformly as possible over the land; spreading over surface water must be avoided.
- After ploughing grassland a nitrogen-fixing crop must be grown. Fertilisation of this crop has to be based on analysis of soil samples taken after ploughing. On sandy and loess soils ploughing of grassland is only permitted in spring.
- After harvesting maize on sandy and loess soils (September-October), a nitrogen-fixing crop must be grown.
- Manure storage has to be sufficient for bridging over a production period of six months (September-February).
- Manure storage has to be covered to reduce ammonia emission.
- The building and installation of pig and poultry stalls must be done in such a way that ammonia losses do not exceed certain standards. By 2010 all existing buildings must be renovated to comply with these standards.

Emission of crop protection products

We will first deal with product use and discuss the emission reductions which have already been reached. Some data on water quality are described as is the Dutch approach on achieving a new balance. Finally, the regulatory measures on crop protection are presented.

Extent and effects of crop protection products

The average use of crop protection products (CPP) in Dutch arable and horticultural farming, measured as active substances, amounts to about 10 kg/ha/yr. This is 2.5 times the average in the European Community (LEI, 2005). This relatively high use is related to the presence of many intensive cultures.

CPP use peaked around 1985, as was the case with nutrients. In 2003, use had fallen by 55% relative to use in 1984-88 (Table 11). The target for 2000 has been realised, but after the actual time specified. The use of soil disinfecting products was reduced substantially. The targeted reduction for insecticides was also attained, but not for herbicides and fungicides.

At the crop specific level, the reduction has been lowered more than at national level, but this gain has been partly lost due to an increase in crops with a relatively high CPP use, such as potatoes, onions and flower bulbs.

Table 11. Use of crop protection products in 1984-2003 and target for 2000 (million kg active substances)

	1984-88	1990	1995	2000	2003	Target 2000*
Soil disinfecting products	10.2	8.9	2,4	1.4	1.2	3.3
Herbicides	4.6	4.1	4.0	3.5	3.3	2.5
Fungicides	4.4	4.7	4.5	5.0	3.5	2.8
Insecticides	0.7	0.8	0,6	0.3	0,3	0.4
Others	1.3	1.7	1.2	1.3	1.4	0.9
Total	21.3	20.3	12.6	11.4	9.6	10.6

Source product use: Nederlandse stichting voor Fytofarmacie (Nefyto) en Plantenziektenkundige Dienst.
Source target: Evaluatie Meerjarenplan Gewasbescherming. Eindevaluatie van de taakstelling voor 1990-2000, LNV, 2001a.

The active substances and their residues form a potential threat for the ecosystem and public health, either through contact or by consumption through breathing, eating, and drinking.

Emissions to ground and surface water can take place by volatilisation, drift, run off, leaching, and drainage. The real environmental effect depends on the concentration and frequency of occurrence, decreasing time and toxicity of the residues.

RIVM (2002) calculated that in 1984-2002 the emission to surface water, expressed in kg active substances, decreased by 78% (Table 12), although the objective was 90% (LNV, 2001). For the emission to groundwater, nearly the same result was reached.

Table 12. Emission of crop protection products to surface water in 1984-2000

	1984-1988	1995	1998-2000
Total emission (10^3 kg/yr)	116	48	25
Fraction of total use (%)	0.58	0.39	0.22

Source: Emissie-evaluatie MJP-G 2000, RIVM, Bilthoven, 2002.

Since 1998, the emission is no longer expressed in kg active substances, but in environmental effects expressed in points (Figure 5). The score depends on the volumes used of different active substances on the one hand and on aquatoxity and sensitivity to leaching to groundwater of the substances in question on the other hand. In 1998-2003, the environmental effects on the surface decreased by 50%; similar effects were observed for ground water but with an even larger 80% drop (MNP-RIVM, 2004b).

Figure 5. Total use of crop protection products (upper line) and calculated environmental effects on surface water (lower line) in 1998-2003 Targets for 2005 and 2010

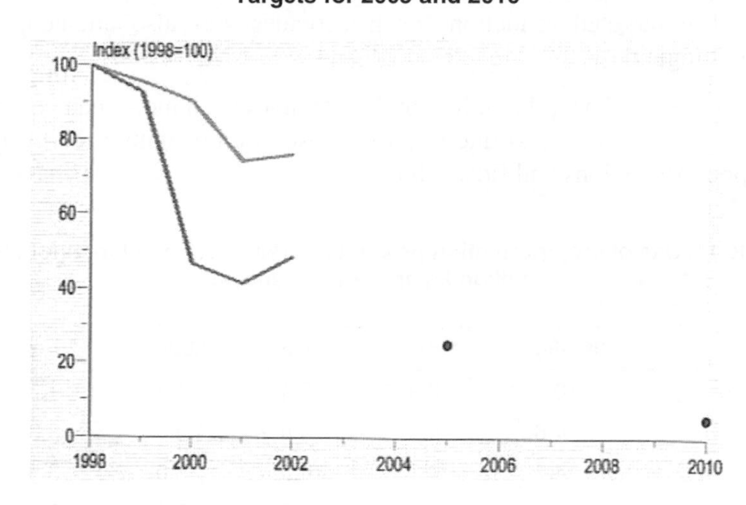

Source: MNP-RIVM, 2004b.

Water quality trends

Concentrations of crop protection products are measured annually in surface water at 400 to 700 locations within agricultural areas. The percentage of locations where the concentration of one substance exceeded the EQS, was about 60% in 1995-98, 30% in 1999 and 45% in 2000 (CIW, 2002). In large rivers the percentage, measured at boundary points, was 76% in 1999 and 100% in 2000. The results observed for saline waters were 10% in 1999 and 34% in 2000 (CIW, 2002).

Residues of crop protection products are still found regularly in surface water consigned for drinking purposes. Only at a few ground water pumping-stations did concentrations exceed the standard of 0.1 μg/l (RIVM, 2004b).

The European Water Framework Directive 2000/60/EG (EU, 2000), which came into force in 2000, will lead to a readjustment of environmental quality standards (EQS) for crop protection products and their residues. The Directive distinguishes between priority substances and other relevant substances. The European Commission has defined 34 priority substances, among which only three crop protection products are admitted into the Netherlands. The EQS for priority substances will be determined by the European Commission. The emission of these substances must be reduced by steps; for the most dangerous ones among them, the emission must be terminated completely within 20 years. With respect to other relevant substances, an EU-country may make its own selection; the EQS must be defined according to an EU-Guidance at latest by 2008.

Dutch approach to reduce emissions of crop protection products

The approach to reduce emissions of crop protection products started in 1990, about five years later than the approach to reduce emissions of nutrients. The approach consists of the two strategic thrusts (LNV, 1991, 2001b, 2004):

- Reducing dependence on crop protection products.
- Introducing products and application techniques which lead to less environmental loading.

Both strategies were incorporated into a covenant concluded by five ministries and eight branch organisations for the period 1990-2000. In 2003 this was followed up by a covenant concluded by two ministries and six branch organisations for the period 2003-2010. The parties collectively have obliged themselves to lower emissions of crop protection products in such a way that, in 2010, environmental quality standards (VROM, 2001) will have been met. Each party has taken responsibility for some tasks. In 2006, the progress will be evaluated and, if necessary, the tasks will be further refined (LNV, 2003). Reaching the current environmental quality standards means reducing environmental loading by 95% relative to loading in 1998 (Figure 5).

In both of the above instances, the branch organisations received assurances that a sufficiently broad set of crop protection products would be maintained as far as EU-directives allowed. Likewise, assurances were made that the relative position of farmers in the Netherlands would not be disadvantaged in comparison with farmers in other EU-countries (level playing field) and that there would be no special levies on crop protection products (LNV, 2003). But it remains the case that, for all crops now grown in the Netherlands, continuity has by no means been guaranteed.

In the period 2000-2003 there was no agreement between the government and the branch organisations, because then the Netherlands ran ahead of the EU-re-evaluation of the admission of crop protection products. The admission of several products was withdrawn before alternatives became available.

Reduction of dependence on crop protection products

Much effort is taken in developing new knowledge and experiences in the field of integrated crop protection. This is based on crop management in the broad sense and has lead to a decision-tree type of framework. In principle, farmers must take the different measures in a prescribed sequence before deciding to apply chemical crop protection products:

- Preventive measures like taking measures on hygiene for food health and safety reasons.
- Crop management measures in the field of crop rotation, choice of crop variety and the like.
- Utilisation of disease and pest infestation forecasting systems.

- Application of non-chemical methods like mechanical weed control.

A change of attitude is asked which is hindered by all kinds of factors such as extra labour needed and uncertainties regarding crop yield and crop quality. Nevertheless, several important breakthroughs have been realised in dealing with pests which attack greenhouse crops.

Within the context of the present covenant, government and branch organisations have developed an extensive research, extension, education and stimulating programme for removing bottlenecks and reaching more breakthroughs. Farmers have interests in it because they also worry about their health and that of their labourers, and consumers and retailers are becoming increasingly aware of food safety and health concerns.

Introduction of new products and application techniques, giving less emission

In 1995 the European Community decided to harmonise the admission procedure for crop protection products and to re-evaluate all products admitted in EU-countries on new, severe criteria. This process will be finished in 2008. At the beginning of the 1990s about 850 substances had been admitted; about 470 of them are now extinct (MNP, 2004b). In the Netherlands the number of substances admitted declined from more than 300 in 1990 to less than 200 in 2004, partly due to being in advance of the 2000-2003 EU-re-evaluation. The crop protection products which disappeared from the market have been substituted by products with a lower environmental loading, partly in the form of newly developed ones. In the Netherlands, a lot of crops have an area too small to cover the costs for developing new products or extending the use of existing products to more crops. Therefore, government and farmer organisations have created a fund for contributing to the costs for solving financial bottlenecks.

Several techniques have been developed for reducing emission by drift; further techniques for avoiding leakage from all kind of apparatus to ground and surface water. Most have been embodied in regulations.

Regulatory measures on crop protection

- Farmers have to draw up yearly a farm plan for crop protection. Thereby the principles of integrated crop protection mentioned above must be followed.
- Farmers have to participate in meetings for improving knowledge and skills on crop protection. Non-attendance can lead to the non-renewal of licences to use crop protection products.
- Farmers have to record amounts of products used. This is done for the sake of analysis at the branch level and to enable benchmarking by farmers themselves.
- No application of crop protection products is permissible in windy weather (wind speed higher than 5 m/s) in order to reduce drift and pollution of surface water.
- No application of crop protection products in zones along water courses. The width varies from 0.25 m for grassland and wheat to 5 m for nurseries where products are sprayed from below. Further application within 14 m zone to water courses must be done with machines equipped with nozzles which reduce drift. Use of nozzles which reduce drift may be substituted by other measures with similar effects on emissions.

- No discharge of washing and condensed water with the remains of crop protection products to surface water is allowed. Washing of machines must take place at a distance of at least 5 m to watercourses. Filling of machines with surface water is only allowed if certain control measures have been taken. Further apparatus for disinfecting bulbs must fulfil requirements which assure that there are no emissions to surface water.
- Use of products for disinfecting soils is only permitted on a low frequency rate (once per five years).

Integrated approach by Dutch greenhouse horticulture

In 1997 a covenant was concluded which focussed on several environmental problems related to greenhouse horticulture (LNV, 1997). Objectives on emission reduction for 1997-2010 have been defined.

Since 2002, every horticultural farmer with more than 2 500 m^2 greenhouses must declare his crops along with their areas. The farmer is then informed about the standards for his farm on the use of fossil energy, nutrients and crop protection products, expressed in physical units and in terms of points. These standards have been further refined. Affected farmers must now measure, record and report their uses. The total amounts must comply with the standards. Points gained by exceeding one standard might be compensated within limits by points lost in not attaining another standard. The results on emission reduction can be certified. With the "license to deliver", farmers obtain a competitive advantage.

In this system, individual accountability and responsibility can yield commercial benefits and successful farmers found they had considerable freedom to fully exploit their entrepreneurship and innovation. Recently, decisions have been taken to rebuild the system and current standards are likely to be replaced by more stringent emission standards (LNV, 2006).

REFERENCES

Boheemen, P.J.M. van (1987), "Extent, Effects and Tackling of a Regional Manure Surplus; a Case-Study for a Dutch Region", *Animal Manure on Grassland and Fodder Crops. Fertiliser or Waste?* Martinus Nijhoff Publishers, Dordrecht/Boston/Lancaster.

Boheemen, P.J.M. van (1990), *Management of Emission and Waste on the Dairy Farm; a European Approach,* Proceedings of the XXIIIrd International Dairy Congress, October 8-12, 1990, Montreal, Canada.

Boheemen, P.J.M. van (1992), Technology and Practices for Reducing Mineral Losses of Dutch Animal Husbandry, OECD Workshop on Sustainable Agriculture, 11-13 February, Paris, France.

CBS/RIVM (2004 and other years), Milieucompendium, Centraal Bureau voor de Statistiek, Voorburg.

CIW (2002), Bestrijdingsmiddelenrapportage, Commissie Integraal Waterbeheer, Den Haag.

CIW (2004), Water in Beeld / Water in Cijfers, Commissie Integraal Waterbeheer, Den Haag.

EC (2000), Richtlijn 2000/60/EG "Kaderrichtlijn Water", Europees Parlement en Raad van Europese Unie, Publicatieblad van de Europese Gemeenschappen, Brussels.

EC (2001), Richtlijn 2001/81/EG van het Europees Parlement van 23 oktober 2001 inzake nationale emissieplafonds voor bepaalde luchtverontreinigende stoffen, Publicatieblad van de Europese Gemeenschappen No L309/22.

EC (2005), Proposal for a Marine Strategy Directive, CEC(2005) COM (2005) 504 final, Brussels.

EC (2006), Towards a Guidance Document on Eutrophication Assessment in the Context of European Water Policies, European Commission, Brussels.

LBOW (2004), Water in Cijfers 2004, Achtergrondinformatie over het waterbeheer in Nederland. Landelijk Bestuur Overleg Water, Rijkswaterstaat, Den Haag.

LEI (2004), Effecten in 2006 en 2009 van Mestakkoord en nieuw EU-Landbouwbeleid, Landbouw economisch Instituut, rapport 30344, Den Haag.

LEI/CBS (2005 and other years), Land- en tuinbouwcijfers. Landbouw Economisch Instituut, Den Haag.

LEI (2005), Landbouw Economisch Bericht 2005, Landbouw Economisch Bericht, Den Haag.

LNV (1991), Meer Jaren Plan Gewasbescherming (MJP-G), Ministerie van Landbouw, Natuur en Voedselkwaliteit, Den Haag.

LNV (1997), Convenant Glastuinbouw en Milieu, Ministerie van Landbouw, Natuur en Voedselkwaliteit, Den Haag.

LNV (2001a), Evaluatie Meerjarenplan Gewasbescherming, Einddocument, Eindevaluatie van de taakstelling 1990-2000, Ministerie van Landbouw, Natuur en Voedselkwaliteit, Expertisecentrum, Ede.

LNV (2001b), Zicht op een gezonde teelt, Gewasbeschermingsbeleid tot 2010, Ministerie van Landbouw, Natuur en Voedselkwaliteit, Den Haag.

LNV (2003), Convenant Duurzame Gewasbescherming, Ministerie van Landbouw, Natuur en Voedselkwaliteit, Den Haag.

LNV (2004), Duurzame gewasbescherming, Beleid voor gewasbescherming tot 2010, Ministerie van landbouw, Natuur en Voedselveiligheid, Den Haag.

LNV (2004), Third Dutch Action Programme (2004-2009) Concerning the Nitrates Directive, 91/676/EEC, Ministerie van Landbouw, Natuur en Voedselkwaliteit, Den Haag.

LNV (2006), Brief aan Tweede Kamer inzake Herijking Glastuinbouw, Ministerie van Landbouw, Natuur en Voedselkwaliteit, kenmerk DL2006/647, Den Haag.

MNP-RIVM (2004a), Milieubalans 2004, Milieu en Natuur Planbureau, Bilthoven.

MNP-RIVM (2004b), Van inzicht naar doorzicht, Beleidsmonitor water, thema chemische kwaliteit van oppervlaktewater, Milieu en Natuur Planbureau, Bilthoven.

MNP-RIVM (2004c), Mineralen beter geregeld, Evaluatie van de werking van de meststoffenwet. Milieu en Natuur Planbureau, Bilthoven.

MNP (2005), Milieubalans 2005, Milieu en Natuur Planbureau, Bilthoven.

MNP (2006a), Nationale Milieuverkenning 6, 2006-2040, Milieu en Natuur Planbureau, Bilthoven.

MNP (2006b), Welke ruimte biedt de Kaderrichtlijn Water? Een quick scan, Milieu en Natuur Planbureau, Bilthoven.

OSPAR (2005), Ecological Quality Objectives for the Greater North Sea with Regard to Nutrients and Eutrophication Effects, Oslo and Paris Convention.

RIVM (2002), Emissie-evaluatie Meer Jaren Plan Gewasbescherming MJP-G 2000, Achtergronden en berekeningen van emissies van gewasbeschermingsmaatregelen, Rijks Instituut voor Volksgezondheid, rapport 716 601 004, Bilthoven.

RIVM (2004a), Agricultural Practice and Water Quality in the Netherlands in the 1992-2002 Period, Background Information for the Third Nitrates Directive Member States Report, Rapport nr. 500003002, Rijks Instituut voor Volksgezondheid, Bilthoven.

RIVM (2004b), De kwaliteit van het drinkwater in Nederland in 2003, Rijks Instituut voor Volksgezondheid, rapport 703719007, Bilthoven.

RIZA (2004), Eutrofiëring van landbouw beïnvloede wateren en meren in Nederland – toestanden en trends, Rijks Instituut voor Integraal Zoetwaterbeheer en Afvalwaterbehandeling, rapport 2004.009, Lelystad.

Schoumans, O.F. (2004), Inventarisatie van de fosfaatverzadiging van landbouwgronden in Nederland, Alterra, Rapport 730.4, Wageningen.

V&W (1989), Derde Nota Waterhuishouding, Ministerie van verkeer en Waterstaat, Den Haag.

V&W (1998), Vierde Nota Waterhuishouding, Ministerie van Verkeer en Waterstaat, Den Haag.

VROM (2001), Vierde Nationaal Milieu Beleidsplan, Ministerie van Volkshuishuisvesting en Ruimtelijke Ordening, Den Haag.

VROM (2002), Vaste waarden, nieuwe vormen, Ministerie van Volkshuishuisvesting en Ruimtelijke Ordening, Den Haag.

Chapter 8.

POLICY ISSUES REGARDING WATER AVAILABILITY AND WATER QUALITY IN AGRICULTURE IN THE UNITED STATES[1]

Dennis Wichelns
Rivers Institute at Hanover College, Hanover, Indiana

Abstract

Most current issues regarding water availability and water quality in agriculture have arisen due to increasing competition for limited resources and changes in public preferences regarding the environment. Demand for water has increased in recent decades with increases in population and income levels. Water supply has remained largely fixed due to natural conditions and a declining rate of investment in new water supply facilities. Investment has slowed as the cost of developing new sites for dams and reservoirs has increased, and as public preferences have shifted toward greater concern for protecting the environment and sustaining the use of natural resources. Public agencies and legislatures have responded to the shift in public preferences by implementing new environmental standards that require changes in agricultural and industrial production methods. The two primary issues that must be resolved by public officials in the future are: determining the best ways to allocate limited water supplies in an era of increasing demands by agriculture, industry, cities, and the environment, and determining the best ways to define and achieve state and national water quality objectives, while not constraining desirable economic growth and development. Many smaller issues are described within the context of these primary issues.

Introduction

Major issues involving water availability and water quality in agriculture have evolved over time in the United States, changing largely as a function of increasing population and wealth, and changes in technology and public preferences. Early in the nation's history there was a vast supply of land, water, and other natural resources, both in aggregate and per person. The eastern states in which the country first developed are humid, with sufficient rainfall in summer to produce most crops without irrigation. As the country expanded westward into the arid states of Kansas, Arizona, Colorado, and Texas, irrigation was required to support commercial agriculture. Dryland farming could support small numbers of residents, but irrigation was needed to generate desirable rates of economic growth.

Public and private agencies were established in the late 1800s and early 1900s to promote irrigation development. Many large dams, reservoirs, and irrigation schemes were constructed in the first half of the 20th century, particularly in states located west of the Mississippi River. Irrigation enabled several arid western states to become major producers of annual and perennial crops. Farmers could achieve high yields of many crops with an assured water supply, abundant sunshine, and affordable labour. The high

1. This paper is Contribution Number 06-08 of the Rivers Institute at Hanover College.

yields ensured low average costs of production that enabled western farmers to compete successfully with farmers in rainfed farming areas in central and eastern states. As profits accumulated, western farmers could adopt new technologies quickly, enabling them to retain and enhance their competitive advantages.

Some of the technological advances have had unintended environmental consequences. The heavy use of nitrogen and phosphorus fertilisers has increased nutrient loading to streams and groundwater throughout the United States. The application of pesticides, growth regulators, and other chemicals has degraded surface water and groundwater quality in many areas. Public knowledge and concern regarding the environmental impacts of agricultural chemicals and nutrients increased substantially in the 1960s and 1970s. Since then, major efforts have been implemented to reduce chemical use and restore water quality.

Public concern regarding water availability has increased also in recent decades, as the demand for high-quality water has increased faster than the supply. The demand for water has increased with population growth and increases in aggregate wealth. Water supply has been limited largely by natural conditions, but also by a declining rate of public investment in new water supply facilities. Public budgets have been under stress for many years, with increasing demands for education, health care, social security, transportation, and other public goods. Water agencies must compete with other public agencies for limited funds. In addition, the cost of building new dams and reservoirs has increased as the cheaper sites already have been developed. New sites generally are more costly to develop and new environmental regulations that require safeguards and mitigation have added substantially to the costs of building dams and reservoirs.

The new environmental regulations reflect changing public preferences regarding water supply and demand in specific, and environmental quality in general. As the nation has become wealthier, many residents have become more concerned about environmental quality. The increase in wealth has provided many residents with more time for leisure and enjoyment of the environment and natural resources. Hence, many residents express concern for maintaining environmental quality, even if safeguards and mitigation generate higher costs of providing goods and services.

Several natural resource agencies have responded to changing public preferences by focusing on water conservation and demand management when water becomes scarce, rather than expanding water supply. The US Bureau of Reclamation, which has constructed most of the large dams, reservoirs, and irrigation schemes in western states, provides an illustrative example. In recent years, the Bureau has publicly changed its focus from water supply development to water resource management. A substantial portion of the Bureau's budget is allocated to supporting water conservation programmes and improvements in water management.

Analytical framework

Most of the current issues involving water availability and water quality in the United States have arisen as a result of two phenomena: 1) Increasing competition for scarce resources, and 2) Changing social preferences regarding the environment and natural resources. The increasing competition for land, water, and environmental quality in the United States has been driven by growth in population and income. The demand for water has increased in agriculture, industry, and cities as population and aggregate wealth have increased. In addition, many residents have expressed greater interest in improving the quality of air, land, and water resources in the country, as their wealth has increased and they have gained more time to enjoy and appreciate environmental amenities.

For many years it was sufficient to list three sectors, or groups of water users, when describing water demand and supply in the United States: Agriculture, Industries, and Cities (Figure 1). From the time the nation was formed and through the 1950s the volume and quality of water resources generally was sufficient to supply all three sectors. In the 1950s, however, the use of modern inputs such as chemical fertilisers, pesticides, irrigation systems, and groundwater pumping increased substantially. Aggregate wealth increased notably in the 1950s as productivity increased with greater use of modern inputs in agriculture and other sectors.

Figure 1. Water supply and demand in the United States, before 1970

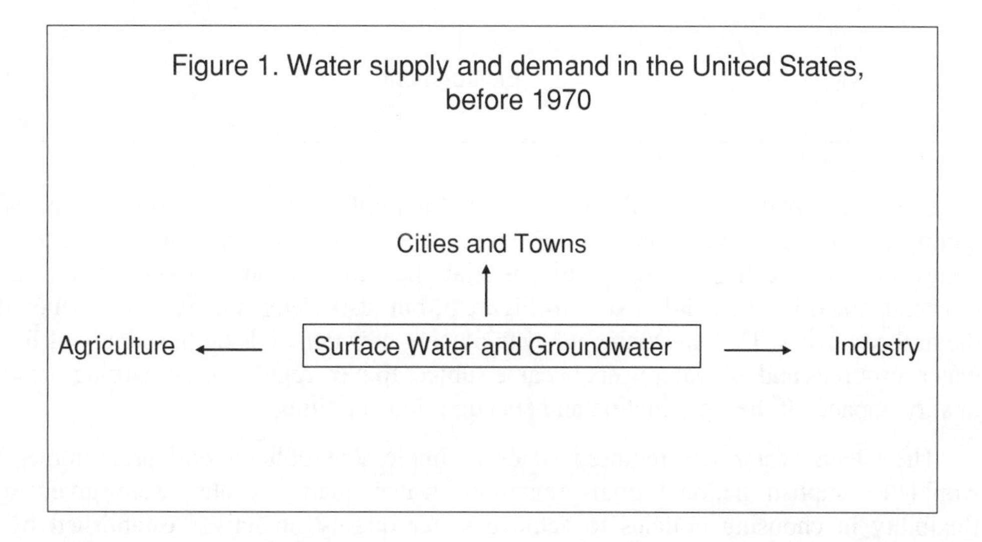

The public became concerned about the long-term impacts of modern inputs in the 1950s and 1960s. Several writers began describing the potential harm that might be caused by the increasing use of chemical fertilisers and pesticides in agriculture, detergents and process chemicals in cities and industries, and fossil fuels in all sectors (Melosi, 2000). The unintended impacts of the productivity gains achieved in the 1950s include the pollution of air, land, and water resources. An "environmental movement" gained substantial support in the 1960s, as many residents began calling for greater concern about the impacts of production activities on environmental quality.

The United States Congress passed the Clean Water Act in the 1970s, due largely to the increasing concern about declining water quality in many areas of the country. The ambitious goals of the Clean Water Act included restoring fishable and swimmable water quality conditions in the nation's navigable rivers and streams within the 1980s and eliminating the discharge of pollutants within the 1990s (Melosi, 2000; Deason *et al.*, 2001). With the passing of the Act, it became helpful to consider the Environment as a fourth water user or sector, in addition to Agriculture, Industries, and Cities. Since the 1970s, public officials have had to consider environmental impacts when designing all policies regarding water supply and water quality. Competition for the nation's supply of surface water and groundwater has included four sectors since the 1970s (Figure 2).

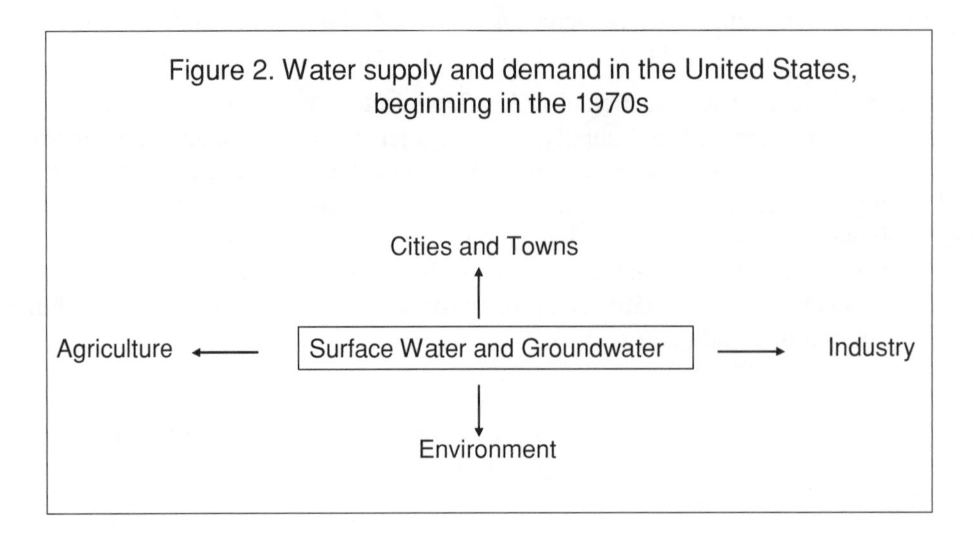

The Clean Water Act made it necessary for public officials to consider the water quality impacts of production and consumption activities in Agriculture, Industry, and Cities and Towns. In a sense, public officials had to consider not only water supply issues, denoted by the solid arrows in Figure 3, but also water quality issues denoted by the dashed arrows. The environment began placing new demands on the supply of limited water resources and all water users became subject to new regulations regarding the water quality impacts of their production and consumption activities.

The Clean Water Act required states to implement policies and programmes that would accomplish national goals regarding water quality. States were given some flexibility in choosing policies to achieve water quality objectives established by the federal government, but the Clean Water Act initially favoured technological solutions (Melosi, 2000). Over time, many states and the federal government have gained appreciation of the role economic incentives can play in achieving environmental objectives. Cost-sharing programmes, water marketing, and tradable emission permit programmes have been implemented in many states in recent years.

The increasing demand for water volume in the United States has not been accompanied by increases in supply infrastructure. Very little surface water storage capacity has been added in the United States since 1980 (Stern, 2003). Reasons for the declining rate of construction include the high cost of developing new reservoir sites, environmental concerns, and the perception among many planners and public officials that improvements in water management will enable better use of existing water supplies at lower cost than developing new supplies. This perception has been sufficiently widespread to transform the goals and operating mode of several state and federal agencies from water development to water management.

Figure 3. Water supply and demand in the United States, beginning in the 1970s

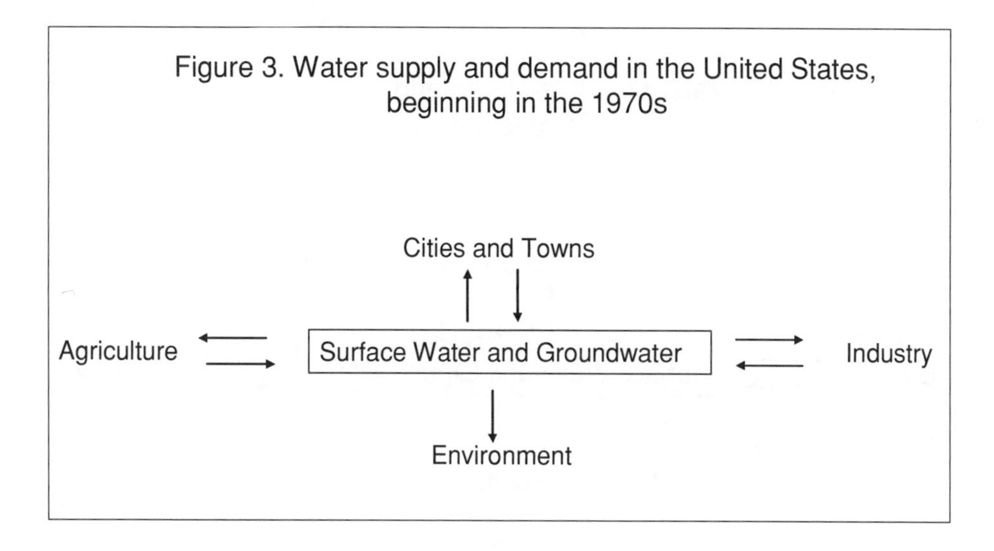

The new focus on water management has created substantial interest in water conservation methods and the transfer of water supplies from agriculture to cities and towns, and to the environment (Figure 4). Particularly in arid states, such as Arizona, California, and Colorado, many farmers have sold or leased a portion of their water supply to cities and towns. In most states farmers are allowed to receive a fair market price for the water they sell or lease. Those prices often are substantially higher than the net returns farmers can earn by using their water to irrigate crops. Hence, farmers choose voluntarily to sell or lease their water supply, and both the buyers and sellers gain net value from the transactions.

Some farmers have received payment for leasing their water supply to public agencies seeking to achieve environmental objectives. In California many farmers lease water rights or contractual allocations to agencies of the state or federal government. Those agencies use the water to maintain minimum flow conditions in rivers to support fish and other aquatic wildlife. Water transfers from agriculture to cities and public agencies represent an innovative approach to increasing the values generated by limited water supplies. In some areas the transfer of water from agriculture also improves environmental quality by reducing surface runoff or deep percolation that carries harmful chemicals and other elements into streams or groundwater.

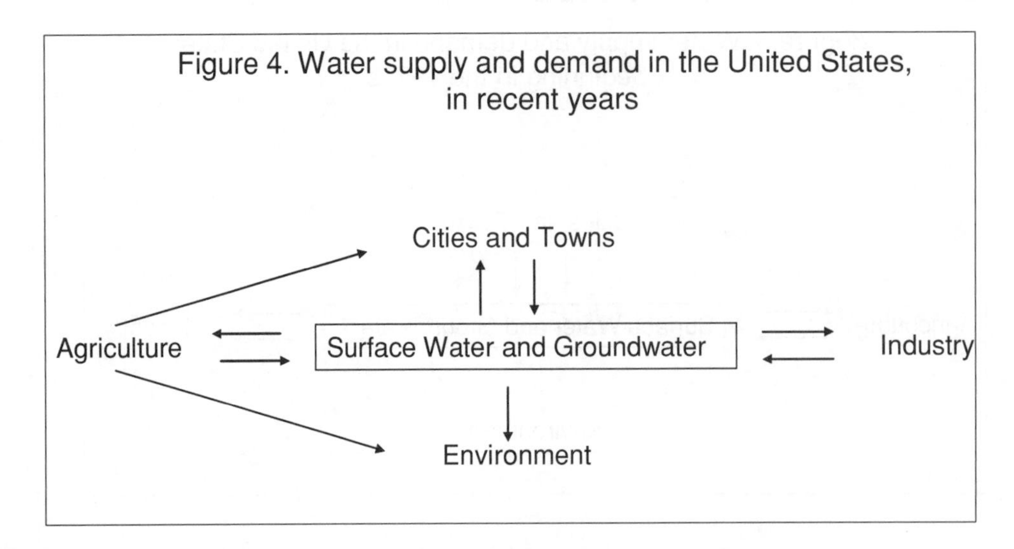

Empirical overview

Irrigation has been the largest use of freshwater in the United States since 1950. Historically, more surface water than groundwater has been used for irrigation, although the portion of total irrigation withdrawals from groundwater has continued to increase, from 23% in 1950 to 42% in 2000 (Hutson *et al.*, 2005). The area irrigated in the United States increased by more than 100% between 1950 and 1980 and then remained somewhat constant before increasing by about 7% between 1995 and 2000. The area irrigated with sprinkler and microirrigation systems has continued to increase and now comprises more than half of the total irrigated area (Hutson *et al.*, 2005). Sprinklers and microirrigation systems were used on an estimated 5 320 hectares of the 10 138 hectares irrigated in 2000 (Table 1).

Sprinklers are used primarily in central states that overly the Ogallala Aquifer and other groundwater formations. Large areas are irrigated with centre pivot irrigation systems in Kansas, Nebraska, Texas, and Colorado. Side-roll sprinkler systems are used extensively in the Pacific northwestern states of Washington, Oregon, and northern California to irrigate alfalfa. Portable and permanent-set sprinkler systems are used throughout California and many other states to irrigate fruits, nuts, and vegetables. Microirrigation systems including microsprinklers and drip irrigation also are used extensively on fruits, nuts, and vegetables.

In 1995, surface water was the source for about 63% of the 185 100 million m^3 of water used for irrigation in the United States (Solley *et al.*, 1998). Groundwater provided about 37%, while wastewater provided a very small portion. In 1995, 61% of the water withdrawn for irrigation was used consumptively by plants, 19% was lost in conveyance, and 20% was returned to surface water or groundwater supplies (Solley *et al.*, 1998). Most irrigation water was used in the arid western states. Five states (California, Idaho, Colorado, Texas, and Montana) accounted for 54% of the nation's irrigation withdrawals. The largest annual withdrawals of groundwater for irrigation occurred in California (14 931 million m^3), Texas (9 033 million m^3), Nebraska (7 996 million m^3), and Kansas (4 368 million m^3). Much of the groundwater withdrawn in Texas, Nebraska, and Kansas is taken from the Ogallala Aquifer, which essentially is a non-renewable source that is being depleted over time (Opie, 1993).

Table 1.
Estimated irrigated areas by technology and irrigation water by source, in selected major irrigation states, 2000

State	Irrigated Land in 1000 Hectares				Withdrawals in Million m^3 per Year, By Source		
	Sprinkler	Micro	Surface	Total	Ground	Surface	Total
Arizona	30.0	2.3	127.6	159.8	4,690	4,538	9,228
Arkansas	103.3	0.0	635.5	738.6	11,101	2,406	13,507
California	271.9	493.0	895.9	1,654.2	19,948	32,130	52,078
Colorado	194.9	0.2	363.6	556.8	3,685	15,837	19,491
Florida	84.3	115.3	137.4	337.4	3,731	3,609	7,324
Georgia	240.8	12.1	0.0	252.2	1,281	668	1,949
Idaho	399.6	0.8	212.9	614.2	6,350	22,841	29,085
Illinois	59.8	0.0	0.0	59.8	256	7	263
Indiana	40.9	0.0	0.0	40.9	95	78	172
Iowa	13.8	0.0	0.0	13.8	35	2	37
Kansas	435.6	0.4	106.0	542.1	5,847	492	6,335
Louisiana	18.0	0.0	135.9	154.0	1,351	397	1,751
Michigan	65.7	1.4	0.8	68.0	219	125	344
Minnesota	89.4	0.0	4.4	93.8	324	63	387
Mississippi	74.5	0.0	158.2	232.6	2,238	169	2,406
Missouri	87.1	0.2	129.7	217.8	2,360	82	2,436
Montana	82.9	0.0	199.8	281.7	142	13,431	13,583
Nebraska	673.1	0.0	607.6	1,280.7	12,669	2,345	15,014
Nevada	31.4	0.0	74.7	106.0	967	2,634	3,594
New Mexico	75.5	1.2	86.8	163.5	2,101	2,787	4,888
Oklahoma	64.2	0.2	18.5	83.0	967	259	1,224
Oregon	190.0	0.7	163.8	355.4	1,351	9,015	10,370
South Dakota	45.2	0.0	12.8	58.0	233	402	637
Texas	656.7	14.6	391.4	1,062.9	11,101	3,639	14,740
Utah	86.1	0.3	144.1	230.9	801	5,786	6,594
Washington	208.0	8.2	41.3	257.1	1,275	3,913	5,177
Wisconsin	58.1	0.0	0.0	58.1	332	3	335
Wyoming	31.1	0.8	157.9	190.0	705	6,974	7,690
All 50 States	4,635	685	4,815	10,138	97,152	136,591	232,982

Note: Figures may not sum to totals because of independent rounding.
Source: Hutson et al., 2005.

Annual withdrawals from surface water and groundwater sources increased substantially in the United States between 1950 and 1980, before declining somewhat in recent decades. The sum of withdrawals from both sources increased from 259 billion m^3 in 1950 to 608 billion m^3 in 1980, an increase of 144% (Table 2). Annual groundwater withdrawals increased by 147% during that period, while surface water withdrawals increased by 141%. The faster rate of increase in groundwater withdrawals was driven in part by the strong demand for agricultural crops in the mid-1970s, particularly in central states where groundwater provides much of the irrigation water supply.

Population in the United States increased from 151 million persons in 1950 to 230 million persons in 1980, an increase of 52% (Table 2). Hence, the sum of annual withdrawals from surface water and groundwater per person in the United States increased from 1 551 m^3 in 1950 to 2 649 m^3 in 1980, an increase of 60%. Surface water withdrawals have been somewhat constant since 1985, at about 450 billion m^3, while groundwater withdrawals have increased to the volume observed in 1980 (Table 1). The annual sum of withdrawals from both sources is about 7% smaller than it was in 1980, while population has grown by 24% since 1980. As a result the sum of withdrawals per person has decreased from 2 649 m^3 to 1 977 m^3, a decline of about 25%. This substantial decline in water withdrawals per person can be attributed to improvements in water management technology, higher energy prices, and lower real prices of agricultural products.

Table 2. Trends in estimated water use in the United States, 1950-2000

Year	Population	Water Withdrawals			Total
		Surface Water	Groundwater	Total	per person
	(million)		(billion m^3)		(m^3)
1950	151	207	47	249	1,651
1955	164	274	66	332	2,023
1960	179	305	70	373	2,081
1965	194	350	84	428	2,211
1970	206	419	95	511	2,484
1975	216	455	115	581	2,683
1980	230	499	116	608	2,649
1985	242	449	102	552	2,275
1990	252	452	111	564	2,235
1995	267	447	107	556	2,080
2000	285	446	117	564	1,977

Source: Hutson et al., 2005. USGS Circular 1268.

Annual water withdrawals for irrigation increased from 123 billion m^3 in 1950 to 207 billion m^3 in 1980, an increase of 69% (Table 3). Since 1985, water withdrawals for irrigation have been relatively constant at about 189 billion m^3 per year. Water withdrawals for industry and public water supply increased even more quickly than water withdrawals for irrigation between 1950 and 1980, with respective rates of increase of 131% and 76% (Table 3). Like agriculture, water withdrawals for industry have been somewhat constant since 1985, while water withdrawals for public supply have continued increasing from 37 billion m^3 in 1985 to 43 billion m^3 per year in 2000. That increase, about 16%, is smaller than the 18% increase in population between 1985 and 2000.

Table 3. Trends in estimated water use in the United States, 1950-2000

Year	Total Withdrawals	Water Uses: Irrigation	Industries	Public Supply	Rural
	(billion m^3)	(billion m^3)			
1950	249	123	106	19	5
1955	332	152	153	17	5
1960	373	152	191	21	5
1965	428	166	243	24	6
1970	511	180	300	27	6
1975	581	194	339	29	7
1980	608	207	352	34	8
1985	552	189	301	37	11
1990	564	189	311	39	13
1995	556	185	303	40	12
2000	564	189	302	43	13

Source: Hutson et al., 2005. USGS Circular 1268.

The empirical information describing irrigation in the United States reflects the rapid increase in irrigated area and irrigation withdrawals that occurred between 1950 and 1980. Irrigated area increased substantially in the early 1970s in response to a sharp rise world prices for food grains. The high prices were not sustained. By the late 1970s, declining real crop prices and rising energy costs reduced the net returns to irrigated agriculture, particularly in central states where farmers irrigated large areas of wheat, corn, and soybeans. Many farmers in central states sold their land and discontinued farming in the early 1980s, contributing to the substantial decline in water withdrawals for irrigation between 1980 and 1985 (Table 2).

Increasing competition for water supplies and greater focus on the environmental impacts of agriculture have motivated farmers and entrepreneurs in other industries to improve water management practices during the 1980s and 1990s. Those improvements have enabled agriculture and industries to continue expanding in the United States, while using about the same volume of water withdrawals that was reported in 1985 (Table 2). Many observers of this empirical record suggest that similar, continuous improvements in water management practices will enable further economic growth without investing in new water supply facilities. The question of whether not to construct new dams and reservoirs is a major policy issue regarding water availability for agriculture and other sectors in the 21st century.

Current issues regarding water availability

The primary, over-arching policy issue regarding water availability for agriculture and other sectors in the United States can be summarised as follows:

- Public officials must determine the best ways to allocate limited water supplies in an era of increasing demands by agriculture, industry, cities and towns, and the environment.

There are many smaller issues that can be described within the context of this primary issue regarding water allocation:

- Establishing systems of water rights or allocations in states where such systems are not yet in existence.
- Modifying systems of water rights and allocations as needed in states where the water supply is not sufficient to satisfy increasing demands.
- Achieving efficient allocation of limited water supplies while not causing undue harm to water users with low-priority water rights or users in downstream or tail end locations.
- Achieving sustainable use of groundwater resources, particularly in areas where natural recharge is much less than current withdrawals, such as the Ogallala Aquifer in the central United States.
- Determining the appropriate balance between public programmes that enhance water supply (such as new dams and reservoirs) and those that modify water demand (such as conservation programmes, water pricing, and other economic incentives).
- Determining the appropriate roles of public agencies and private sector firms in providing water service to cities, towns, and industries.
- Determining how to pay for maintaining existing water supply facilities and for investments in new facilities that will be needed to meet increasing demands for water in future.
- Determining the appropriate role of regional conjunctive use projects in augmenting water supplies, particularly in arid states.
- Predicting future demands for water in agriculture, which currently uses the largest portion of developed water supplies.
- Predicting the impact of rising energy prices on the demand and supply of ethanol and other biofuels, and implications for water use in agriculture.

- Predicting the long-term impacts of global climate change on rainfall and surface runoff, with particular emphasis on water supply implications.
- Determining the appropriate importance to place on constructing new water supply facilities, given the large number of demands placed on public agencies for investments in education, health care, transportation, and other public goods.
- Determining whether historical agreements involving the flow of rivers between states need to be re-negotiated, due to changes in population and economic growth that have occurred since the original agreements were enacted.
- Determining whether international agreements with Mexico and Canada involving transborder rivers and aquifers need to be re-negotiated, due to changes in population and economic growth that have occurred since the original agreements were enacted.

Each of these issues is being considered by staff members of one or more state and federal agencies. Technical specialists and attorneys spend large amounts of time developing information, drafting plans, and preparing arguments in favour of a desired resolution of a water allocation decision. The substantial public and private cost of the policy process regarding water allocation issues has become an issue of its own. Some observers of the process in western states such as Colorado and California have suggested that the transaction costs or resolving water allocation issues can exceed the benefits gained by individuals and communities involved in the process.

Selected water availability issues

Several states in the eastern and central United States are struggling with the issue of allocating scarce water supplies among competing users. Water laws and allocation rules are not well developed in many eastern and central states because water resources generally have been available in sufficient volume during the 19th and 20th centuries. Increasing populations and income levels are changing the balance between water demands and supplies, causing many states to seek assistance in developing new water laws and allocation policies. In some states the imbalance occurs only during extended dry periods in summer when rainfall is less than normal and river flows decline below the levels needed to support agriculture, cities, and recreational users. The challenge in those states is to develop policies that are sufficiently detailed and comprehensive to achieve water allocation goals, while also being fair to existing water users.

Water laws and allocation rules are better defined and more established in the arid, western states, where water allocation policies have been required for many years. Perhaps the two most challenging policy issues in western states involve the ongoing conversion of agricultural land to municipal and residential uses, with implications for water supplies, and the long-term impacts of global climate change. Agriculture is seen by many observers in the west as an important source of water needed by real estate developers to continue expanding the supply of homes and businesses. Already substantial volumes of water have been transferred voluntarily from farmers to cities and towns. The likelihood that such transfers will continue in future has generated concern regarding the potential impacts on employment and livelihoods in agricultural communities.

Some states and counties have considered or implemented laws to limit the transfer of water, to prevent the decline of agricultural activity. It is not yet clear if such efforts will be successful in limiting water transfers or if the impacts on local communities will be notable when water supplies are moved away to a distant city or residential development. Public officials are considering also programmes that require developers or other purchasers of water rights to make payments that might be used to train or re-locate residents who become unemployed when water is transferred away from an agricultural area. Such programmes might enhance the public perception and acceptance of water transfer programmes in agricultural communities.

Current issues regarding water quality

The primary, over-arching policy issue regarding water quality in agriculture and other sectors in the United States can be summarised as follows:

- Public officials must determine the best ways to define and achieve state and national water quality objectives, while not constraining desirable economic growth and development.

The following specific issues can be described within the primary water quality issue of defining and achieving water quality goals at affordable cost:

- Determining the appropriate water quality objectives for a large number of constituents that include nutrients, chemicals, naturally occurring elements, and synthetic organic compounds.
- Determining the appropriate balance between efforts to improve water quality and impacts on economic activities such as production, consumption, employment, and growth.
- Determining how to achieve water quality objectives in small cities and towns that have limited financial resources.
- Determining the best mix of policies and programmes to motivate reductions in point source and nonpoint source pollution.
- Evaluating interactions involving agricultural programmes and water quality objectives.
- Determining optimal strategies for restoring groundwater quality or treating groundwater in areas where it is used as a source of drinking water.
- Determining the potential health and environmental impacts of synthetic organic compounds in surface water and groundwater, and establishing appropriate policies and programmes.
- Predicting the likely long-term impacts of constituents that degrade surface water and groundwater quality, and selecting appropriate policies to minimise those impacts.
- Predicting the likely long-term impacts of global climate change on efforts to maintain water quality in arid and humid regions.

Selected water quality issues

One of the most challenging policy issue regarding water quality and agriculture involves public efforts to reduce nonpoint source pollution of rivers, streams, lakes, and groundwater. Since the passing of the Clean Water Act in the 1970s, substantial progress has been made in reducing pollution of surface waters from point sources. Most of the remaining pollution of surface waters is from nonpoint sources including agriculture and cities (Carpenter *et al.*, 1998).

Nitrogen inputs to the environment from human uses doubled in the United States between 1961 and 1997, with most of the increase occurring in the 1960s and 1970s (Howarth *et al.*, 2002). Reducing nonpoint source pollution remains a challenging problem for state and federal agencies, given the difficulty of measuring discharges by individuals (Melosi, 2000). Policies and incentives that have worked well in reducing point source pollution, such as effluent charges and tradable emission permits, have been less effective in reducing nonpoint source pollution.

The issue of groundwater quality has nationwide significance because groundwater is the primary source of drinking water for about half the nation's population (Kolpin *et al.*, 1998; Melosi, 2000). More than 95% of individual water supplies involve groundwater and most of that water is not treated (Solley *et al.*, 1998). Groundwater quality is degraded by pollution from both point and nonpoint sources and by naturally occurring elements, such as arsenic and selenium (Korte, 1991; Lemly, 1994). Many of the chemicals and nutrients applied on farms are found in groundwater, often at concentrations that require treatment before the water can be consumed (Hallberg, 1989). Pesticides are found in groundwater in both urban and agricultural settings (Kolpin *et al.*, 2000).

Several researchers have begun to measure and assess the concentrations of synthetic organic compounds such as pharmaceuticals and hormones in surface water resources (Kolpin *et al.*, 2002; Squillace *et al.*, 2002). Further study is needed to determine the potential health and environmental impacts of those constituents. Hexavalent chromium and perchlorate have gained attention in recent years with new knowledge of potential impacts on drinking water quality and improved methods of measuring concentrations in surface water and groundwater. More than 10 states have sites where effluent contaminated with perchlorate has been discharged into sewage streams, natural waterways, or groundwater (Urbansky and Schock, 1999).

Public officials in several western states have been seeking cost-effective solutions to extensive salinity and drainage problems for many years. In the mid-1980s, elevated concentrations of selenium at the Kesterson National Wildlife Refuge were attributed to agricultural drainage water that had been used there as a source of water supply (Letey *et al.*, 1986; National Research Council, 1989; Posnikoff and Knapp, 1997; Ohlendorf, 2002). Selenium occurs naturally in soils on the west side of the San Joaquin Valley and it is mobilised when the soils are irrigated. An extensive programme of drainage water reduction was implemented throughout a large portion of the Valley after the discovery at Kesterson. In addition, scientists began investigating the occurrence of selenium in surface water in irrigated areas in other western states (Lemly, 1997, 2004).

It is not yet clear that irrigated agriculture can be sustained in areas where selenium is found in agricultural drainage water. Subsurface drainage systems are required in arid regions with impervious soil layers that restrict the downward movement of water, to prevent salt accumulation in the root zone. The cost of treating drainage water to remove selenium before the drainage water is discharged to rivers and streams is greater than the net returns earned in crop production. Improvements in farm-level water management can reduce drainage water volume substantially and the remaining volume can be used to irrigate salt-tolerant crops (Oster and Wichelns, 2003). Agricultural land retirement also will reduce drainage water volume, while making water available for use in agriculture, industry, or cities in other areas (Wichelns and Cone, in press).

Summary

Most of the current issues involving water availability and water quality in agriculture in the United States have been generated either by increasing demands for limited water supplies or changes in public preferences regarding environmental quality. In a sense, even these two over-arching issues might be summarised in a single statement: As population and aggregate wealth have increased in the United States, many residents have asked for greater service from the nation's limited surface water and groundwater resources, while also pledging to provide better care of those resources. It is no longer sufficient that water is used only to produce agricultural goods, power industrial growth, and provide drinking water for cities and towns. Water must also be used to protect and enhance environmental amenities such as wetlands, fisheries, wildlife habitat, and other ecosystem services.

Increasing competition for limited supplies is placing pressure on existing water laws, allocation rules, and interstate compacts in western states, while generating the need for new institutions in eastern and central states. Substantial public and private efforts are being invested in the process of revising or defining water rights and other allocation mechanisms. At the same time, individuals and communities are being required to spend large sums to improve water quality by reducing pollution from point and nonpoint sources. The larger challenge involves nonpoint sources including agriculture and municipal storm water systems. In some areas, the combination of increasing demands for water supplies and the need to reduce environmental impacts from irrigation are motivating farmers to sell a portion of their water supply to cities, towns, and real estate developers. Agriculture's role in providing water to enable continuous economic growth, particularly in arid states, likely will continue in future.

All of the policy issues involving water availability and water quality in agriculture become even more challenging when considering the potential impacts of global climate change. New water supply facilities might be needed in some states if rainfall patterns change or if the timing of snowmelt in spring is altered in ways that reduce the effectiveness of existing dams and reservoirs. Water quality in rivers and streams also might be impacted by structural changes in the timing, frequency, or magnitude of rainfall events. Public officials will require helpful input from scientists in assessing the likely impacts of global climate change as they evaluate near-term investment strategies that will minimise the potential social costs of changes in climate that might not occur for many years. The policy process will be time consuming and expensive, but the cost of not planning for climate change and not investing wisely in programmes to ensure water availability and protect water quality in future might be much larger.

REFERENCES

Carpenter, S.R., N.F. Caraco, D.L. Correll, R.W. Howarth, A.N. Sharpley, and V.H Smith (1998), "Nonpoint Pollution of Surface Waters with Phosphorus and Nitrogen", *Ecological Applications,* Vol. 8, No. 3, pp. 559-568.

Deason, J.P., T.M. Schad, G.W. Sherk (2001), "Water Policy in the United States: A Perspective", *Water Policy,* Vol. 3, No. 3, pp. 175-192.

Hallberg, G.R. (1989), "Pesticide Pollution of Groundwater in the Humid United States", *Agriculture, Ecosystems and Environment*, Vol. 26, No. 3-4, pp. 299-367.

Howarth, R.W., E.W. Boyer, W.J. Pabich, and J.N. Galloway (2002), "Nitrogen use in the United States from 1961-2000 and Potential Future Trends", *Ambio,* Vol. 31, No. 2, pp. 88-96.

Hutson, S.S. N.L. Barber, J.F. Kenny, K.S. Linsey, D.S. Lumia, and M.A. Maupin (2005), "Estimated Use of Water in the United States in 2000", Circular Number 1268, *U.S. Geological Survey*, Denver, Colorado.

Kolpin, D.W., J.E. Barbash, and R.J. Gilliom (1998), "Occurrence of Pesticides in Shallow Groundwater of the United States: Initial Results from the National Water-Quality Assessment Program", *Environmental Science & Technology,* Vol. 32, No. 5, pp. 558-566.

Kolpin, D.W., J.E. Barbash, and R.J. Gilliom (2000), "Pesticides in Ground Water of the United States, 1992-1996", *Ground Water,* Vol. 38, No. 6, pp. 858-863.

Kolpin, D.W., E.T. Furlong, M.T. Meyer, E.M. Thurman, S.D. Zaugg, L.B. Barber, and H.T. Buxton (2002), "Pharmaceuticals, Hormones, and Other Organic Wastewater Contaminants in U.S. Streams, 1999-2000: A National Reconnaissance", *Environmental Science & Technology,* Vol. 36, No. 6, pp. 1202-1211.

Korte, N. (1991), "Naturally Occurring Arsenic in Groundwaters of the Midwestern United States", *Environmental Geology,* Vol. 18, No. 2, pp. 137-141.

Lemly, A.D. (1994), "Agriculture and Wildlife: Ecological Implications of Subsurface Irrigation Drainage", *Journal of Arid Environments,* Vol. 28, No. 2, pp. 85-94.

Lemly, A.D. (1997), "Environmental Hazard of Selenium in the Animas La Plata Water Development Project", *Ecotoxicology and Environmental Safety,* Vol. 37, No. 1, pp. 92-96.

Lemly, A.D. (2004), "Aquatic Selenium Pollution is a Global Environmental Safety Issue", *Ecotoxicology and Environmental Safety,* Vol. 59, No. 1, pp. 44-56.

Letey, J., Roberts, C., Penberth, M., Vasek, C. (1986), "An Agricultural Dilemma: Drainage Water and Toxics Disposal in the San Joaquin Valley", Special Publication 3319, Division of Agriculture and Natural Resources, University of California, 56 pp.

Melosi, M.V. (2000), "Pure and Plentiful: the Development of Modern Waterworks in the United States: 1801-2000", *Water Policy,* Vol. 2, No. 4-5, pp. 243-265.

National Research Council (1989), *Irrigation-Induced Water Quality Problems*, National Academy Press, Washington, DC, 157 pp.

Ohlendorf, H.M. (2002), "The Birds of Kesterson Reservoir: a Historical Perspective", *Aquatic Toxicology,* Vol. 57, No. 1-2, pp. 1-10.

Opie, J. (1993), *Ogallala: Water for a Dry Land,* University of Nebraska Press, Lincoln.

Oster, J.D. and D. Wichelns (2003), "Economic and Agronomic Strategies to Achieve Sustainable Irrigation", *Irrigation Science,* Vol. 22, No. 3-4, pp. 107-120.

Posnikoff, J.E. and K.C. Knapp (1997), "Farm-Level Management of Deep Percolation Emissions in Irrigated Agriculture", *Journal of the American Water Resources Association,* Vol. 33, No. 2, pp. 375-386.

Solley, W.B. R.R. Pierce, and H.A. Perlman (1998), "Estimated Use of Water in the United States in 1995", Circular 1200, *U.S. Geological Survey*, Denver, Colorado.

Squillace, P.J., J.C. Scott, M.J. Moran, B.T. Nolan, and D.W. Kolpin (2002), "VOCS, Pesticides, Nitrate, and their Mixtures in Groundwater Used for Drinking Water in the United States", *Environmental Science & Technology,* Vol. 36, No. 9, pp. 1923-1930.

Stern, A. (2003), "Storage Capacity and Water Use in the 21 Water-Resource Regions of the United States Geological Survey", *International Journal of Production Economics,* Vol. 81-82, pp. 1-12.

Urbansky, E.T. and M.R. Schock (1999), "Issues in Managing the Risks Associated with Perchlorate in Drinking Water", *Journal of Environmental Management,* Vol. 56, No. 1, pp. 79-95.

Welch, A.H., D.B. Westjohn, D.R. Helsel, and R.B. Wanty (2000), "Arsenic in Ground Water of the United States: Occurrence and Geochemistry", *Ground Water,* Vol. 38, No. 4, pp. 589-604.

Wichelns, D. and D. Cone, in press, A Water Transfer and Agricultural Land Retirement in a Drainage Problem Area, *Irrigation and Drainage Systems*, forthcoming.

Chapter 9.

DECISION SUPPORT TOOLS TO AID POLICY DESIGN AND IMPLEMENTATION FOR SUSTAINABLE RESOURCE USE IN AGRICULTURE

Kevin Parris
Agriculture Directorate, OECD

Abstract

Given the growing importance of agri-environmental policies in many countries, underpinning policy-making with better analysis is a matter of some urgency to enhance policy monitoring, evaluation and future scenario analysis. In this regard the OECD has made a significant contribution in terms of: identifying and measuring the environmental performance of agriculture; building an inventory of policy measures addressing environmental issues in agriculture; and, analysing the linkages between policies and environmental outcomes. This paper provides examples of the OECD activities across these areas and identifies future directions of work to strengthen decision support tools that can improve policy design and implementation for sustainable resource use in agriculture.

Introduction and background

A key objective of this Workshop is to determine the best policy mix in addressing concerns related to agriculture and environment. A crucial step toward meeting this objective is to establish and utilise a range of decision support tools that can provide information to aid policy design and implementation including:

- Identifying and measuring the environmental performance of agriculture.
- Building an inventory of policy measures addressing environmental issues in agriculture.
- Analysing the linkages between policies and environmental outcomes.

Policy decision-making in the environmental and agri-environmental domain is a challenging undertaking. Most countries are addressing a wide range of environmental issues in agriculture. But in addressing these issues policy makers face a number of uncertainties and knowledge gaps including: the complexity of the linkages involved between policies and environmental outcomes; a lack of scientific knowledge of many agri-environmental interactions; incomplete data to undertake policy analysis; and other knowledge gaps, for example the impact of the environment on agriculture and the monetary valuation of environmental costs and benefits.

Given the growing importance of agri-environmental policies in the wider context of agricultural and environmental policies in many countries, underpinning policy-making with better analysis is a matter of some urgency to enhance policy monitoring, evaluation and scenario analysis. This requires improving the information and decision support tools as a prerequisite for efficient and effective policy decision making, as revealed in efforts underway in the OECD and its member countries.

Identifying and measuring the environmental performance of agriculture

Identifying and measuring the environmental performance of agriculture can help answer a broad range of questions that many policy makers are having to address:

- What are the impacts of agriculture on the environment, positive and negative, so that policy makers can target the most important impacts?
- What are the environmental impacts of different agricultural policy instruments, such as commodity price support, input subsidies (*e.g.* pesticides, fertilisers, water, energy) and direct payments for environmental services (*e.g.* protection of biodiversity)?
- What might be the environmental impacts of extending current policies and farming practices into the future, including the implications of climate change on agriculture?
- What are the trade-offs between meeting environmental targets and trade liberalisation commitments?

In helping to answer such questions the OECD and many of its member countries are developing a set of agri-environmental indicators (AEIs), drawing on the ***Driving Force–State–Response (DSR)*** model as an organising framework to establish indicators for policy purposes (OECD, 1997; 2003). This framework takes into account the specific characteristics of agriculture and its relation to the environment and the consideration of agriculture in the broader context of sustainable development (Figure 1). The DSR framework addresses a set of questions, including:

- What is causing environmental conditions in agriculture to change (*driving force*)?
- What effect is this having on the state or condition of the environment in agriculture (*state*)?
- What actions are being taken to respond to changes in the state of the environment in agriculture (*response*)?

In addressing these questions the DSR framework can provide a flexible framework in which analysis can help to improve understanding of the complexity of linkages and feedbacks between the causes and effects of agriculture's impact on the environment. It can also shed light on the responses by farmers, policy makers and society to changes in agri-environmental conditions and, identify indicators to explain and quantify these linkages and feedbacks.

Indicators to monitor changes in the environmental conditions in agriculture

In order to help select and develop indicators to monitor sustainable agriculture, OECD countries agree that they should possess a number of attributes (OECD, 1997), namely that they are:

- ***Policy-relevant***, that is they should be demand (issue) rather than supply (data) driven, and address the environmental issues faced by governments and others in the agriculture sector.
- ***Analytically sound***, based on sound science, but recognising that their development involves successive stages of improvement.
- ***Measurable***, that is feasible in terms of current or planned data availability and cost effective in terms of data collection, processing and dissemination.
- ***Easy to interpret*** and communicate essential information to policy makers and other stakeholders.

Within the context of the DSR framework and building on previous OECD work on indicators this has led to considerable progress in both the identification and specification of policy-relevant indicators as listed in Annex 1. In summary, the indicators are being developed in terms of agriculture's role in:

- ***Protecting the stock of natural resources impacted by agriculture***: Agriculture plays a critical role in the protection (or depletion) of the stock of natural resources used for production, notably soil and water resources, because for many countries agriculture accounts for the major share in the use of these resources. Farming activities also impact on the quality and quantity of natural plant and animal resources (*i.e.* biodiversity) and cultural landscapes, both on and off-farm.

- ***Reducing environmental pollution from agriculture***: Flows of materials into water (*e.g.* nutrients, pesticides) and emissions into the atmosphere (*e.g.* ammonia, greenhouse gases) are an inevitable part of agricultural production systems. Reducing the flows of these materials and emissions to an 'acceptable' level of risk in terms of human and environmental health is a priority for policy.

- ***Improving agri-environmental management practices and resource use efficiency***: The quantity of agricultural production is affected by the financial resources available to agriculture (both returns from the market and government support), the incentives and disincentives facing farming, and the kinds of management practices and technologies adopted by farmers. These practices and technologies impact on the productivity of the natural resources (*e.g.* soil) and purchased inputs (*e.g.* fertilisers) used by farmers. Depending on the management and productivity of agriculture's use of resources and inputs this will affect the rate of depletion and degradation of soils and water; the flows of harmful emissions into soils, water, air and the atmosphere; and the quantity and quality of plant and animal resources and landscape features.

Figure 1. The Driving Force - State - Response Framework to Address Agri-Environmental Linkages

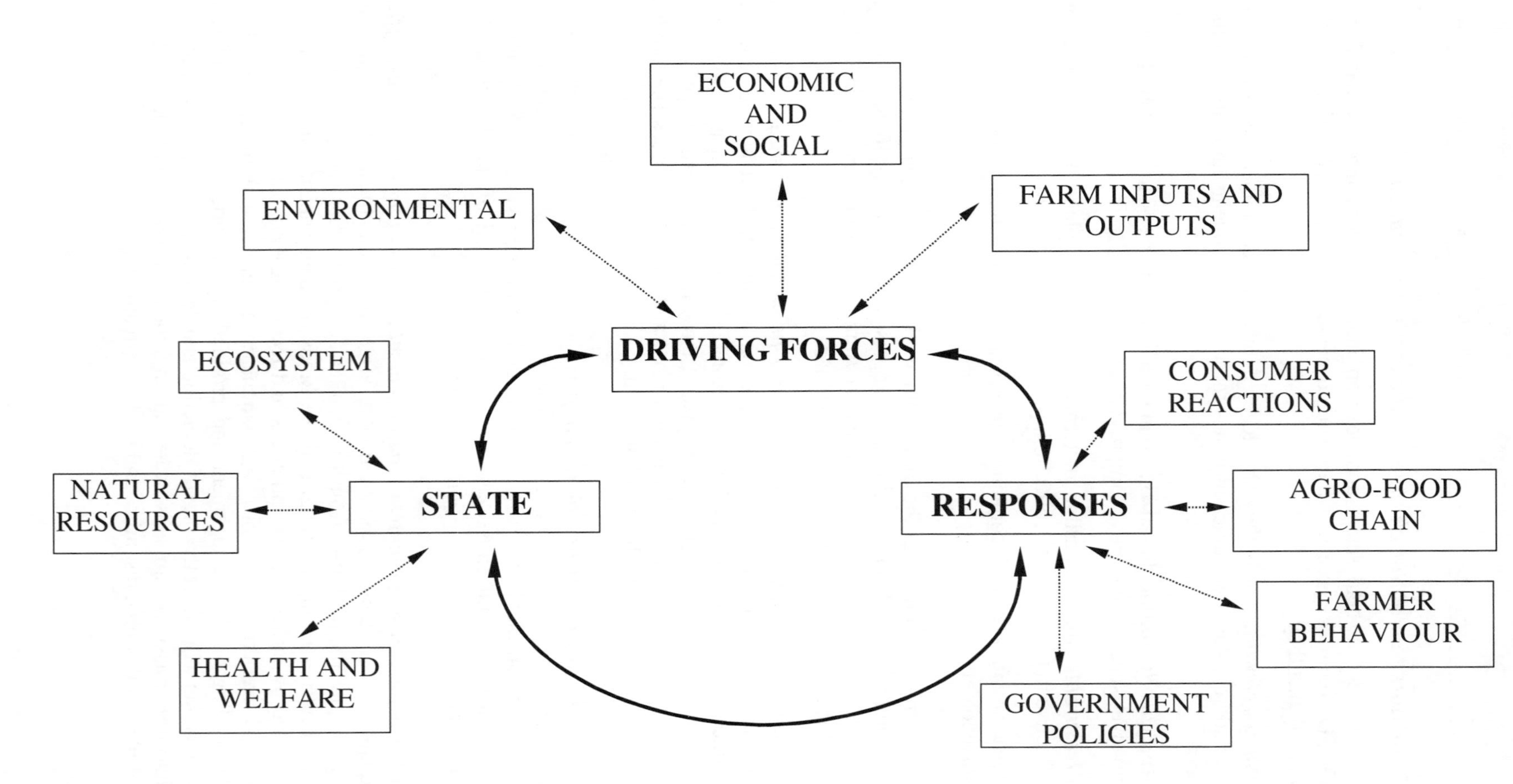

Source: OECD (1997).

Selected illustrations of OECD agri-environmental indicators

This section provides some illustrative examples of OECD agri-environmental indicators which, in terms of methodology and data requirements to calculate the indicators, could be extended to non-member OECD countries.

Nutrient balances

Inputs of nutrients, such as nitrogen and phosphorus, are important in farming systems as they are critical in raising crop and forage productivity, and a nutrient deficiency can impair soil fertility and crop yields. A build up of surplus nutrients in excess of immediate crop and forage needs, however, can lead to nutrient losses representing not only a possible cause of economic inefficiency in nutrient use by farmers, but especially a source of potential harm to the environment. This can occur in terms of water pollution (*e.g.* eutrophication of surface water caused by nutrient runoff and groundwater pollution by leaching), and air pollution, notably ammonia, as well as greenhouse gas emissions.

There are a complex range of physical processes that affect nutrient supplies in an agricultural system, illustrated by nutrient cycles. The extent to which these processes can harm the environment will depend on the: type of nutrients applied to crops; efficiency of crop nutrient use; type of crop and livestock systems; environmental assimilative capacity of an agro-ecosystem; farming practices; and economic and policy drivers (*e.g.* fertiliser prices and crop subsidies).

The OECD ***gross nutrient balances*** are calculated as the difference between the total quantity of nutrient inputs entering an agricultural system, and the quantity of nutrient outputs leaving the system (Figure 2). This calculation can be used as a proxy to reveal the status of environmental pressures, such as declining soil fertility in the case of a nutrient deficit, or for a nutrient surplus the risk of polluting soil, water and air. The methodology has been jointly developed by OECD country nutrient experts and the OECD and Eurostat Secretariats (OECD, 2005a; 2005b; 2006a forthcoming).

Figure 2. The main elements in the OECD gross nutrient (nitrogen and phosphorous) balance calculation

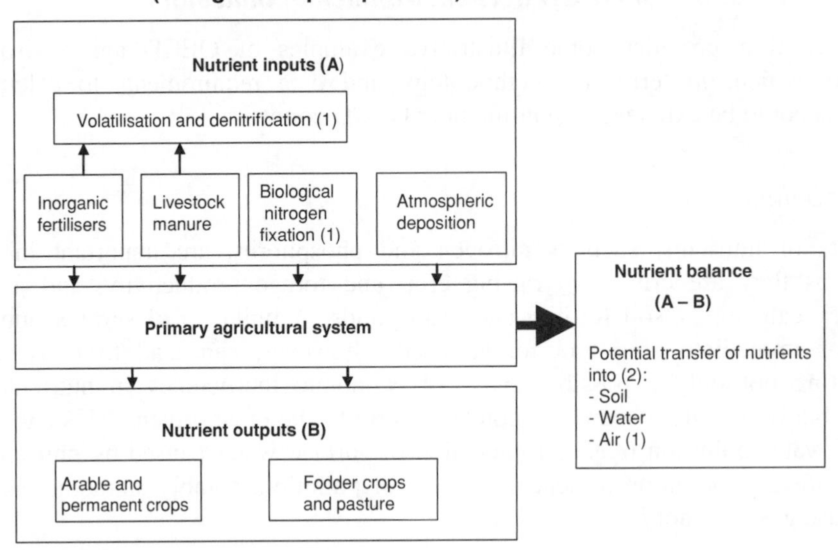

Notes:
1. Applies to the nitrogen balance only.
2. In most situations an excess of nutrients are transported into the environment, potentially polluting soils, water and air, but a deficit of nutrients in soils can also occur to the detriment of soil fertility and crop productivity.
Source: OECD (2005*a*; 2005*b*).

The nutrient balance indicator can be expressed in terms of the kilograms of nutrient surplus (deficit) per hectare of agricultural land per annum (Annex 2). This expression facilitates the comparison of the relative intensity of nutrients in agricultural systems between countries (*e.g.* very high in the Netherlands and very low in Australia, Annex 2) and also helps describe the main sources of nutrient inputs and outputs. In addition, the nutrient balances are expressed in terms of changes in the physical quantities of nutrient surpluses (deficits), which provide an indication of the trend and level of potential physical pressure of nutrient surpluses into the environment. While the nutrient balances are calculated at the national level, the same methodology can be used to estimate regional (sub-national) balances. This is important given the significant spatial variation in balances around national average values.

It should be stressed that the methodology is a gross balance calculation which takes account of all the total potential, not effective, losses of nutrients into the environment (*i.e.* soil, water and air). This includes for the *nitrogen balance* ammonia, volatilisation during the process of manure accumulation and manure storage and nitrogen losses from the soil (leaching, denitrification, and ammonia volatilisation). Denitrification, which is the conversion of soil nitrate to nitrogen gases, mainly occurs under anaerobic conditions (*e.g.* wet paddy rice and wet soil conditions). This process leads to the release of both dinitrogen gas (N_2) which is not harmful to the environment or human health, and also nitrous oxide which although released in small amounts is a very potent greenhouse gas. The components of the *phosphorus balance* are similar to the nitrogen balance, but exclude volatilisation and biological nitrogen fixation.

Agricultural water balances, groundwater use and irrigation

In many OECD countries there is growing competition for water resources between industry, household consumers, agriculture and the environment (*i.e.* aquatic ecosystems). The demand for water is also affecting aquatic ecosystems particularly where water extraction is in excess of minimum environmental needs for rivers, lakes and wetland habitats. Some OECD countries possess abundant water resources and, as a result, do not consider water availability to be a significant environmental issue in terms of resource protection. There are also important social issues concerning water, such as access for the poor in rural areas, while in some societies water has a significant cultural value, for example, for the indigenous peoples of Australia and New Zealand, and in Korea and Japan (OECD, 2006b forthcoming). The OECD agricultural water use indicators include the following showing the change of the:

- Quantity and share of agricultural water use in total national water utilisation.
- Quantity and share of agriculture's use of groundwater in total national groundwater utilisation.
- Area and share of the irrigated land in total agricultural land area.

Water use indicators, drawing on the concept of the water cycle shown in Figure 3, provide information on the trends in agricultural water use, and the importance of the sector in total national water use. Calculations of water use indicators are complex and not all OECD countries use the same data collection methods, which is a limitation in using these indicators. A further limitation is that water use balances are not usually calculated annually, but derived from 5 or even 10 year surveys. Moreover, the extent of groundwater reserves and their rate of depletion are also not easily measured, and cross country time series data are lacking. An additional complication is that under some systems, agriculture has the potential to recharge groundwater.

As environmental *driving forces* the agricultural water use and irrigated area indicators are linked to the *state* of (changes in) groundwater reserves and competition over water resources with other major users. *Responses* to these changes in the sustainability of water use are revealed through, for example, indicators of the uptake of more efficient irrigation management technologies and practices. Recent trends in the OECD agricultural water use indicators, are shown in Annex 3.

Interpretation and limitations of agri-environmental indicators

While indicators provide a basis on which policy makers can obtain an overview of trends that may require action and as a tool for monitoring and analysing the impact of farming and policies on the environment, they need to be interpreted and used with caution (OECD, 2001). For the OECD agri-environmental indicators the following caveats are important:

- *Definitions and methodologies for calculating indicators* are mostly standardised, but in particular further work is required for biodiversity and farm management indicators. Moreover, for some indicators, such as greenhouse gas emissions (GHGs), work is ongoing toward their further improvement (*e.g.* by incorporating agricultural carbon sequestration into a net GHG balance).

Figure 3. The interaction between the hydraulic cycle, water resources and water use

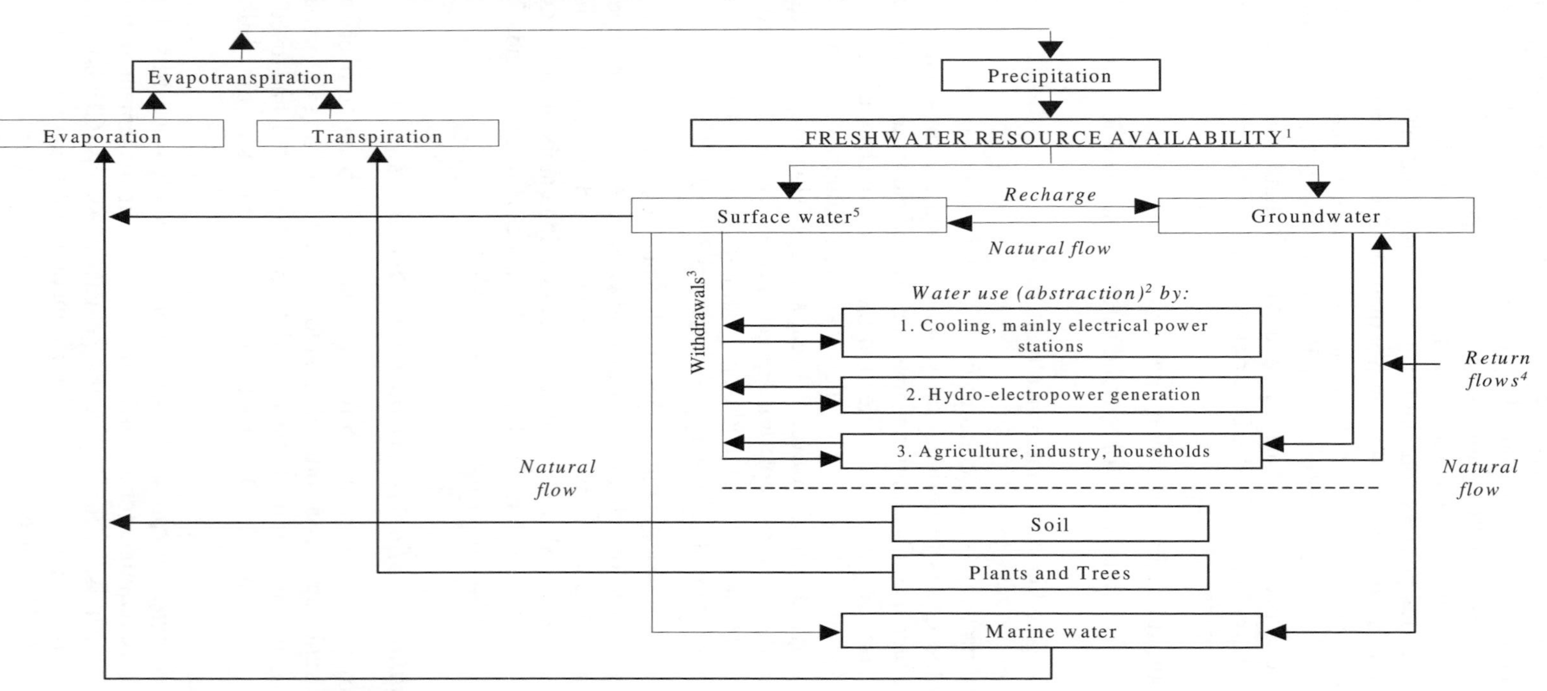

Notes:

1. Mean annual precipitation + transborder water flows - mean annual evapotranspiration. Evapotranspiration is the addition of evaporation (i.e. water loss from mainly surface and marine water, but also soil) and transpiration (i.e. water loss from plants and trees). Overexploitation of groundwater resources was not included in the calculation.
2. Water use (abstraction) defined as the amount of water drawn from surface or groundwater and conveyed to place of use.
3. Water withdrawals defined as water use from surface and groundwater resources and conveyed to main users, plus water returned to surface water and used by a downstream user, for example, from irrigation systems.
4. Return flows in the form of water infiltration into soil and drainage.
5. Rivers and lakes.

Source: OECD Secretariat.

- *Data quality and comparability* are as far as possible consistent and harmonised across the various indicators, but deficiencies remain such as the absence of data series (*e.g.* pesticide risks, biodiversity), variability in data coverage (*e.g.* pesticide use and energy consumption), and differences related to how the data was collected (*e.g.* surveys for farm management, census for land use, and models for water use).

- *Spatial aggregation* of indicators is given at the national level, but for certain indicators (*e.g.* water quality) this can mask significant variations at the regional level.

- *Trends and ranges in indicators* rather than absolute levels are important for comparative purposes across countries for many indicator areas, especially as local site specific conditions can vary considerably within and across countries. But absolute levels are of significance where: limits are defined by governments on the basis of scientific evidence (*e.g.* nitrates in water); targets agreed under national and international agreements (*e.g.* ammonia emissions); or where the contribution to global pollution is important (*e.g.* greenhouse gases).

- *Agriculture's contribution to specific environmental impacts* is sometimes difficult to isolate, especially for areas such as soil and water quality, where the impact of other economic activities is important (*e.g.* forestry) or the 'natural' state of the environment itself contributes to pollutant loadings (*e.g.* water may contain high levels of naturally occurring salts).

- *Environmental improvement or deterioration* is in most cases clearly revealed by the direction of change in the indicators (*e.g.* soil erosion, greenhouses gases), but for some indicators changes can be ambiguous. For example, changes in farm practices, such as no or minimum till agriculture can lower soil erosion rates, but at the same time may result in an increase in the use of herbicides with potential deterioration in adjacent aquatic ecosystems.

- *Baselines, threshold levels or targets for indicators* are generally not used to assess OECD indicator trends as these may vary between countries and regions due to difference in environmental and climatic conditions, as well as national regulations. But for some indicators threshold levels can be used to assess indicator change (*e.g.* drinking water standards) or targets compared against indicators trends (*e.g.* ammonia emissions and methyl bromide use).

The caveats to the interpretation of the indicators need to be viewed in a broader context, as in many cases they also apply to other indicators regularly used by policy makers. For example, there can be wide variations around national averages of socio-economic indicators (*e.g.* employment), and methodological and data deficiency problems are not uncommon (*e.g.* wealth distribution). Work on establishing agri-environmental indicators is relatively recent compared with the much longer history of developing economic indicators, such as gross domestic product. Measuring the linkages between the biophysical environment and human activities through indicators is often more complex than monitoring trends in socio-economic phenomena, given that many agri-environmental effects are not valued in markets or are not easily measured in physical terms (*e.g.* biodiversity).

Building an inventory of policy measures addressing environmental issues in agriculture

Over recent years there has been an increase in the range and extent of policy measures and approaches to address environmental issues in agriculture. As the significance of agri-environmental and environmental policy measures has grown, it is important to evaluate them in terms of their efficiency in achieving the environmental outcomes for which they were designed, and their effects on agricultural production and trade. Against the background of these policy developments OECD is establishing a policy inventory in the context of work on agricultural policy reform and sustainable development. The inventory covers a broad range of policy measures addressing environmental issues in agriculture, namely:

- Agricultural policy measures where environment outcomes are the primary objective.
- Mechanisms tying general agricultural support programmes to environmental conditions.
- General environmental policy measures which have a significant impact on agriculture.

The inventory provides a complement to the OECD databases on agri-environmental indicators, described above, and agricultural support policies. Support policies cover the broader set of agricultural and economy-wide policy measures that influence the overall impact (positive or negative) of agriculture on environment - either directly or indirectly. While these policy measures are beyond the scope of the agri-environmental policy inventory they are relevant for the evaluation of policies.

Agricultural support policies are measured by OECD on an annual basis through the Producer Support Estimate (PSE) and derived indicators, and are the principal tools used to monitor and evaluate agricultural policy developments. It is important to distinguish between support provided to producers and its impact on individual production decisions, and support provided to general services for the agricultural sector as a whole. Policy measures within the PSE are classified in terms of how policies are implemented, and are regularly measured and analysed in the annual OECD *Monitoring and Evaluation of Agricultural Policies* (OECD, 2005c), and recently in the OECD study of agricultural policies in China (OECD, 2005d).

The choice of the way in which to classify policy measures addressing environmental issues in agriculture depends on the purpose or intended use of the inventory, in particular, such as to:

- Gather information on the different policy approaches used, then this would include collecting information on the different ***types of instruments*** used (economic instruments, command and control, voluntary agreements, and information, see Annex 4).
- Evaluate the effectiveness in achieving stated environmental objectives, this would suggest a categorisation of policies based on their ***environmental objectives,*** based on the agri-environmental indicators (Annex 2).
- Assess the impacts of the measures on farmers' production practices and incomes, this would involve a classification based on ***methods of implementation.***

To be helpful for policy analysis and evaluation the inventory has to cover all three concerns and needs to be undertaken using a multi-tiered approach. To provide the ***data*** necessary for policy analysis and evaluation, the main outcomes of each policy measure are classified in terms of the:

- ***Standard description*** of the policy measure (Annex 5).
- ***Monetary transfers***, including implicit and explicit *budgetary transfers (costs)*; implicit monetary transfers associated with, for example, higher water or electricity consumer prices due to water pollution or methane produced from animal waste - *consumer transfers (costs)*; and the opportunity costs incurred by farmers - *compliance costs.*
- ***Programme coverage,*** including number of participants (individuals, communities or regions), number of contracts or farms assisted, type of land eligible and area covered.

Access to the OECD Policy Inventory can be obtained through the website at: www.oecd.org/agr/env under *Inventory of Policy Measures Addressing Environmental Issues in Agriculture,* and as an example for New Zealand of an agri-environmental policy inventory (Ministry of Agriculture and Forestry, 2005).

Analysing the linkages between policies and environmental outcomes

As the Driving Force-State-Response model (Figure 1) recognises agri-environmental linkages are highly complex, and disentangling the effects of different policies measures and mixes on environmental outcomes to better inform policy makers is a significant challenge for a number of reasons outlined below.

- Linkages between policies and environmental outcomes occur through changes in farm practice decisions, use of inputs and natural resources, which impact on production. While the linkages between policies and production have been reasonably well quantified in the OECD and elsewhere, this is not the case for agri-environmental linkages.
- There are multiple environmental outcomes as a result of many factors (driving forces), not only policies, which are rarely implemented in isolation, but as a mix of policy measures, and it is difficult to separate "causality" from "correlation".
- Some environmental outcomes are complementary with production (more production is associated with higher environmental outcomes, whether those outcomes are harmful or beneficial), while for others the outcomes are substitutes (more or less production is associated with less or more of the environmental outcomes).
- A given environmental outcome from a driving force will differ depending on the ecological carrying capacity in a particular site (spatial effects) and are unlikely to be linear. Also some environmental outcomes take a considerable period of time to become evident, such as leaching of phosphorus into rivers and groundwater. Hence, from a policy perspective targeting practices will be more efficient in some cases than policies directed at environmental outcomes.
- Environmental outcomes are rarely quantified in a common, monetary unit of measurement, while physical values (indicators) often refer to total and not marginal changes.

There are three broad questions to be addressed by the quantitative analysis of policy-environmental linkages; namely to examine the:

- Relative effects of different *agricultural and agri-environmental policy measures* on production and the environment, hence, the resulting environmental effects (as measured for example by agri-environmental indicators) from a given change to different policy measure (as measured for example by a 10% increase in support estimates) are quantified, which throws light on understanding the environmental implications of meeting agricultural policy objectives.
- Cost-effectiveness of different policy measures in achieving specified *environmental outcomes*. In this case, to achieve a given change in an environmental outcome (as measured for example by a 10% decrease in nutrient surpluses or greenhouse gas emissions), the costs of alternative policy measures (as measured for example by changes in support estimates) are quantified, which helps to better understand the agricultural implications of meeting environmental policy objectives.
- Implications of future projections and policy scenarios, drawing on agricultural and environmental projection models to provide information to policy makers of likely future developments.

OECD is using two main modelling tools to help address these questions, SAPIM and AGLINK, which are briefly outlined below. A discussion of other modelling approaches that can (and have) been used to address these questions can be found in OECD, 1999.

SAPIM: Stylised Agri-environmental Policy Impact Model

OECD is developing a cause-effect model for analysing the general linkages between agricultural and agri-environmental policies and their environmental effects. The main purpose of the model is to provide illustrations of how different types of stylised policy interventions affect the environment through changes in intensive (input use) and extensive (land use) margins as well as those related to entry and exit of agricultural land. The Stylised Agri-environmental Policy Impact Model (SAPIM) adopts integrated economic and bio-physical modelling at a disaggregate level necessary to capture the heterogeneity of the environment and economic behaviour. To this end SAPIM captures these cause-effect linkages in a field parcel and extrapolate them to provide insight at more spatially aggregate levels.

SAPIM is conceptually grounded in farm-level modelling, however, it can also be interpreted at more aggregate land use levels to infer what the impact of policies would be at the extensive and intensive margins. Several caveats apply to the model used in this manner. First, commodity prices are exogenous because the demand for the commodities produced is not taken into consideration. Policies may affect prices by increasing or decreasing supply, but currently the model does not capture these feedback effects. Second, the land use specification is very simplified, as it is assumed there is no transfer of environmental effects between field parcels.

The model has been developed to facilitate its application across countries by allowing for a diverse set of environmental issues. However, the model has been defined as "stylised" because it is meant to capture qualitative differences in agri-environmental systems of OECD member states, not to obtain a detailed characterisation of agriculture with full OECD coverage.

AGLINK: medium term agricultural projections model

The OECD's AGLINK is an econometrically estimated dynamic supply-demand model of world agriculture, developed by the OECD Secretariat in close co-operation with member countries. It represents annual supply, demand and prices for the principal agricultural commodities produced, consumed and traded in member countries. The overall design of the model focuses particular attention to the potential influence of agricultural policy on agricultural markets in the medium term.

AGLINK is a partial equilibrium model including over thirty of the most important products, in terms of output and trade in the OECD area, with wheat, coarse grains, rice, oilseeds, oilseed meals and oils, key dairy products, milk, beef, pork, sheep, poultry meats, eggs and wool being a concise representation of commodities covered. The model consists of complete modules for ten OECD countries/regions – Australia, Canada, the European Union, Hungary, Japan, Korea, Mexico, New Zealand, Poland and the United States, as well as three non-OECD countries/regions – Argentina, China and the Rest of the World.

Illustrating the structure of supply and demand specifications for the crop functions, crop production is expressed in AGLINK as the product of area harvested and yield per unit area. Area harvested and yield are represented separately and each may be influenced by relative prices and government policy variables. Competition among alternative crops for land is represented in the model by cross-price effects in the area equations. Although land is included in the production process, it is a very simplified representation based on average yield, and other factors of production are not included.

A modelling simulation drawing on AGLINK was used in the *OECD Environmental Outlook* (OECD, 2001b) to compare the effects of removing agricultural subsidies and the application of *ad valorem* tax on chemical input use in OECD regions to 2020 compared to a baseline 'Reference' scenario. The policy simulation found significant environmental benefits from removal of agricultural subsidies compared with the Reference Scenario (Table 1). The reductions in irrigation water usage and nitrogen loading from agriculture are expected to be particularly significant, especially in the regions of Western Europe, Japan, Korea, Central and Eastern Europe. The simulation which examined the application of an *ad valorem* tax on chemical input use in OECD agriculture found significant reductions in nitrogen loading from agriculture compared with the Reference Scenario (Table 1).

Table 1. Environmental effects of OECD agricultural subsidy removal and a tax on agrochemical use

(% change from Reference Scenario in 2020 for OECD regions)

	Irrigation water use	Agricultural methane emissions	Agricultural nitrogen loading
Agricultural subsidies removal[1]	-10%	-1%	-7%
Tax on agrochemicals use[2]	-2%	-1%	-21%

1. This policy simulation explores the impacts of eliminating all subsidies on agricultural products used as inputs of final demand, all agricultural producer and export subsidies, all subsidies on inputs to agriculture, and market price support to agriculture in OECD regions. A proxy was used to simulate the removal of OECD market price support, combining a tax on agricultural production equal to market price support estimates with an equivalent subsidy to household consumption of agricultural products.
2. This policy simulation explores the impacts of applying an *ad valorem* annually increasing tax of 2% (*i.e.* reaching a total tax levy of 50% by 2020) on the use of all chemical inputs to OECD agricultural production.

Challenges and limitations in quantifying agri-environmental linkages

In using the SAPIM and AGLINK models to quantify agri-environmental linkages there are a number of challenges and limitations which need to be taken into account by policy makers discussed here.

The more removed a policy instrument is from an observed environmental outcome the more difficult it is to asses the specific policy's contribution to the outcome. For example, the role of a conservation tillage incentive payment in an observed reduction in soil erosion is likely to be easier to assess than the role of agricultural trade liberalisation in such a change. However, a word of caution is necessary when assessing the impact of even highly targeted agri-environmental payments, and that is whether farmers are providing an environmental benefit that they would not otherwise have provided.

An added difficulty in disentangling the role of policies in environmental change is that many of the recent changes in farming that have led to environmental impacts may be linked to technological developments driven by competition in agricultural markets. The role of policy in influencing these trends is not always apparent. For example, while application rates of fertilisers and pesticides vary considerably because of site specific agro-ecological and climatic conditions, the use of different technologies and farm management practices can alter application rates considerably.

There is very limited opportunity for models to make a policy-on versus policy-off comparison, in examining policy effects, thus many modelling exercises do not consider the counterfactual situations. One such counterfactual, but one often overlooked in agri-environmental analysis, is that at a global level food production is essential and that the very act of production will inevitably lead to some environmental disturbance. Perhaps all agri-environmental analysis might be accompanied with a reminder of this, although where and how food is produced to minimise environmental harm is a key issue (Pearce, 2001).

Another counterfactual example, is not so obvious (Pearce, 2001). While there is frequently a focus on pesticides and fertilisers pollution and the overexploitation of soil and water resources, it is important to understand that these inputs and resources contribute to higher agricultural productivity. Without that extra productivity on existing land, the race to convert existing non-agricultural land, especially forest land, will be even faster than it is. Hence, it is necessary to take account of the counterfactual, hard as it may be to identify and measure it, of the land area not converted because of increased agricultural productivity. The foregone land conversion allows time toward making agriculture and the environment more compatible.

There are many examples in quantifying links between policies and the environment of assumed links without any clear rationale to explain these assumptions. Hence, there is value in understanding the underlying cause and effect relationships, not just between, say, nitrogen runoff and eutrophication of water, but why the nitrogen runoff occurs in the first place. This might be for a variety of reasons not directly related to a policy impact, such as the farmers managerial training in handling fertilisers or managing livestock nutrient waste, and the financial resources available to enact change, such as purchasing better livestock waste storage facilities that reduce nutrient leakages into the environment.

Due to the difficulties in gathering empirical data and establishing policy counterfactuals, it is often the case that an ex-ante assessment of policies is performed using a combination of economic and bio-physical models. For example, site-specific changes in-field soil erosion due to particular soil conservation practices can be estimated using the Universal Soil Loss Equation and the Wind Erosion Equation. These models provide reasonably accurate results and require only minimal data describing climate, topography, soil, and cropping information at the field level. In contrast, models of nutrient and pesticide runoff are far more complex, simulating multiple environmental effects from the transport and fate of multiple pollutants into environmental sinks.

Even with elaborate process models, the results are unlikely to match real world observations because farming practices are not the only factors that affect environmental quality. Weather variability and non-agricultural pollutants may have a considerable impact on the variation encountered in physical measurements. Thus, models can provide guidance on the expected environmental outcome, but do not substitute an ex-post analysis of an impact of a specific policy.

Conclusions

There are many encouraging signs that policy makers are increasingly recognising and addressing the environmental costs and benefits associated with agricultural activities. This paper has highlighted, based on the OECD experience, how the policy decision making process is being better informed through the development of a set of agri-environmental indicators and an agri-environmental policy inventory as building blocks in quantifying the causes and effects of policies on environmental outcomes in agriculture.

To further advance work on the information and analytical requirements to underpin efficient and effective policy making for sustainable resource use in agriculture, particularly at the regional and national level of decision making, OECD work suggests that it is necessary to address a number of issues:

- Identifying and agreeing a recognised set of agri-environmental indicators. These should be developed by multi-disciplinary teams that will use the indicators in model development and to convey to a wider public the state and trends of environmental conditions in agriculture.
- Determining the most appropriate and highest priority investments in monitoring and reporting capacity given limited resources and the large array of 'policy relevant' environmental issues that governments typically need to address. This relates to moving from physical measures of agricultural impacts on the environment to a set of economic or monetary measures of these impacts. This would enable comparison and evaluation of different environmental issues on a common basis, which is not feasible with physical measures, for example, the relative of importance of water pollution compared with soil erosion. While there are numerous difficulties in measuring the economic costs and benefits of agriculture on the environment, especially deriving estimates of the benefits which often have no market value (*e.g.* biodiversity), a number of countries have begun to develop work in this area (EFTEC and IEEP, 2004).
- Building an inventory of agri-environmental policies and measures of agricultural support helps to clearly identify the scope of environmental objectives and targets and the range and extent of support provided to agriculture, and is essential information for quantifying the linkages between policies, agriculture and environmental outcomes.

- Developing models should proceed in conjunction with data development. Data sources should be reviewed in order to identify appropriate proxies that will be used in the modelling process. Both economic and bio-physical data should be collected and analysed together in order that the links between the economic and environmental activities are identified.
- Establishing better feedback mechanisms to develop economic decision-making models and bio-physical models and vice versa, which will increase the integration between economic and bio-physical relationships.
- Examining how modelling efforts can be developed at various levels of aggregation: macroeconomic, national and ecological dimensions, with model development designed to interface with one another in order to capture the full range of environmental impacts.

Annex 1.

List of Main OECD Agri-Environmental Indicators

Theme	Indicator title	Indicator definition (trends over time for all indicators)
I. Soil	**i. Soil erosion**	1. Area and share of agricultural land affected by water erosion in terms of different classes of erosion, *i.e.* tolerable, low, moderate, high and severe.
		2. Area and share of agricultural land affected by wind erosion in terms of different classes of erosion, *i.e.* tolerable, low, moderate, high and severe.
II. Water	**ii. Water use**	3. Quantity and share of agricultural water use in total national water utilisation.
		4. Quantity and share of agriculture's use of groundwater in total national groundwater utilisation.
		5. Area and share of the irrigated land in total agricultural land area.
	iii. Water quality	6. Nitrate concentrations in surface water and groundwater in excess of national water threshold values in agricultural areas.
		7. Phosphate concentrations in surface water in excess of national water threshold values in agricultural areas.
		8. Share of nitrate and phosphate contamination derived from agriculture in surface water, groundwater and coastal waters.
		9. Pesticide concentrations in surface water and groundwater in excess of national water threshold values in agricultural areas.
		10. Share of monitoring sites in agricultural areas where one or more pesticides are present in surface water and groundwater
III. Air	**iv. Ammonia emissions, acidification and eutrophication**	11. Quantity and share of agricultural ammonia emissions in national total ammonia emissions
	v. Methyl bromide use and ozone depletion	12. Quantity of methyl bromide use in terms of tonnes of ozone depleting substance equivalents.
	vi. Greenhouse gas emissions and climate change	13. Gross total agricultural greenhouse gas emissions (carbon dioxide, methane and nitrous oxide, expressed in carbon dioxide equivalents) and their share in total greenhouse gas emissions.

IV. Biodiversity	**vii. Genetic diversity**	14. Number of plant varieties registered and certified for marketing for the main crop categories (*i.e.* cereals, oilcrops, pulses and beans, root crops, fruit, vegetables and forage).
		15. Share of the dominant (one to five) crop varieties in total marketed production for selected crops (*i.e.* wheat, barley, maize, oats, rapeseed, field peas and soyabeans).
		16. Area and share of land under transgenic crops in total agricultural land.
		17. Number of livestock breeds registered and certified for marketing for the main livestock categories (*i.e.* cattle, pigs, poultry, sheep and goats).
		18. Share of the three dominant livestock breeds in total livestock numbers for the main livestock categories (*i.e.* cattle, pigs, poultry, sheep and goats).
		19. Total number of livestock (*i.e.* cattle, pigs, poultry and sheep) in endangered and critical risk status categories and under conservation programmes.
		20. Status of plant and livestock genetic resources under in situ and ex situ national conservation programmes.
	viii. Wild species diversity	21. Share of wild species that use agricultural land as primary habitat.
		22. Populations of a selected group of breeding bird species that are dependent on agricultural land for nesting or breeding.
	ix. Ecosystem diversity	23. Conversion of agricultural land area to (land exits)/from (land entries) with other land uses (*i.e.* forest land; built-up land, wetlands, and other rural land).
		24. Area and share of agricultural semi-natural habitats (*i.e.* fallow land, farm woodlands) in the total agricultural land area.
		25. Share of national important bird habitat areas where intensive agricultural practices are identified as either posing a serious threat or a high impact on the area's ecological function.

V. Farm management	**x. Nutrient management**	26. Number (area) and share of farms (agricultural land area) under environmental farm management plans.
		27. Share of farms using soil nutrient testing (agricultural land regularly sampled and analysed for nutrient content).
	xi. Pest management	28. Share of agricultural land area under non-chemical pest control methods.
		29. Share of arable and permanent crop area under integrated pest management
	xii. Soil management	30. Share of arable land area under soil conservation practices.
		31. Share of agricultural land area under vegetative cover over a year.
	xiii. Water management	32. Share of irrigated land area using different irrigation technology systems.
	xiv. Biodiversity management	33. Share of agricultural land area under biodiversity management plans.
	xv. Organic management	34. Share of agricultural land area under certified organic farm management (or in the process of conversion to an organic system).
VI. Agri-cultural inputs	**xvi. Nutrients**	35. Gross balance between the quantities of **nitrogen** (N) inputs (*e.g.* fertilisers, manure) into, and outputs (*e.g.* crops, pasture) from farming and those per hectare of agricultural land.
		36. Gross balance between the quantities of **phosphorus** (P) inputs (*e.g.* fertilisers, manure) into, and outputs (*e.g.* crops, pasture) from farming and those per hectare of agricultural land.
	xvii. Pesticides	37. Quantity of pesticide use (or sales) in terms of tonnes of active ingredients.
		38. Index of the risk of damage to terrestrial and aquatic environments, and human health from pesticide toxicity and exposure.
	xviii. Energy	39. Quantity and share of direct agricultural energy consumption in national total energy consumption.

Annex 2.

OECD Nitrogen balance estimates: 1990-92 to 2002-04

Change in the nitrogen balance (tonnes N)

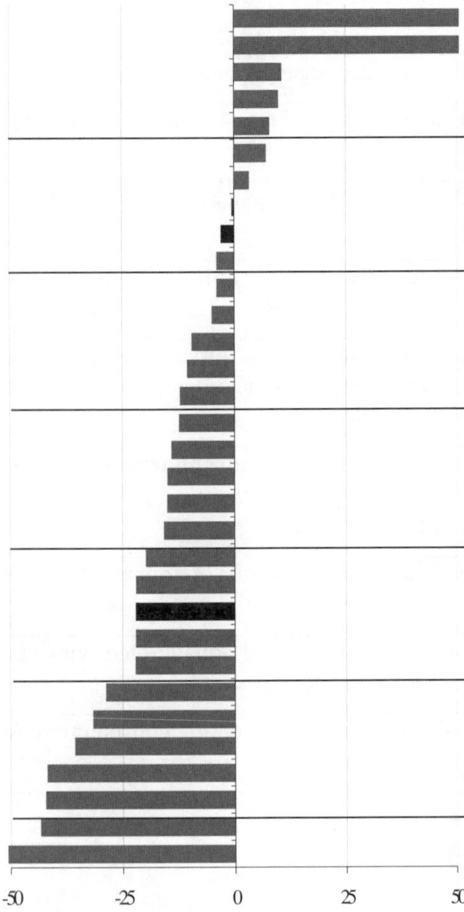

	Balance expressed as tonnes of nitrogen (N)				Balance expressed as kg nitrogen per hectare of total agricultural land		
	Average•	Average•	Change		Average•	Average•	Change
	1990-92	2002-04	1990-92 to 2002-04•		•1990-92	2002-04	1990-92 to 2002-04•
	(000) tonnes N	(000) tonnes N	(000) tonnes N	%	kgN/ha	kgN/ha	%
Canada (2) (3)	1073	1960	887	83	15	29	92
New Zealand (2) (3)	307	501	195	63	22	35	60
Spain (2)	713	790	77	11	24	27	14
Hungary	227	250	23	10	36	43	20
United States (2)	13419	14541	1122	8	31	35	12
Ireland (4)	331	355	24	7	75	81	8
Portugal (2)	173	179	6	3	43	47	8
Korea (2)	465	462	-3	-1	213	240	13
OECD	39784	38545	-1239	-3	89	76	-15
Switzerland	120	116	-5	-4	76	76	0
Australia (2)	7581	7284	-297	-4	16	16	-1
Iceland (4)	9	8	0	-5	4	3	-5
Czech Republic	332	300	-31	-9	77	70	-9
Mexico	2843	2545	-298	-10	28	24	-13
Poland	862	758	-104	-12	46	46	-1
Norway	92	81	-11	-12	92	77	-16
Sweden (2)	195	167	-27	-14	58	55	-5
Japan (2)	935	795	-139	-15	180	166	-8
Luxembourg	30	25	-4	-15	238	199	-16
Italy (2)	835	703	-131	-16	48	46	-4
France (2)	1832	1470	-363	-20	60	50	-18
Belgium (4)	344	269	-75	-22	255	193	-24
EU-15	9830	7668	-2162	-22	114	89	-22
Turkey (2)	1493	1164	-329	-22	37	29	-21
Germany (2)	2624	2043	-581	-22	151	120	-21
Austria	226	161	-65	-29	66	48	-27
Denmark	493	338	-156	-32	178	127	-29
Netherlands	688	443	-245	-36	345	229	-34
Finland	211	123	-88	-42	83	55	-34
United Kingdom (2)	702	407	-296	-42	39	25	-36
Slovak Republic	197	111	-85	-43	80	46	-43
Greece (4)	276	106	-169	-61	32	13	-61

-50 -25 0 25 50 %

1. The gross nitrogen balance calculates the difference between the nitrogen inputs entering a farming system (*i.e.* mainly livestock manure and fertilisers) and the nitrogen outputs leaving the system (*i.e* the uptake of nutrients for crop and pasture production).
2. Average for the period 2002-04 refer to the average 2000-02.
3. For Canada, change in the nitrogen balance is 83%.
   For New Zealand, change in the nitrogen balance is 63%.
   For Greece, change in the nitrogen balance is -61%.
4. Average for the period 2002-04 refer to the average 2001-03.

Source: OECD Secretariat (2006).

Annex 3.

Figure 1. Total agricultural water use1: 1990-92 to 2001-03

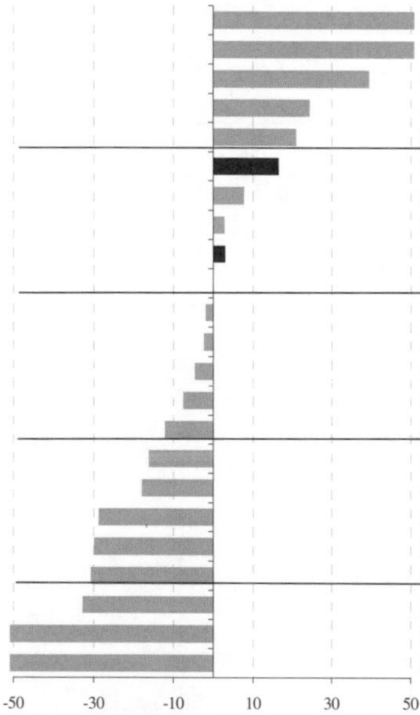

% Change in total agriculture water use 1990-92 to 2001-03

	Total agriculture water use 1990-92 (million m³)	Total agriculture water use 2001-03 (million m³)	Change in total agriculture water use 1990-92 to 2001-03 %	Change in total water use[2] 1990-92 to 2001-03 %	Share of agriculture in total water use[2] 2001-03 %
Greece (3)	4 600	8 941	94	24	87
Turkey (4)	18 812	31 000	65	28	78
United Kingdom (5)	1 347	1 880	40	5	15
Australia (6)	13 384	16 660	24	12	75
Portugal (7)	5 100	6 178	21	-2	73
EU-15 (8)	41 817	48 791	17	-9	34
Spain (9)	23 700	25 538	8	4	66
Canada (10)	3 991	4 104	3	-6	10
OECD (11)	405 850	416 237	3	-1	42
Iceland (12)	5	5	0	-7	3
United States (13)	195 200	191 555	-2	2	40
Japan (14)	58 630	57 240	-2	-2	66
France	4 901	4 676	-5	-14	14
Mexico (15)	62 500	57 763	-8	-2	80
Sweden	169	148	-12	-9	6
Korea (16)	8 671	7 272	-16	8	46
Austria (17)	100	82	-18	-49	5
Germany (18)	1 600	1 140	-29	-21	3
Poland	1 527	1 070	-30	-18	9
Denmark (19)	300	208	-31	-17	28
Hungary	1 032	694	-33	-21	13
Slovak Republic (20)	188	68	-64	-39	6
Czech Republic (21)	93	15	-84	-45	1

-50 -30 -10 10 30 50 %

1. Agricultural water use is defined as water for irrigation and other agriculture uses such as for livestock operations. It includes water abstracted from surface and groundwater, and return flows from irrigation but excludes precipitation directly onto agricultural land.
2. Total water use are the total water abstractions for public water supply + irrigation + manufacturing industry no cooling + electrical cooling, and EU15 and OECD include the total of the countries shown in this figure.
3. Data for the period 1990-92 and 2001-03 refer to the year 1985 and 2001. Data for irrigation are used because data for agriculture water use are not available. Share of agriculture in total water use is for 1997. For Greece, change in total agriculture water use is 94%.
4. Data for the period 2001-03 refer to the year 2001. Data for irrigation are used because data for agriculture water use are not available. For Turkey, change in total agriculture water use is 65%.
5. Data for the period 2001-03 refer to the year 2000.
6. Average 1990-92 = Average 1993-95, Average 2001-03 = (2000).
7. Data for the period 1990-92 and 2001-03 refer to the year 1991 and 2002. Data for irrigation water (year 1991) are used because data for agriculture water use are not available.
8. EU15 excludes: Belgium, Finland, Ireland, Italy, Luxembourg, Netherlands.
9. Data for the period 1990-92 and 2001-03 refer to the year 1991 and 2001. Data for irrigation (year 1991) are used because data for agriculture water use are not available.
10. Data for the period 1990-92 and 2001-03 refer to the year 1991 and 1996.
11. OECD excludes: Belgium, Finland, Ireland, Italy, Luxembourg, Netherlands, New Zealand, Norway, Switzerland.
12. Data for the period 1990-92 refer to the year 1992.
13. Data for the period 1990-92 and 2001-03 refer to the year 1990 and 2000.
14. Data for the period 1990-92 and 2001-03 refer to the year 1990 and 2000.
15. Data for the period 1990-92 and 2001-03 refer to the year 1995 and 2000.
16. Data for the period 1990-92 and 2001-03 refer to the year 1990 and 2000.
17. Data for the period 2001-03 refer to the year 2003.
18. Data for the period 2001-03 refer to the year 2001. Data for irrigation are used because data for agriculture water use are not available.
19. Data for the period 1990-92 and 2001-03 refer to the year 1991 and 1998.
20. For Slovak Republic, change in total agriculture water use is -64%.
21. For Czech Republic, change in total agriculture water use is -84%.

Source: OECD Environmental Data Compendium 2004; OECD estimation; National data for Australia, Austria, Denmark and Hungary.

Figure 2. Change in irrigated area and share of area irrigated in total cultivated and total agricultural area: 1990-92 to 2001-03

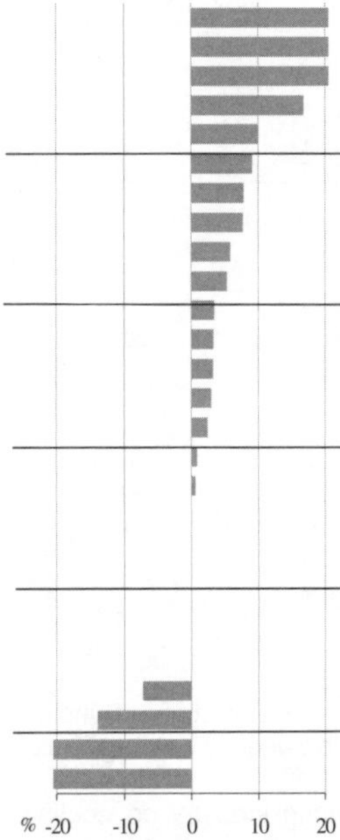

	Irrigated area ('000 hectares)		Change in irrigated area ('000 hectares)	Change in irrigated area %	Change in total agriculture area %	Share of irrigated area in arable and permanent crop area %	Share of irrigated[1] area in total agricultural area. %	Share of irrigation[1] water use in total agriculture water use. %	Irrigation water application rates Megaliters per hectare of irrigated land		
	1990-92	2001-03	1990-92 to 2001-03	1990-92 to 2001-03	1990-92 to 2001-03[2]	2001-03	2001-03	2001-03	1990-92	2001-03[5]	Change
New Zealand[3]	250	475	225	90	-11	14	3	75	..	..	..
Belgium[4]	24	40	16	67	2	5	3	..	0.1	0.1	-10
France[5]	2150	2632	482	22	-4	17	9	..	..	..	..
Australia	2057	2402	345	17	-6	5	1	43	8.7	4.3	-50
Spain	3388	3727	339	10	-3	21	13	60	7.0	6.2	-11
Canada	719	785	66	9	0	2	1	9	4.4	5.2	17
United States	20900	22543	1643	8	-3	13	5	40	9.0	8.4	-7
EU15[6]	12033	12965	932	8	-3	17	10	..	6.5	6.6	2
OECD[7]	50284	53207	2923	6	-4	12	4	..	8.9	8.5	-4
Turkey	3329	3506	177	5	-2	12	8	78	5.7	8.8	56
Greece	1383	1431	48	3	-1	37	17	87	5.5	6.2	14
Denmark	433	448	14	3	-5	20	17	25	0.7	0.4	-48
United Kingdom	165	170	5	3	-6	3	1	1	1.0	0.6	-40
Portugal	631	650	19	3	-6	25	17	79	8.1	13.5	67
Mexico	6170	6320	150	2	1	23	6	78	9.9	8.9	-10
Netherlands	560	565	5	1	-1	60	29	1	0.3	0.1	-59
Germany	482	485	3	1	-1	4	3	3	3.3	2.4	-29
Switzerland	25	25	0	0	-3	6	2	..	..	..	..
Sweden	115	115	0	0	-7	4	4	4	0.9	0.9	3
Poland	100	100	0	0	-7	1	1	1	3.7	0.9	-77
Italy	2698	2698	0	0	-1	24	17	46	9.4	9.6	2
Czech Republic	24	24	0	0	0	1	1	1	1.3	0.5	-64
Austria	4	4	0	0	-7	0	0	5	12.5	2.5	-80
Japan	2846	2641	-205	-7	-8	63	55	66	20.4	21.5	5
Korea	1327	1142	-185	-14	-12	61	59	..	..	..	..
Hungary	205	126	-79	-39	-8	3	2	3	2.1	1.2	-44
Slovak Republic	299	153	-146	-49	0	10	6	5	0.5	0.4	-29

% -20 -10 0 10 20

.. Not available.

1. To be consistent, the years used for the average calculations are the same for irrigation water use and total agricultural water use, irrigated area and total agricultural area.
2. For some countries, data in brackets are used to replace the average due to missing data:
   - Australia: 1990-92(1996), 2001-03(2003)
   - Czech Republic: 1990-92(1993)
   - New Zealand: 1990-92(1985), 2001-03(2003)
   - Slovak Republic: 1990-92(1993)
3. New Zealand, share of irrigation water in total agriculture water use, for 2002, see Chapter 3. Change in irrigated area is 90%.
4. For Belgium, change in irrigated area is 67%.
5. For France, change in irrigated area is 22%.
6. EU15 excludes: Finland, Ireland, Luxembourg.
7. OECD excludes: Finland, Iceland, Ireland, Luxembourg, Norway, Switzerland.

Source: FAOSTAT 2006, OECD 2nd Agri-Environmental Indicators Questionnaire, 2005. OECD Environmental Data Compendium 2004.

Figure 3. Share of agricultural groundwater use in total groundwater use and total groundwater use in total water use: 2002

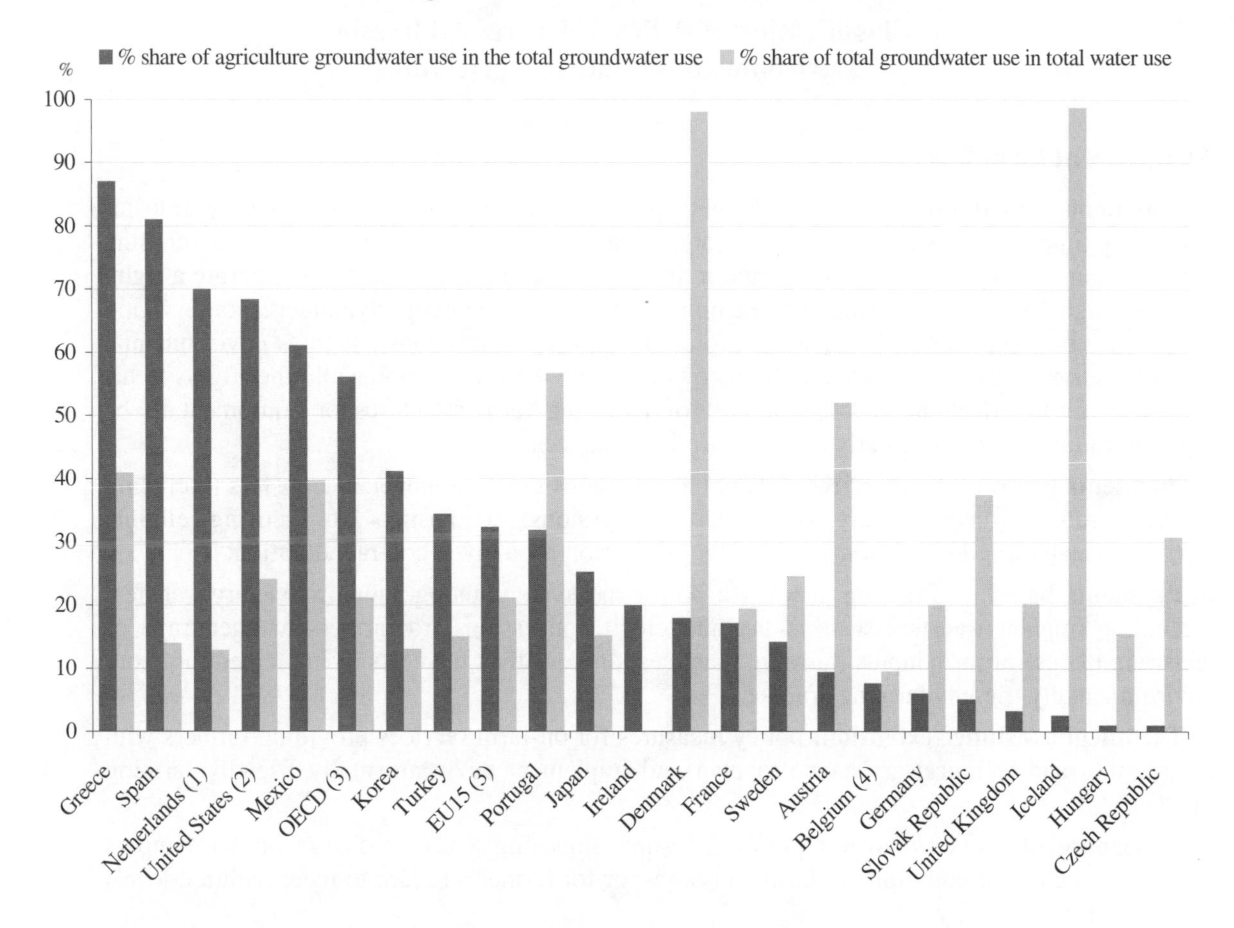

For the relevant time series see Annex Tables on the Volume 4 website.
Data of 1994 are used to replace missing data of 2002 for: France, Ireland, Portugal.
Data of 1995 are used to replace missing data of 2002 for: Germany, Netherlands, Spain and Sweden.
Data of 1997 are used to replace missing data of 2002 for: Czech Rep., Greece, Iceland, Mexico, Slovak Rep. and Turkey.
Data of 2000 are used to replace missing data of 2002 for: United Kingdom.
1. Source. Chapter 3, Netherlands country section.
2. United States: groundwater for irrigation is used as data on total agricultural groundwater are unavailable.
3. The EU15 and OECD data must be interpreted with caution as they consist of totals using different years across countries, and do not include all member countries. EU15 excludes: Finland, Italy, Luxembourg.
OECD excludes: Australia, Canada, Finland, Italy, Luxembourg, New Zealand, Norway, Poland, Switzerland.
4. Year 2000.
Sources: OECD Environmental Data Compendium 2004; OECD 2nd Agri-Environmental Indicators Questionnaire, 2004.

Annex 4.

Classification of Policy Measures Addressing Environmental Issues in Agriculture

Measures at farm level

1. Payments based on farm fixed assets: policy measures granting a monetary transfer (including implicit transfers such as tax and credit concessions) to farmers for purchasing conservation easements, not including land retirement (an easement is the act of creating a right, such as a right of way, to a person to make limited use of another's property, under a conservation easement a person acquires the right to enforce the contract restrictions , such as environmental conditions, on the use of the property by its owner, while the owner retains all other rights to his property), or to offset the investment cost of adjusting farm structures or equipment to be adjusted to more environmentally friendly farming practices.

2. Payments based on land retirement: policy measures granting monetary transfers (including implicit transfers such as tax and credit concessions) to farmers for retiring eligible environmentally fragile land from commodity production for a given contract duration.

3. Payments based on farming practices: policy measures granting annual monetary transfers (including implicit transfers such as tax and credit concessions) to farmers to encourage or constrain the use of farm inputs (farming practices) and/or offset the costs of implementing more environmentally friendly farming practices.

4. Technical assistance/extension: policy measures for on-farm services providing farmers with information and technical assistance to plan and implement environmentally friendly farming practices.

5. Environmental taxes/charges: policy measures imposing a tax or charge on farm inputs which are sources of pollution, on farm emissions, or for farmer's failure to meet required levels of environmental quality.

6. Cross-compliance mechanisms: policy measures imposing environmentally friendly farming practices or levels of environmental performance to farmers participating on a voluntary or mandatory basis on specific agricultural support programmes.

7. Labelling standards/certification: voluntary participation measures defining standards that have to be met by private goods for certification according to specific eco-labelling.

8. Regulatory measures: involuntary participation measures imposing regulatory requirements to achieve specific environmental quality, including environmental quotas, permits, restrictions and bans, maximum rights or minimum obligations assigned to economic agents that may take the form of *transferable or tradable permits.*

Measures at sector level

9. Community-based measures: measures granting support to public agencies or community-based associations (*e.g.* Landcare groups, conservation clubs, environmental co-operatives) to implement collective projects to improve environmental quality in agriculture.

10. Research/education: measures granting support to institutional services to improve environmental performance of agriculture through promoting research and education on environmentally friendly production technologies, pollution prevention, quality control management systems, and green marketing.

11. Inspection/control: measures granting support to institutional services controlling the environment associated with agriculture, including monitoring and enforcement of policy measures addressing the environmental effects of agriculture (administration costs).

Annex 5.

Standard Description of Policy Measures

Policy	The title of the policy measure.
Agency	The agency through which the policy measure is implemented.
Relevant legislation	The legislation relevant to the policy measure.
Overall objective	A description of the overall objective(s) of the policy measure.
Delivery mechanism	A description of how the policy measures works (*e.g.* details regarding how direct payments are implemented, enforcement mechanisms for regulatory requirements, etc.), based on the classification in Annex 4.
Targets	Any specific targets (*e.g.* benchmarks, or thresholds levels etc.).
When applied	The years in which the policy measure is applied.
Coverage/eligibility	The coverage of the policy measure - *e.g.* whether it is applied at a national, regional, local level; and whether participation is compulsory or voluntary. Also the relevant eligibility criteria – *i.e.* to all or a specific group.
Costs	The available information on costs – *e.g.* fiscal/budgetary costs in the case of direct payments, or costs to farmers in the case of regulatory requirements.
How the programme is monitored/evaluated	A description of the process and methods of monitoring and evaluation.
Participation/results	Any quantitative information on participation in the programme is available (*e.g.* number of farms, area covered, etc.)
Sources/further information	The source of information, particularly any web-links to sites where further information is available.

REFERENCES

EFTEC and IEEP (2004), *Framework for Environmental Accounts for Agriculture,* Economics for the Environment Consultancy (EFTEC) in association with Institute for European Environmental Policy (IEEP), report submitted to the Department for Environment, Food and Rural Affairs, London, United Kingdom, http://statistics.defra.gov.uk/esg/reports/env.asp

Ministry of Agriculture and Forestry (2005), *Policy Measures Addressing Environmental Issues in New Zealand Agriculture*, MAF Technical Paper No: 2005/05, Wellington, New Zealand, http://202.78.129.207/mafnet/publications/techpap.html

OECD (1997), *Environmental Indicators for Agriculture Volume 1: Concepts and Frameworks*, Paris, France, www.oecd.org/agr/env/indicators.htm

OECD (1999), *Environmental Indicators for Agriculture Volume 2: Issues and Design*, Paris, France, www.oecd.org/agr/env/indicators.htm

OECD (2001a), *Environmental Indicators for Agriculture Volume 3: Methods and Results*, Paris, France, www.oecd.org/agr/env/indicators.htm

OECD (2001b), *OECD Environmental Outlook*, Paris, France, www.oecd.org/env

OECD (2003), *OECD-China Seminar on Environmental Indicators: Proceedings,* Paris, France, http://www.oecd.org/dataoecd/28/3/34760501.pdf

OECD (2005*a*), *OECD Nitrogen Balance Handbook*, jointly published with Eurostat, Paris, France, www.oecd.org/agr/env/indicators.htm

OECD (2005*b*), *OECD Phosphorus Balance Handbook*, jointly published with Eurostat, Paris, France, www.oecd.org/agr/env/indicators.htm

OECD (2005c), *Agricultural Policies in OECD Countries Monitoring and Evaluation 2005,* Paris, France, www.oecd.org/agr

OECD (2005d), *OECD Review of Agricultural Policies: China,* Paris, France, www.oecd.org/agr

OECD (2006a forthcoming), *Environmental Indicators for Agriculture Volume 4*, Paris, France, www.oecd.org/agr/env/indicators.htm

OECD (2006b forthcoming), *Water and Agriculture: Sustainability, Markets and Policies,* Paris, France, www.oecd.org/agr/env

Pearce, D. (1999), "Measuring Sustainable Development: Implications for Agri-Environmental Indicators", in OECD, *Environmental Indicators for Agriculture Volume 2: Issues and Design*, Paris, France, www.oecd.org/agr/env/indicators.htm

PART III.

POLICY OPTIONS FOR CHINA

Chapter 10.

FERTILISER USE IN CHINESE AGRICULTURE

Chen Mengshan
Department of Crop Production, Ministry of Agriculture, China

Abstract

This paper presents a retrospective analysis of the history of fertiliser use in China, discusses the significance of fertilisers in the development of the national agricultural sector, and assesses the present situation and existing problems. We also consider measures to improve fertiliser resource utilisation rates with the purpose of promoting the development of sustainable agriculture in China.

Introduction

Fertilisers are an important element of the means of agricultural production, and make a significant contribution to development in the national agricultural sector and to safeguarding the nation's food security. In order to bring about a scientific approach to development and build a conservation-oriented, environmentally friendly society, China's government has proposed new approaches to fostering economic growth and improving the overall productive capacities of the agricultural sector. In this respect, initiatives include the promotion of water-saving techniques in irrigation, the scientific use of fertilisers, pesticides and similar resources, and vigorously supporting sustainable agriculture.

China's Ministry of Agriculture and related agencies have also adopted practical and effective measures to implement, on a nationwide basis, soil testing programmes to devise fertiliser formulas specific to local conditions. Technical subsidies are available for these programmes and related services are provided to farmers free of charge. Farmers are also strongly encouraged to use organic fertilisers in the nation's efforts to resolve the current low fertiliser utilisation rates and put in place a scientific system for fertiliser use that is adapted to China's own situation.

Historical overview of fertiliser use in China

China boasts an agricultural tradition that stretches to the most remote antiquity, with the essence of that tradition to be found in meticulous farming methods and the understanding of soil nutrients. As long ago as the Western Zhou dynasty (c. 1000 – 771 BCE) the *Zhou Li* texts advocated that "*marshland grasses be killed to plant wheat*", while the contemporaneous *Book of Odes (Shi Jing)* records the use of organic fertilisers in the verse, "*These weeds being decayed, the millets grow luxuriantly.*" In the "Spring & Autumn" and "Warring States" periods, the use of organic fertilisers became more widespread, while the *Fuguo* chapter of the philosophical treatise *Xunzi* advocates "*generous manuring of the fields*" and "*the necessity of manures and irrigation to draw harvests from the fields.*"

The *Chen Fu Book of Agriculture* of the Song Dynasty (960+ CE) notes that *"adding new fertility to the soil is done with manures"* so that the "*powers of the earth are ever renewed.*" An agronomical treatise of the Yuan Dynasty (1279+ CE), meanwhile, goes so far as to advocate "*treasuring manure as you treasure gold*", which raises the concepts of fertilising and nurturing the soil to unprecedented levels of importance. Later, the *Zhiben Tigang* of the Qing Dynasty (1644+) describes the linkages of fertiliser use to advances in crop productivity and soil quality, observations that descend directly from the "*earth becomes manure, manure becomes earth*" dialectic of the *Zhou Li Situ* chapters of 2 000 years earlier.

These principles and practical measures have guaranteed the growth, prosperity, expansion and social progress of China's peoples and had an important positive influence on the enduring Chinese civilisation. Chemical nitrogen fertilisers were first introduced to Chinese agriculture in 1901 when the Japanese occupiers of Taiwan of the time began to experiment with their applications to sugarcane farming. Use of such fertilisers gradually spread inland from the coastal areas over the following years but, as there were no domestic manufacturers at the time, all supplies needed to be imported. In view of this and various other (primarily economic) factors, the speed of advance was slow and quantities used small.

After the founding of the People's Republic, the government of China attached great importance to fertilisers and their use along scientific lines. In 1955, Comrade Mao Zedong drew on the agricultural experiences of various regions of the country to personally draft a *National Plan for Agricultural Development* in 40 articles that proposed 12 measures for improving crop productivity. These included increasing the use of fertilisers. When promulgated in 1956, the draft *National Plan for Agricultural Development* put particular emphasis on deep tillage and soil improvement, and the increased use of organic fertilisers. In 1957, the National Chemical Fertiliser Experimentation Network was set up, and the State published the *Guidelines for Deep Tillage and Soil Improvement* and *Guidelines regarding Fertiliser Problems*, which stressed the important effect increased use of organic fertilisers in soil improvement schemes would exert on crop productivity. In 1959, Comrade Mao Zedong made the further recommendation that "China's first source of fertilisers should be the raising of pigs and cattle. If we can raise one pig per person, one pig per *mu*, the main source of fertiliser is settled. A pig is a small organic fertiliser factory. Fertilisers are the foodstuffs of plants, plants are the foodstuffs of animals and animals are the foods of humans." This succinctly stated the relationships that link the soil, plants and animals in ecological cycles.

The Third Plan for National Economic and Social Development proposed by Comrade Deng Xiaoping in 1963 aimed to establish in first priority 500 million *mu* of stable, high-productivity farmland treated primarily by increased applications of organic fertilisers in conjunction with chemical fertilisers.[1] The government's great interest in organic fertilisers then led to the green-manure campaigns of the 1960s and the organic fertiliser movement of the 1970s, which encouraged among the people a widespread and positive concern for improving the quality of farmed land. At the same time, development in China's chemical fertiliser industry fed a trend toward rapid and sustained increases in the volumes of chemical fertiliser being used.

1. The *mu* is a traditional unit of land measure equal to 1/15 ha (0.0667 ha)

Since the beginning of the 1980s, three major adjustments have occurred with respect to fertiliser use. In the first half of that decade, to begin with, the single-minded objective to raise productivity led to widespread applications of nitrogen, phosphorous and potassium fertilisers, and continual growth in fertiliser use efficiency. The second phase came in the latter half of the 1980s when emphasis in the agricultural sector turned to high productivity, high efficiency and high quality products. Models for the use of compound fertilisers were promoted, high-density single-ingredient fertilisers were developed and the overall quantities of chemical fertilisers used increased very quickly. The third major adjustment has come about since the turn of the century as goals in agriculture have shifted once more, to emphasise ecology and safety as well as enhanced quality, productivity and efficiency. Fertiliser use has thus entered a new phase in which eco-environmental protection and agricultural production are viewed with equal importance.

Current situation and problems with fertiliser use in China

Volumes of chemical fertiliser use, and product structures

Since the mid-20th century, chemical fertilisers have been viewed with increasing importance in China's efforts to develop its agricultural sector. Over the past 20 years, applications of these fertilisers have increased by 1.57 million tonnes (of pure nutrient – this applies also to figures below) on average every year. While in 1984, the gross volume of fertiliser used was 14.82 million tonnes, this figure had grown to 47.66 million tonnes in 2005 and was the highest in the world. At a cost exceeding 200 billion Yuan, these chemical fertilisers represent 25% of total agricultural production costs (material expenses plus labour). Some 390 kilograms of chemical fertiliser are thus used per hectare of cultivated land, somewhat higher than the world average.

There is a higher proportion of single-ingredient and low-grade fertilisers produced than of high-grade or complex fertilisers. Utilisation of complex fertilisers reached 13.04 million tonnes in 2005, representing some 27% of all chemical fertilisers used. Meanwhile, most complex (or compound) fertilisers on the domestic market are general-use products. Fewer specialised complex (or compound) fertilisers are manufactured according to formulas that have been developed on the basis of soil testing or in-field performance experiments. In terms of fertiliser use, meanwhile, there is insufficient development and application of organic fertilisers and an unreasonable preponderance of chemical fertilisers. In 2004, the relative ratios of nitrogen, phosphorous and potassium used were 1: 0.47: 0.10, which indicates a slight insufficiency of potassium use.

Organic fertiliser resources and utilisation

China has an abundance of organic fertiliser resources of a great variety of types. In addition to manures and urines, composts, crop straws, green manures and general farmyard manures, there are cake fertilisers, marine fertilisers, urban wastes with agricultural applications, and biogas wastes. It can be deduced from available figures that China's livestock and poultry manures amount to 2 billion tonnes, composts to 2 billion tonnes as well, crop straws to 700 million tonnes, cake manures to 20 million tonnes and green manures to 100 million tonnes. These organic resources contain great quantities of nitrogen, phosphorous and potassium and trace elements, amounting to some 70 million tonnes of (pure) nutrients, or 1.46 times the amount of chemical fertilisers utilised nationwide. Clearly, the potential for their utilisation is great.

Practical evidence shows that a scientific and rational approach to using organic fertilisers provides an abundance of nutrients to food crops, improving both their external appearance and internal quality, thus bringing down production costs and enhancing the competitive profile of farm products on the market. In addition, these fertilisers are irreplaceable in terms of their salutary effects in soil improvement, building soil fertility, enhancing the overall productive capacity of the agricultural sector and maintaining farmland in a "constantly renewed" condition. At the present moment, the government of China is implementing a programme to increase the organic content of farmland soils, advocating in particular the return of crop straws to the soil, farmyard waste composting, the planting of green manures and industrialising the manufacture of organic fertilisers. These various measures aim to utilise to their best advantage the available organic resources as fertilisers.

Research, development and promotion of new fertilisers

Since the 1980s, China has conducted extensive research into foliar, microbiological and slow-release or controlled-release fertilisers, as well as the industrial treatment of organic wastes. Many favourable results have been obtained. With respect to foliar fertilisers, multi-nutrient types have been developed, including new types containing organic or amino acids that have proven to be very effective in leaf vegetable and fruit-tree applications. Initial work with microbiological fertilisers, meanwhile, has resulted in the foundation of a manufacturing industry that now services more than 100 million *mu* of farmland, while also regularly producing new types of these fertilisers. Technical innovations have led to breakthroughs in coatings for slow-release and controlled-release fertilisers, and certain new technologies are already being applied to their production and use. With respect to the treatment and utilisation of organic wastes, meanwhile, a variety of biotechnological techniques and products such as stalk and refuse rotting preparations are already in wide use. Significant advances have been achieved in the utilisation of manure wastes from large and medium-sized poultry and livestock breeding operations, and the technologies and facilities for treatment of these wastes are at a mature stage of development.

Fundamental research in soil fertilisers and their applications

China now has a large corps of soil and fertiliser technologists of more than 17 000 individuals, and agricultural agencies at all levels operate soil chemistry laboratories. The past twenty years of reform and development have produced manpower with the technical understanding, operational know-how and management skills to achieve noteworthy results in research and development while also acquiring valuable experience in technical outreach programmes.

A series of national technical standards and manuals have been drawn up that put in place "National land-type zones and grading of cultivable land", "National medium and low-yield farmland zones and technical improvement standards", "Technical specifications for farmland soil-testing technology", "Principles for the reasonable use of fertilisers", "Technical specifications for fertiliser performance experimentation and evaluation" and "Assessment standards for investments in infrastructure development to improve medium and low-yield farmland," among others. Other technical issues have been addressed in such documents as the "Technical specifications for fertiliser formulation", the "Technical standards for soil analysis", the "Technical specifications for the national soil quality survey," and the "Handbook of national standards for water-saving agricultural technology and guidelines for investment estimates."

At the same time, such initiatives as the "County Farmland Resources Information System", the "Information Compilation System for Soil Testing and Fertiliser Formulation" and a dictionary of standards have been realised and are now in use in farmland surveys and assessments across China.

Recommendations for promotion of the scientific use of fertilisers

The scientific use of fertilisers has been incorporated into the 11th 5-Year Plan for National Economic and Social Development and funding for soil testing and fertiliser formulation work has been brought within the scope and jurisdiction of public financial organs. The Ministry of Agriculture will coordinate with relevant agencies the various major strategies discussed below so as to ensure that a scientific approach to fertiliser use receives its due emphasis.

Promotion of soil testing and fertiliser formulation technologies

Soil testing and fertiliser formulation is generally referred to internationally as balanced fertilisation. It is an advanced agricultural technology widely promoted by the United Nations. In general terms, it involves the five stages of soil testing, making a formula, developing a fertiliser to match the particular actual and desired conditions, then supplying and applying such fertilisers. China's Ministry of Agriculture began a country wide promotion of soil testing and fertiliser formulation technology in 2005, a programme that has been well received by the farming community and the general public alike. Practical experience in the early stages of the programme demonstrates that it serves as an entry point for the promotion of fertiliser technologies. It has proven to be an important factor in boosting overall productivity in the agricultural sector as it leads to increased yields, higher farmer incomes and greater efficiency.

The Ministry of Agriculture has also drawn up its recommendations for fertiliser work in the context of the 11th 5-Year Plan. With the guidance and financial support of the government, a farmer-oriented system for fertiliser use will be progressively put in place that is based on scientific research, aims for comprehensive promotion, and incorporates the resources of industry. The orderly system for soil testing and fertiliser formulation that will result will give the rural population the benefits of stable and enduring access to the needed fertilisers and supporting services.

Substantive quality improvements to cultivated land

Tillable land is the foundation stone of the agricultural industry and the ultimate source of wealth. Only when land is usable on a sustainable basis can sustainable agriculture be realised. It follows that at the same time as China promotes a scientific approach to fertiliser use, equal attention must be paid to improving the quality of cultivated land. Indeed, work in this regard is a great responsibility of the current generation on behalf of those that will follow.

With respect to the improvement of the nation's farmland, the government has set in motion new soil-fertility programmes which include initiatives to greatly increase production of organic fertilisers, increase the organic content of soils, and put in place a dynamic monitoring and warning system to track soil quality. At the same time, efforts will continue to improve the production standards of quality grains at the field level and to build up large commodity-grain production bases. Model programmes demonstrating conservation farming techniques will be set up throughout the country and improvements will be made to support water-saving infrastructure in arid areas. In short, every effort will be made to ensure that farmland remains strong and constantly renewed with enhanced overall productive capacity, and that a solid foundation of healthy, fertile soils is in place to support sustainable agriculture.

Improving farmer understanding of fertilisation techniques

Farmers are the key decision makers with respect to fertiliser use, and only by improving understanding among this community of relevant techniques can a truly comprehensive scientific approach to fertilisation be achieved. The government is making great efforts to upgrade the standards of technical knowledge among farmers. To this end, it has set up training programmes and encourages technical personnel to get out onto the farms proper and familiarise themselves with specific issues and challenges facing farmers. This will make these specialists better able to give farmers face-to-face instructions and advice. Meanwhile, national agricultural technology promotion agencies are actively engaged in setting up model or demonstration programmes that bring farmers in for hands-on learning opportunities featuring live presentations of practical, relevant information. These various initiatives have a very positive effect in raising the levels of scientific understanding of fertilisers among the rural populace.

Adjustments to fertiliser product structures

The government also encourages the private sector to introduce advanced production techniques so as to improve product structures and develop, right at the manufacturing level, the high-density complex fertilisers for special uses that favour balanced inputs of fertiliser nutrients. In step with market dynamics, meanwhile, fertiliser manufacturers will be encouraged to involve themselves in the operational mechanisms of the soil testing and fertiliser formulation field, and will be supported in such efforts. The production of formula fertilisers will be organised in cooperation with national technical outreach agencies, transforming manufacturing models to produce the complex and compound fertilisers specific to the conditions that testing indicates. Such an approach will favour the early and widespread adoption of these technologies.

Encouraging increased use of organic fertilisers

Increased inputs of organic fertilisers are an effective means of improving land. China is a country with massive organic matter resources available for use as fertilisers, yet only about one-third of these resources are currently exploited. The government intends to organise and guide the production by farmers of organic fertilisers, drawing on all available resources to significantly increase the amounts produced and utilised. Research, education and outreach programmes will be put in place to instruct farmers in the composting and other production processes for these fertilisers, while developing new techniques will ensure that the most advanced approaches are being used. These approaches will be expanded: (1) through government guidance, advice and financial support, (2) through market-oriented management and fair competition, and (3) based on the principles of favouring the diffusion of technology. Every effort will be made to promote the scientific use of formula fertilisers and increase the income of rural residents. Great emphasis will be placed on providing the appropriate models, adapted to local conditions, for the efficient implementation of the soil testing and formula fertiliser model.

Legal measures for the monitoring and management of fertiliser quality

The agricultural agencies of China's various levels of government are currently engaged in tightening laws and regulations that deal with fertiliser use and, where necessary, drawing up new statutes. These various measures will provide added levels of regulation to the production, marketing, utilisation and management of fertilisers. Through amendments to and coordination of the various statutes governing farmland and general land management, and with respect to the maintenance of cultivated land, the responsibilities, rights and obligations of land users as well as of the various levels of People's governments will be set out in clearer terms. Meanwhile, additional measures will be taken to enhance monitoring of fertiliser product labelling, and any type of false labelling or misleading advertising, or any other such activity detrimental to the rural population, will be severely dealt with. In short, various measures to combat product piracy and enhance product quality in this important sector will be taken. This will serve to promote improved quality in the fertilisers made available, and ensure the enforcement of industrial standards in fertiliser manufacture, and regulated, orderly procedures for the entire fertiliser market.

Chapter 11.

CONSERVING AGRICULTURAL BIODIVERSITY THROUGH WATER MARKETS IN CHINA: LESSONS FROM THE MILLENNIUM ECOSYSTEM ASSESSMENT

Jeffrey A. McNeely
IUCN, Gland, Switzerland

Abstract

The concept of eco-compensation measures is being developed in China as an important means of providing a more diverse flow of benefits to rural people. Compensating up-stream landowners for managing their land in ways that maintain downstream water quality is particularly important for China. While biodiversity itself is difficult to value, it can be linked to other markets, such as certification in the case of organic agriculture. Drawing on the findings of the recently-released Millennium Ecosystem Assessment, this paper expands on some of the markets for agriculture-related ecosystem services (especially watershed protection), identifies relevant sources of information, and highlights some of the initiatives linking such markets to poverty alleviation. Making markets work for ecosystem services requires an appropriate policy framework, government support, operational institutional support, and innovation at scales from the site to the country.

Introduction

China is in the midst of rapid changes. Demographic changes, income growth, climate changes, changes in diet, new sources of energy, growth in international trade, expansion of invasive alien species, and many others contribute to a highly dynamic set of challenges. Maintaining agricultural biodiversity will contribute to China's capacity to adapt to such changing conditions. Agricultural biodiversity can also be an essential element in enabling the shift to a more sustainable, and biologically richer, future for all.

China has already made substantial contributions to ecological agriculture. For example, some 51 eco-agriculture counties have been identified in China, each based on a series of factors such as environmental monitoring and conservation, cultivation and breeding techniques, system design and analysis techniques, techniques for controlling diseases and pests, and techniques for managing wastes (Li, 2001).

It is well known that China is able to support over 22% of the world's population on less than 7% of the world's arable land. More worrying is the fact that some of the best agricultural land is being lost to urbanisation, transportation infrastructure, dams, factories and other components of China's rapidly expanding economy. Such trade-offs need to be carefully considered, and ecological approaches to agriculture will be an essential component of such efforts. With the Millennium Development Goals firmly in mind, it is especially important to address the needs of the rural poor. Such people have a strong tendency to live in the counties suffering from severe soil erosion; improving the productivity of such lands requires an ecosystem-based approach beginning with land restoration and often involving diversification of crops, farming systems and supporting vegetation (such as trees and bamboo).

The challenges to conserving biodiversity in agricultural ecosystems are certain to increase over the next few decades, as demand for food and fibre in China is expected to grow by at least 50% by 2030. Increasing human populations and changing consumption patterns as people become wealthier will demand increases in food, forest, and fisheries production along with the equally important need to protect wild plant and animal species and the ecosystem services upon which all life depends. The ecological footprint of agriculture on our planet is already substantial, as indicated by the findings of the Millennium Ecosystem Assessment (Hassan *et al*., 2005). For example:

- Nearly half of all temperate broadleaf forest and tropical and subtropical dry forest, and a third of temperate grass and shrubland, have been converted to agricultural use; conversion rates are especially high in Asia and Europe.
- Over 250 million hectares are irrigated, using over 70% of all freshwater (89% in low-income countries), often diverting water resources needed for other purposes.
- Over half of the world's wetlands have been converted to agriculture, losing their important functions in the hydrological cycle (including flood control).
- Farming has led to significant soil degradation on 16% of all crop, pasture and forestland worldwide, and half of all land within the agricultural extent, thereby affecting the diversity of soil micro-organisms and their role in maintaining productive soils.
- Excessive use and poor management of crop nutrients, pesticides, and penned livestock wastes are a major cause of pollution that can reduce the delivery of other ecosystem services.
- A significant majority of the world's estimated 104 000 protected areas contain substantial amounts of land used for agriculture; over half of the planet's biodiversity hotspots contain large human populations whose livelihoods depend on farming, forestry, herding, or fisheries, and many of these people are plagued by chronic poverty and hunger.

Unsustainable fishing practices threaten both freshwater and coastal and marine fisheries, as the top predators are systematically removed and replaced by less valuable fish. Fishing methods such as trawling have led to bycatch amounting to about 27 million tonnes per year. The declines of keystone top predator species are leading to ecosystem changes whose effects are difficult to predict. For example, the large predatory fish in the oceans have suffered a 90% decline since pre-industrial times (Myers and Worm, 2003). It is hard to avoid the conclusion that biodiversity continues to be depleted in ways that threaten food production and the sustainability of agroecosystems.

Continuing loss of biodiversity is of considerable concern to farmers, who draw on both wild and domestic genetic resources to maintain productivity in their crops, depend on pollinators (often from the wild), use water from watersheds whose productivity is maintained at least partly through biodiversity, and so forth. But how can farmers be encouraged to conserve biodiversity actively as part of their daily work? In fact, such conservation is already being widely practiced (McNeely and Scherr, 2003; Imhoff, 2003; Swaminathan, 2001; Jackson and Jackson, 2002), and one can find considerable support for the idea that much of today's biodiversity is the result of the actions of farmers over the past 10 000 years (Williams, 2003).

On the other hand, agriculture is facing new challenges in meeting expanding production needs within growing environmental constraints. Declining environmental conditions will make it more difficult to provide more food. Loss of biodiversity, deforestation, water shortages, desertification, soil erosion, climate change, salinisation of irrigated lands, and various other dynamic factors make it increasingly difficult to increase productivity. Irrigation historically has been an important means of increasing productivity. The area of irrigated cropland increased by an average of 2.8% per year between 1950 and 1980, but has now declined to only 1.2% per year. On a per capita basis, irrigated lands declined by 6% between 1978 and 1990 and are expected to contract by a further 12% per capita by the year 2010. Even worse, nearly 10% of irrigated lands have become so salinised as to reduce crop yields (Myers, 1999). Agricultural land is becoming increasingly a limited resource, and per capita arable land has declined by an average of 1.9% per year since 1984; some agricultural lands are being so abused that they lose much of their agricultural value, while other areas are falling victim to urban sprawl and the spread of factories into farming areas.

In recent decades, the development of new agricultural technologies, such as genetic engineering and expanding use of chemicals, has been driven by the private sector and economic integration of agricultural processes and products. These technologies have helped to support the quadrupling of the human population over the past 100 years. Famines, when they occurred, were not the result of food shortage but of lack of economic access to food supplies. Feeding hungry people continues to receive lower priority in the current food system than does the substantial profit to be made by global trade in "luxury foods", that ironically may have negative impacts on both human and ecosystem health.

The Global Biodiversity Assessment (Heywood and Watson, 1992), in an extensive review of the literature, concluded that "overwhelming evidence leads to the conclusion that modern commercial agriculture has had a direct negative impact on biodiversity at all levels: ecosystem, species and genetic; and on natural and domestic diversity. On the other hand, the same modern intensive agriculture has made it possible for the ever-increasing human population to be fed without extensive destruction of habitat". Thus agriculture has both positive and negative impacts, and depends on biodiversity for its continued existence. This diversity is currently being threatened by the very activities that depend on it, with habitat conversion being the most serious threat to biodiversity (IUCN, 2003; Hassan *et al.*, 2005).

China urgently needs a means of reshaping its food system to emphasise sustainability and adaptability, based on nurturing biodiversity. This paper will suggest some means of doing so.

Ecosystem services

The Millennium Ecosystem Assessment (MEA), offers a productive framework for communicating the importance of agricultural biodiversity more effectively to decision-makers, through a broader consideration of the benefits of ecosystems for people (MEA, 2005). These so-called "ecosystem services" include:

- **Provisioning services:** Goods produced or provided by ecosystems, such as food, freshwater, fuelwood, and genetic resources.
- **Regulating services:** The benefits obtained from regulation of ecosystem processes, including pollinators, climate, diseases, nutrients and extreme natural events.
- **Cultural services:** The non-material benefits from ecosystems, including spiritual, recreational, aesthetic, inspirational, and educational benefits. In many ways, these cultural services help to define who we are as citizens of our respective countries.
- **Supporting services:** The services necessary for the production of the other ecosystem services, including soil formation, nutrient cycling, primary production, carbon sequestration, and so forth.

The approach taken by the MEA implies that ecosystem services have value to people, which in turn implies that these ecosystem services have an *economic* value which can be internalised in economic policy and the market system. Some of these services - like food production - are relatively easy to quantify, which facilitates the estimation of their economic value and the development of appropriate market incentives. Others are more abstract, but are nonetheless valuable; for example, developing markets for watershed protection, soil formation, pollination, or conservation of wild genetic resources for agriculture, can be extremely challenging, especially when lack of resource tenure discourages people from caring about biodiversity. Current markets often are distorted, so this paper will describe some new approaches to building efficient markets for such ecosystem services, based on the findings of the MEA.

All ecosystem services are supported by biodiversity, which includes genes, populations, species, communities, and ecosystems. The MEA did not consider biodiversity conservation to be an ecosystem service on its own, because it is a structural feature of all ecosystems. Nonetheless, conserving biodiversity provides many values because genes, species, habitats and ecosystems support the provision of all of the other services, such as producing crops, enabling genetic resources to continue evolving, and providing clean water for growing crops. The multiple relationships between biodiversity and ecosystem services remain only partially understood and this is an area of active research.

The study of ecosystems can provide numerous insights into agriculture. For example, systems with one or several species performing a given ecological function (known as "redundancy") enhances the stability of the ecosystem and enables it to respond better to perturbations. While the cause and effect relationships between complexity and stability are still only poorly understood, such functional complexity does seem to enhance stability (McCann, 2000; Hassan *et al.,* 2005; Tilman *et al.*, 2006).

Together, the ecosystem services contribute to the constituents of human well-being, which include security, basic material for a good life, health, good social relations, and the ability to make choices on how to live one's life. This model demonstrates to decision makers how important ecosystem services, and the biodiversity that supports them, are for all aspects of human development. Ecosystem services also underlie virtually all of the Millennium Development Goals approved by the governments of the world at the 2000 Millennium Summit (UN Millennium Project, 2005), though this link has not yet been stated as clearly as it needs to be.

The concept of ecosystem services also implies that those who are providing the services (in the past, often as a public good) deserve to be compensated when they manage ecosystems to deliver more services to others. Payment of conservation incentives (called "eco-compensation measures" in China) can reward farmers for being good stewards of the land, and ensure that payments are made by those who are receiving benefits. Similarly, those who degrade ecosystems and reduce the supply of ecosystem services should pay an appropriate level of compensation for the damage they cause, following the Polluter Pays Principle.

Farmers know better than anyone that a healthy, resilient farm is essential for a productive and profitable ecosystem. Basing the conservation of ecosystem services on economic incentives recognises the capacity of farmers to care for the land, and it supports practices that may not provide the greatest short-term financial return, but pay off well in the longer term. With appropriate incentives provided by government, farmers can become land managers as well as commodity producers, ensuring that areas under their control are sustainably managed to provide multiple ecosystem benefits.

The high-input agriculture that characterises much of Europe, North America, and parts of Asia involves using a wide range of physical, chemical and biological inputs, including irrigation, tractors, fertilisers, pesticides, high-yield varieties, genetic modification, and development of fast-growing cultivars that allow double and even triple cropping. The ecosystem service of providing food often competes against other ecosystem services, with high associated costs in terms of water consumption, pollution of soil and water, soil erosion, and development pest resistance to chemical control measures.

Such stark trade-offs between services are not inevitable, and alternative agricultural practices can on the contrary incorporate other ecosystem services such as nutrient recycling, nitrogen fixation, and pest-predator relationships. They can reduce off-farm inputs that may damage some services, instead making greater use of the biological and genetic potential of plant and animal species. Matching cropping patterns with the potential and limitations of the physical resource base can lead to improving farm management in terms of the conservation of soil, water, energy, and biological resources.

By basing agriculture on a wider range of ecosystem services, China will be better able to obtain the three essentials goals of sustainable agriculture, namely food security; employment and income generation; and natural resource conservation and environmental protection (Li, 2001).

Values of ecosystem services

A market value: pollination services

Assessing the economic values of ecosystem services remains very much a work in progress (Boyd and Banzhaf, 2005). However, some detailed estimates have been made, of which one is presented here.

Support to pollinators is especially important, given their economic value to farmers. The value of pollination services has not been estimated at a global level, but some indications are available. Kevan and Phillips (2001) have provided an approach to assessing the economic consequences of pollinator declines. For example, the value of pollination to alfalfa seed growers in the Canadian prairies is estimated to be 35% of annual crop production (Blawat and Fingler, 1994). This amounts to a value of about USD 8 million per year. The value of native pollinators to the agricultural economy of the US is estimated to be on the order of at least USD 4.1 billion per year (Southwick and

Southwick, 1992). In Costa Rica, forest-based pollinators increased coffee yields by 20% within one kilometre of forest, and improved coffee quality as well. Pollination services from two forest fragments of 46 and 111 ha yielded a benefit of USD 60 000 per year for one Costa Rican farm (Ricketts *et al.*, 2004). The value of honeybee pollination to agriculture in Australia is estimated at over USD 1 billion per year (Gordon and Davis, 2003).

Most ecosystem services have been seen as public goods that benefit large groups of people and resist ownership. A major challenge for eco-compensation is to align private incentives with the public interest. For detailed references on payments for ecosystem services, see Pennington (2005). A useful valuation website is www.naturevaluation.org.html.

A non-marketed value: protection against extreme natural events

Recent human disasters caused by extreme natural events, including the 2004 Indian Ocean tsunami and the 2005 Kashmir earthquake, have demonstrated the value of intact ecosystems in reducing the impact of such extreme natural events on human wellbeing. In the case of the tsunami, intact coral reefs and mangroves greatly reduced the negative impact of the tsunami on people (Danielsen *et al.,* 2005); and in Kashmir, slopes that remained forest-covered suffered far less landslide damage than those where forests had been wilfully over-exploited.

The value of ecosystem services to protect human well-being against the implications of such extreme natural events is seldom quantified as no market exists for them. However, the implications in terms of human fatalities, economic disruptions, and social disruptions carry a very real cost: in the two events mentioned above, human fatalities totalled over 300 000 and the economic costs of restoration exceed USD 5 billion. Such costs need to be better quantified and incorporated in decision-making that affects ecosystem functioning. These costs were externalised in Kashmir and along the coasts of the Indian Ocean, to the great detriment of the people living there. One element in the payment for ecosystem services, therefore, is to avoid expenditures that lead to ecosystem destruction or degradation. China has its own long history of human disasters caused by extreme natural events, especially floods, earthquakes, and - to a lesser extent - dust storms. Healthy ecosystems and the services they provide can ameliorate many such disasters.

Markets for ecosystem services

Over the past ten years or so, markets and other payments for ecosystem services have emerged in many parts of the world (Wunder, 2005; Pagiola *et al.*, 2005). Landell-Mills and Porras (2002) identified 287 initiatives for payments for ecosystem services, including 61 examples specifically associated with watersheds. The emergence of these markets has been driven by frustration with traditional government regulatory approaches, the demands of society for ecologically sound products, and the need of rural people to find additional revenue sources to remain competitive. The expectation is that such markets can contribute to watershed protection and restoration and become a sustainable source of new income for the rural poor who have few other income options (Scherr *et al.*, 2005).

This paper discusses three categories of market and payment schemes:

- Eco-labelling of farm products, an indirect form of payment for ecosystem services.
- Open trading under a regulatory cap or floor, such as carbon trading or mitigation banking.
- Public payment schemes to farmers to maintain or enhance ecosystem services, such as "conservation banking" and watershed protection.

One way to use markets to support biodiversity is to provide a premium for agricultural commodities that are grown in sustainable agroecosystems. The most important instrument that has been designed to achieve this has been producer certification. In Austria, the European country where organic production has become most important, 10% of the food consumed is now organic. The World Organic Commodity Exchange (WOCS, www.wocx.net) represents over 2 500 organic products, including textiles, furniture, cosmetics, wine, vegetables, fruits, dog food, baby food, and ice cream. The indication of public interest in such products is high and growing, often mainly in response to human health concerns, but increasingly because of environmental concerns as well.

Organic products have long been labelled, and the organic movement, through its International Federation of Organic Agriculture Movements (IFOAM), is seeking to ensure that organic farming is also biodiversity-friendly (www.ifoam.org). The global organic market was worth USD 27.8 billion in 2004 and is expected to reach USD 133.7 billion by 2012, with the greatest growth in China (though credible certification remains a limitation). Other eco-friendly labels are also being used; for example, shade-grown coffee has a market of USD 5 billion in the US alone.

Certification of biodiversity impacts may become a consideration in financial markets, as "green" mutual funds seek agroindustries that contribute actively to "sustainable development" (Daily and Walker, 2000). Large companies traded on stock exchanges around the world are judged by potential investors according to a variety of criteria. Increasingly, some of those criteria relate to environmental sustainability; many mutual funds exclusively invest in environment-friendly companies. These companies can achieve a competitive advantage by marketing their products as sustainably produced and packaged, and by advertising their environmental responsibility in managing corporate land, water, and forest resources. With further efforts to educate and animate both investors and the public, their performance as stewards of biodiversity might also be rewarded.

The most well studied and widespread of the marketed ecosystem services is carbon sequestration. Forests, grasslands, and other ecosystems remove carbon dioxide from the atmosphere through the storage of carbon. A reasonably prosperous industry has been established in trading "certified emission reductions" within the Clean Development Mechanism (CDM) of the Kyoto Protocol or "verified carbon emission reductions" (CERs) outside of the Kyoto regime (see, for example, Swingland, 2003). The carbon market is substantial, with 64 million metric tonnes of carbon dioxide equivalent exchanged through projects (most transactions intended for compliance with the Kyoto Protocol) from January to May 2004, nearly as much as during the whole year 2003 (78 million tonnes) (Lecoq, 2003). Japanese companies are the largest market buyers, with 41% of the 2003-2004 market, and Asia is the largest seller of emission reduction projects, accounting for 51% of the volume supplied. The total value of the market was about USD 11 billion in 2005.

The Kyoto-compliant carbon emission offset market is expected to grow to a minimum of 15 million tonnes of carbon dioxide in 2008-2012 (Scherr *et al.*, 2005). The European Union Emissions Trading Scheme began in 2005, with futures and spot contracts trading on several exchanges across Europe; it is used mostly by high-emission power and steel sectors. The European carbon market is now being linked to CDM projects in Asia, including Asia Carbon Global activities in China. The International Emissions Trading Association (www.ieta.org) is probably the best source of information on these issues.

Watershed protection

Water is an especially important limiting factor for agriculture in China. The per capita fresh water reserve in China is only 2300 cubic meters per person, well below the world average. As Li (2001) points out, "if water demand in the first decade of the 21st century is to increase by 2-3% annually, as estimated, then the total annual water demand by 2010 will be as high as 720 million cubic meters. This means that the increment of water demand in 10 years will be over 100 billion cubic meters. It will be necessary to increase the water supply capacity by 120 billion cubic meters before 2010. By the turn of the century, North China, Shandong Province, Northwest China, mid-south Liaoning Province and dozens of coastal cities will be acutely hit by freshwater deficits". Li also indicates that of the 1 200 Chinese rivers being monitored at present, 850 are severely polluted. Some major lakes are suffering from eutrophication, coastal areas are suffering from seawater intrusion because of inadequate river flow, and coastal waters are suffering from various forms of pollution.

Watershed protection is therefore an essential service to farmers. Watershed services are far more numerous and complex than is usually appreciated, and provide numerous kinds of benefits to people, including the rural poor (Dyson *et al.*, 2003). A partial list includes:

- Provision of water for consumptive uses, such as drinking water, agriculture, domestic uses, and industrial uses.
- Non-consumptive uses such as hydropower generation, cooling water, and navigation.
- Water storage in soils, wetlands and flood plains to buffer floods and droughts.
- Control of erosion and sedimentation, which can have effects on productive aquatic systems.
- Maintain a flow of water required to enable river dynamism, riparian habitats, fisheries, and water management systems for rice cultivation and fertilisation of flood plains.
- Maintenance of mangroves, estuaries, and other coastal ecosystems that may require fresh water infusions.
- Control of the level of groundwater tables, avoiding adverse effects on agriculture by keeping salinity well below the surface.
- Maintenance of water quality that may have been reduced through inputs of nutrients, pathogens, pesticides, fertilisers, heavy metals, or salinity.

- Support for cultural values including aesthetic qualities that support tourism and recreationally uses as well as supporting traditional ways of life and providing opportunities for adapting to changing conditions.

The services provided by forests protecting watersheds overlap with many other ecosystem services, indicating the synergies that can be realised through improved management of forest systems. Many of these services have market values, while others have non-market values that are nonetheless significant.

Many countries in various parts of the world are developing mechanisms for collecting payments for watershed protection. Of just a few that could be quoted:

- **Brazil**: A water utility in Sao Paulo pays 1% of total revenues for the restoration and conservation of the Corumbatai watershed. The funds collected are used to establish tree nurseries and to support reforestation along riverbanks.
- **Costa Rica:** A hydropower company pays USD 10 per ha/year to a local conservation NGO for hydrological service in the Peñas Blancas watershed. In the town of Heredia, the drinking water company earmarks a portion of water sales revenue for reforestation and forest conservation.
- **Ecuador:** Municipal water companies in Quito, Cuenca, and Pimampiro impose levies on water sales, which are invested in the conservation of upstream areas and payments to forest owners (Landell-Mills and Porras, 2002).
- **Lao PDR:** The Phou Khao Khouay Protected Area currently receives 1% of the gross revenues from a downstream hydropower dam, and the proposed Nam Theun 2 hydropower project is expected to prove over USD 1 million per year for the management of the Nakai-Nam Theun Protected Area.
- **Japan**: the Kanagawa Prefectural Assembly adopted an ordinance in October 2005 that will impose an additional residence tax to be used exclusively for protecting water sources, with the funds going to projects aimed at conserving and restoring forests and rivers. The new tax will be introduced in April 2007 and continue for five years.

  www.japanfs.org/db/database.cgi?cmd=dp&num=1253&dp=data_e.html
- **Colombia**: In the Cauca valley, water user associations have assessed themselves additional charges and used the revenue to finance conservation activities in their watershed areas (Echevarria, 2002). Watershed management in the country is partly funded through a 6% tax on the revenue of large hydroelectric plants (Tognetti *et al.*, 2003).

Maintaining the value of watershed services will depend on:

- Maintaining the integrity of ecosystem functions or processes that support the watershed protection service.
- The scale at which the benefits from watershed protection have economic significance.
- The effectiveness of the institutional arrangements that have been put in place to ensure provision and access, including such issues as land secure tenure (Tognetti *et al.*, 2003).

Payments for watershed services are often politically popular, as the value of water is well recognised. Information on recent developments in this field is available from www.flowsonline.net. Linking watershed protection services with improved livelihoods is the objective of a project carried out by IIED in London www.iied.org/forestry/research/projects/water.html.

Conservation payments to farmers

Conventional wisdom holds that modern farming is largely incompatible with conserving wild biodiversity. Thus policies to protect wild species and ecosystems typically rely on land use segregation, establishing protected areas from which agriculture is excluded (at least legally). Those promoting this view of wildlife conservation see farmers as sources of problems. However, it is becoming apparent that farming systems can make important contributions to biodiversity conservation, with forms of land use that support the objectives of biodiversity conservation rather than conflict with them (McNeely and Scherr, 2003). These contributions can be enhanced by new approaches to resource management, supported by technical and policy research.

An essential strategy for conserving wild biodiversity, especially that found in highly populated, poor rural areas around the world, is to convert agriculture that is destructive of biodiversity into a new type of agriculture: "ecoagriculture" (McNeely and Scherr, 2003). Ecoagriculture, which builds on the concept of "ecosystem management", refers to land-use systems that are managed both to produce food and to protect wild biodiversity. For ecoagriculture, enhancing rural livelihoods through more productive and profitable farming systems becomes a core strategy for both agricultural development and conservation of biodiversity. Farmers working to maintain biodiversity are receiving payments for those services in a growing number of countries.

Given that farmers - like most other people - seek rewarding financial opportunities, some governments are using economic incentives to encourage them to conserve wild or domestic biodiversity, or otherwise maintain healthy environmental conditions on their farms. In Europe, the controversial subsidies of agricultural production are increasingly being converted to subsidies for improved environmental conditions. Such subsidies have increased from USD 60 million per year in 1993 to about USD 350 million in 2003. Farmers can earn an additional USD 40 per ha by joining a basic environmental scheme, such as leaving the edges of fields unploughed in order to support insects and birds that feed in such areas. It also appears that some farmers will simply allow at least part of their land to lie fallow, collecting the subsidy that requires them to keep the land in "good agricultural and environmental condition", but forgoing the income from crops.

Roughly 20% of the farmland in the European Union is now under some form of agri-environment scheme to counteract the negative environmental impacts of modern agriculture, at a cost of about USD 1.5 billion (about 4% of the EU expenditure on the Common Agricultural Policy). The Netherlands has been implementing management agreements designed to conserve biodiversity on farms since 1981, often obliging farmers to postpone agricultural activities on individual fields until a set date that will allow certain species of birds to safely hatch their chicks. Other management agreements are designed to conserve species-rich vegetation in grasslands, restricting the use of fertiliser or postponing the first mowing or grazing date. These arrangements are not always effective in contributing to biodiversity conservation. Kleijn *et al.* (2001) evaluated the contribution of such schemes to the protection of biodiversity in intensively used Dutch agricultural landscapes, finding no positive effects on plant and bird species diversity. On average, edges accounted for 96% of the total plant species richness of a field, and 66%

of uncounted species were never found in the field centre. Some of the management activities may have had perverse effects. For example, postponing the first mowing or grazing date forced farmers to reduce the input of fertiliser, which may have adversely affected the abundance of soil animals that certain bird species use for food. On the other hand, management agreements appear to have a positive effect on reproductive success of birds. This can lead to an "ecological trap" where the cues that individual birds use to select their nesting habitat (for example food availability) are decoupled from the main factor that determines their reproductive success (delayed mowing/grazing). Since the primary concern of farmers is necessarily to secure their income, conserving biodiversity will be of secondary importance to them, especially in the context of a farming system that is driven by economic pressures to increase its intensity. These results are not surprising, and indicate that such agri-environment schemes must be accompanied by scientifically sound evaluation plans and are carefully designed to be ecologically appropriate.

In Central America, researchers are developing modified systems of shaded coffee with domesticated native shade tree species, be maintain coffee yields while also diversifying income sources and conserving wild biodiversity. Farmer adoption of these systems has been promoted through changes in public coffee policy to favour shade systems, technical assistance, and in some cases price premiums in international markets for certified "biodiversity-friendly" coffee (Giovannucci, 2001).

In order to conserve the diversity of crop varieties in Yunnan Province, China, Zhu *et al.* (2003) promoted crop diversity by mixed planting (intercropping) of traditional and hybrid rice varieties. Since the adoption of this form of crop diversity management in 1997, the number of traditional rice varieties in cultivation has increased dramatically and now includes some varieties that formerly were locally extinct. The cultivated area of traditional varieties has also been greatly expanded. They point out that this form of management is easy to implement and links the economic concerns of the farmers to conservation. Management for crop diversity can promote on-farm conservation practices in a feasible and sustainable way.

Incorporating wild animals into agricultural systems enables additional benefit to be delivered to the agricultural fields. For example, the Chinese forest frog (*Ranatemporaria chensinensis*) lives in north eastern China and is an effective predator on crop pests, consuming at least 150 insects per day or a total of 18 000 crop and forest pests during a single summer. It therefore has a high economic value simply in terms of its pest control characteristics. But in addition, the oil that is derived from the frog is a rare nourishing medicine sold in markets both domestically and abroad (Bai, 1988).

Building markets for ecosystem services

In New South Wales (Australia), the State Forest Department has initiated an Environmental Services Scheme that compensates landowners through credits for multiple benefits of forests. Such benefits include biodiversity, carbon sequestration, soil conservation, and protection of water quality that offsets the rise in salinity levels and supports downstream farmers (Oliver *et al.,* 2005).

In support of the implementation of the Millennium Development Goals, the World Bank and OECD have promoted environmental fiscal reform (EFR), stressing that poverty reduction and improved environmental management go hand-in-hand. They advocate a range of taxation or pricing instruments that can raise revenue while simultaneously furthering environmental goals. This is achieved by providing economic incentives to correct market failure in the management of natural resources and the control of pollution (World Bank, 2005). They believe that EFR can mobilise revenue for governments, improve environmental management practices, conserve resources, and reduce poverty. EFR includes a wide range of economic instruments, including:

- Natural resource use taxes (for example, on forestry and fisheries) that will reduce the inefficient exploitation of publicly owned or controlled natural resources that results from operators paying a price that does not reflects the full value of the resources they extract.
- User charges or fees and subsidy reform that will improve the provision and quality of basic services such as water, while providing incentives to reduce any unintentional negative environmental effects arising from inefficient use.
- Environmental taxes that will make polluters pay for the "external costs of their activities and encourage them to reduce these activities to a more socially desirable level."

Payment for ecosystem services may also have some hidden dangers. For example, by becoming commonplace, payments for ecosystem services may risk eroding the sense of an environmental duty of caring for natural resources and managing them sustainably. They may even discourage private investment in the environment by creating the impression that environmental stewardship is the duty of governments rather than individuals (Salzman, 2005). Other potential dangers to consider include rent-seeking behaviour, where certain individuals may exaggerate their potentially negative impacts on ecosystem services in hopes of gaining greater compensation. At least some subsidies may pay the farmers for precisely the behaviour that the subsidies are seeking to overturn. Payments for ecosystem services also need to be provided equitably. Those who are already providing an ecosystem service (such as maintaining many varieties of crops on their land) need to be compensated appropriately, as well as those who are expected to change their behaviour to come into conformity with the provision of a service (for example, watershed protection). But in any case, the establishment of an appropriate system of payments for ecosystem services will certainly change the perception of farmers about how they should manage their land.

The issue of payment for ecosystem services (eco-compensation) is still in its infancy, and further experimentation and research is required. This research should involve inter-disciplinary teams of economists, ecologists and entrepreneurs to determine what ecosystem functions support the provision of particular benefits, how their key parameters can be measured or estimated, and how efficient economic incentives can be created to encourage the sustainable supply of ecosystem services.

Capturing the willingness to pay

Payments for ecosystem services will not work unless governments provide an enabling framework. Farmers or forest managers provide the ecosystem services discussed above. The markets for these ecosystem services often work through an intermediary who issues certificates for the ecosystem services, with a verifier who controls and monitors the sustainable management of the ecosystem providing the services. The buyer of certificates from the intermediary provides financial resources to the system. The intermediary plays a critical role in managing the transaction, though of course it is also possible for the farmer to provide the services directly to the buyer and to receive the funding directly.

However, formal legislation is not always necessary. For example, most certification for organic products is voluntary yet it seems to work relatively well and meets a market demand. And in the case of carbon, at least, the Kyoto Protocol provides a supporting policy framework.

The certificates that are issued can represent units such as hectares of the ecosystem that is providing the service, tonnes of carbon being sequestered, area of crops being pollinated, cubic meters of clean water being provided, or amount of certified organic crops being produced. A system of certificates for ecosystem services may enable them to be traded, as carbon sequestration certificates are now on the market in many parts of the world.

The commercial sector may also purchase many ecosystem services. Their motivations for investing their funds in this way might include the following:

- The service creates direct financial income for the firm.
- The service reduces costs for the firm.
- The ecosystem services ensure that the firm receives necessary supplies of natural resources.
- The firm is active in this field due to the demand of clients (for example, consumer demand for organic agriculture).
- The firm may need to provide mandatory compensation due to legal compliance (offsets).
- The firm may wish to compensate for its impacts on a voluntary basis, often to improve its public image.

Conclusions

Approaches being developed under many biodiversity-related international agreements and programmes call for ecosystems to be managed to meet multiple national objectives. These may include (but are not limited to) providing timber, forage, fibre, and energy, retaining options for future economic use, carrying out various ecosystem services, providing ethical and aesthetic values, and supplying that nation's share of global benefits (MEA, 2003). Achieving these sometimes-conflicting objectives in a time of rising expectations and tight government budgets will require new approaches such as those outlined in this paper.

These approaches are designed to encourage the development of technologies and practices that increase productivity and reduce degradation; reclaim, rehabilitate, restore, and enhance biodiversity, especially that which is relevant to agricultural development; and monitor adverse effects of agriculture on biodiversity. Organic farming, ecoagriculture, integrated pest management, biological control, multi-cropping, crop rotation, and agro-forestry are all approaches that can contribute further in this regard.

An essential component of any effort at sustainable landscape management is the economic viability of the various enterprises that are involved. While growing a crop is the most obvious money-earner from good agricultural land, many crops and methods of growing them are possible, with variable implications for biodiversity. Further, if local people can benefit financially from enterprises that depend on the biodiversity of the surrounding non-domestic land within which they live, then they might reasonably be expected to support the conservation and sustainable use of the ecosystem.

Ecosystem services related to agriculture and based on biodiversity have four major market characteristics:

- The total value of direct ecosystem service payments is still modest, but payments have grown dramatically over the past decade and are especially significant to low-income producers. Some ecosystem services are not yet linked to significant commodities, but instead support niche markets for products of special value to a narrow range of buyers. Scherr *et al.* (2005) estimate the annual value of direct payments through ecosystem markets in tropical countries is on the order of hundreds of millions of US dollars, while indirect payments via eco-labelled products such as organic crops generates over USD 25 billion per year (Halweil, 2001).
- Markets for ecosystem services are expected to grow quickly over the next 20 years. The potential for increased demand for watershed services is immense, providing significant opportunities for increased payments. The growth of these markets can generate new forms of financing and open up new opportunities for non-extractive management regimes for forest ecosystems that provide benefits to downstream farmers.
- Governments play a critical role as the direct buyers of many ecosystem services and encourage many private sector direct payment schemes. Since many ecosystem services are public goods, government intervention may be required to establish a market. This may entail directly paying for a service, establishing property rights, or establishing regulations that set caps and govern trading schemes.
- Ecosystem service payments will usually cover only a relatively modest share of the costs of good farm management, but this contribution can be important in improving the way forests are managed. The prices of ecosystem services are not yet sufficient to justify forest conservation in areas with moderate to high opportunity costs for the land. Even so, these payments can have a disproportionate catalytic effect on forest establishment and management, with subsequent benefits for the agricultural lands in the region (Scherr *et al.,* 2005).

While programmes of payments for ecosystem services are not necessarily designed explicitly for poverty reduction, the poor can receive some important benefits when the programme is designed with that objective and the local conditions are favourable (Pagiola *et al.,* 2005). Possible adverse effects can result in cases where property rights are insecure, the poor are forced to sell their land through economic pressures, or if the programme of payments for services encourages practices that are less labour-intensive (and hence lead to less employment for the rural poor) (Figure 1).

Figure 1. Can payments for environmental services help reduce poverty?

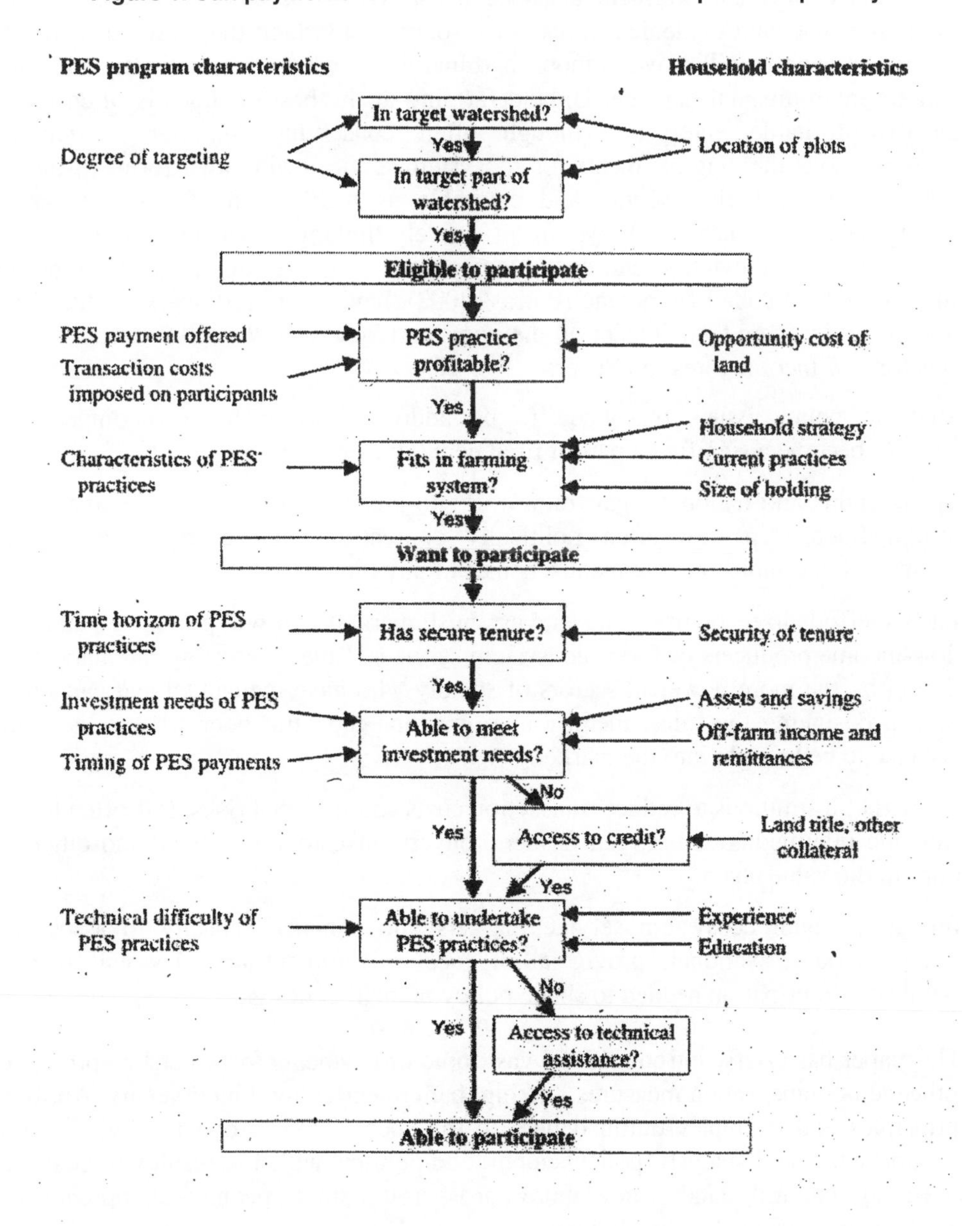

Factors that affect household participation in payments for environmental services programmes.
Source: Pagiola *et al.*, 2005.

One important forest-based initiative to address poverty explicitly is a project known as "Rewarding Upland Poor for Environmental Services" (RUPES), run by the World Agroforestry Centre (www.worldagroforestry.org/sea/Networks/RUPES/). The goal of the programme is to enhance the livelihoods and reduce poverty of the upland poor while supporting environmental conservation on biodiversity protection, watershed management, carbon sequestration and landscape beauty at local and global levels. It has project sites in Nepal, the Philippines, and Indonesia, with areas of work and interest in China, India, Thailand, and Vietnam. The Regional Community Forestry Training Centre for Asia and the Pacific (www.recoftc.org) also does useful work on this topic.

Payments to upstream farmers is especially justified, because if they are simply removed from the land or denied access to resources on which they have traditionally relied, they frequently will move to more marginal lands, where their activities can cause even more environmental damage. Upland farming at the best of times is often only a marginally profitable enterprise (though often conserving important agricultural biodiversity), so combining payments for ecosystem services with other forms of income generation may be both essential and equitable as well as an effective means of addressing poverty concerns. Ways of effectively linking payments for ecosystem services to poverty alleviation and ecosystem conservation are presented by Pfund and Robinson (2004), Landell-Mills and Porras (2002) have reviewed markets for forest ecosystem services and their impact on the poor, and Scherr *et al.* (2004) have reviewed markets for low-income forest producers.

Strategic policy issues that need to be addressed to enable eco-compensation measures to become a significant part of rural economies include:

- Property rights and national legal frameworks to enable ecosystem service markets to develop. Such steps are often politically contentious and costly, yet they are fundamental to establishing payment schemes of any type.
- Proactive efforts to recognise rights and establish markets that will provide equal access to low-income producers of forest ecosystem services. Rules governing the market tend to be set by the more powerful sectors of society who have the capital and capacity to invest in designing the rules, thereby marginalising the rural poor who most require assistance to be brought into the market.
- New market institutions to reduce transaction costs and financial risks. It is often helpful to provide intermediaries between farmers, buyers, investors, certifiers, and other key groups in the value chain.
- Information about ecosystem service markets and enhance capacity to assess and develop markets. National, provincial or local government agencies will need to generate the information needed to shape policy on market design.

This paper has briefly introduced the vast topic of payments for watershed protection and other eco-compensation measures, linking them to agricultural biodiversity. Applying the principles and examples outlined here to the specific needs of China will require better information and analysis, policy support and political will. The result will be better-managed agricultural lands and more prosperous rural people: comprehensive, harmonious, and sustainable development.

REFERENCES

Bai, X.M. (1988), *The Natural Resources Development and Eco-environment Protection in Chanbai Mountainous Areas*, Jilin Science and Technology Press, Jilin.

Blawat, P. and B. Fingler (1994), *Guidelines for Estimating Cost of Production: Alfalfa Seed*, Manitoba Agriculture, Winnipeg CA.

Boyd, James and H.S. Banzhaf (2005), "Ecosystem Services and Government Accountability: The Need for a New Way of Judging Nature's Value", *Resources*, Summer.

Daily, Gretchen C. *et al.* (1997), "Ecosystem Services: Benefits Supplied to Human Societies by Natural Ecosystems", *Issues in Ecology* 2:1-16.

Dyson, Megan, G. Bergkamp and J. Scanlon (eds.) (2003), *Flows: The Essential of Environmental Flows*, IUCN, Gland, Switzerland.

Echevarria, M. (2002), *Water User Associations in the Cauca Valley: A Voluntary Mechanism to Promote Upstream-downstream Cooperation in the Protection of Rural Watersheds*, FAO, Rome.

Giovannucci, Daniele (2001), *Sustainable Coffee Survey of the North American Speciality Coffee Industry*, World Bank, Washington, DC.

Gordon, Jenny and Lee Davis (2003), *Valuing Honeybee Pollination,* Rural Industries Research and Development Corporation, Canberra.

Halweil, B. (2001), "Organic Gold Rush", *Worldwatch,* May/June.

Hassan, Rashid, Robert Scholes and Neville Ash (eds.) (2005), *Ecosystems and Human Well-being: Current State and Trends*, Volume 1, Island Press, Washington, DC.

Heywood, V.H. and R.T. Watson (1995), *Global Biodiversity Assessment*, UNEP, University Press, Cambridge, UK.

Imhoff, Daniel (2003), *Farming with the Wild: Enhancing Biodiversity on Farms and Ranches*, Sierra Club Books, San Francisco, CA.

IUCN (2003), "2003 Red List of Threatened Species", IUCN, Gland, Switzerland. http://www.redlist.org

Jackson, Dana L. and Laura L. Jackson (2002), *The Farm as Natural Habitat: Reconnecting Food Systems with Ecosystems*, Island Press, Washington, DC.

Kevan, T.P. and T.P. Phillips (2001), "The Economic Impacts of Pollinator Declines: An Approach to Assessing the Consequences", *Conservation Ecology* 5(1):8. www.ecologyandsociety.org/articles/272.html (accessed 22 April 2006).

Landell-Mills, N. and I.T. Porras (2001), *Silver Bullet or Fools' Gold?: A Global Review of Markets for Forest Environmental Services and Their Impact on the Poor*, IIED, London.

Lecoq, Franck (2003), "Pioneering Transactions, Catalyzing Markets and Building Capacity: the Prototype Carbon Fund Contributions to Climate Policies", *American Journal of Agricultural Economics* 85(3): 703-707.

Li, Wenhua (ed.) (2001), *Agro-Ecological Farming Systems in China*, UNESCO-MAB, Paris.

McCann, Kevin S. (2000), "The Diversity-Stability Debate", *Nature* 405: 228-233.

McNeely, J.A. (1999), *Mobilizing Broader Support for Asia's Biodiversity: How Civil Society Can Contribute to Protected Area Management*, Asian Development Bank, Manila.

McNeely, Jeffrey A. and Sara J. Scherr (2003), *Ecoagriculture: Strategies for Feeding the World and Conserving Wild Biodiversity*, Island Press, Washington, DC.

Millennium Ecosystem Assessment (MEA) (2003), *Ecosystems and Human Well-being: A Framework for Assessment*, Island Press, Washington, DC.

Millennium Ecosystem Assessment (2005), *Synthesis Report*, Kuala Lumpur, Malaysia, Also available electronically at www.maweb.org.

Myers, Norman (1999), "Pushed to the Edge", *Natural History* 108 (2): 20-22.

Myers, Ransom A. and Boris Worm (2003), "Rapid Worldwide Depletion of Predatory Fish Communities", *Nature* 423: 280-284.

Oliver, Ian, A. Ede, W. Hawes and A. Grieve (2005), "The NSW Environmental Services Scheme: Results for the Biodiversity Benefits Index, Lessons Learned and the Way Forward", *Ecological Management and Restoration* 6:197-206.

Pagiola, Stefano, Agustin Arcenas and Gunars Paltais (2005), "Can Payments for Ecosystem Services Help Reduce Poverty? An Exploration of the Issues and the Evidence to Date from Latin America", *World Development* 33(2): 237-253.

Pfund, Jean-Laurent and Patrick Robinson (2004), *Non-timber Forest Products: Between Poverty Alleviation and Market Forces,* Intercooperation, Bern.

Ricketts, Taylor H. (2004), "Tropical Forest Fragments Enhance Pollinator Activity in Nearby Coffee Crops", *Conservation Biology* 18 (5): 1262-1271.

Salzman, Jim (2005), "The Promise and Perils of Payments for Ecosystem Services", *International Journal of Innovation and Sustainable Development* 1(1/2): 5-20.

Scherr, Sara, Andy White and David Kaimowitz (2004), *A New Agenda for Forest Conservation and Poverty Reduction: Making Markets Work for Low-income Producers*, Forest Trends, Washington, DC.

Scherr, Sara, Andy White and Arvind Kahare (2005), *For Services Rendered: The Current Status and Future Potential of Markets for the Ecosystem Services Provided by Tropical Forests*, Forest Trends, Washington, DC.

Southwick, E.E. and L. Southwick Jr. (1992), "Estimating the Economic Value of Honeybees (Hymenoptera: Apidae) as Agricultural Pollinators in the United States", *Journal of Economic Entomology* 85:621-633.

Swaminathan, M.S. (2001), *From Rio de Janeiro to Johannesburg: Action Today and Not Just Promises for Tomorrow*, East West Books, Madras.

Swingland, Ian R. (ed.) (2003), *Capturing Carbon and Conserving Biodiversity: The Market Approach,* EarchScan, London.

Tilman, David, Peter Reich and Johannes Knops (2006), "Biodiversity and Ecosystem Stability in a Decade-long Grassland Experiment", *Nature* 441: 629-632.

Tognetti, S.S., G. Mendoza, D. Southgate, B. Aylward and L. Garcia (2003), *Assessing the Effectiveness of Payment Arrangements for Watershed Environmental Services*, Presented at the Third Latin American Congress on Watershed Management, Arequipa, Peru.

UN Millennium Project (2005), *Investing in Development: A Practical Plan to Achieve the Millennium Development Goals,* United Nations, New York.

Williams, Michael (2003), *Deforesting the Earth: From Prehistory to Global Crisis*, The University of Chicago Press, Chicago.

World Bank (2005), *Environmental Fiscal Reform: What Should be Done and How to Achieve it*, IBRD, Washington, DC.

Wunder, Sven (2005), "Payments for Ecosystem Services: Some Nuts and Bolts", *CIFOR Occasional Paper* 42:1-24, CIFOR, Bogor, Indonesia.

Chapter 12.

A RESOURCE UTILISATION APPROACH TO RESOLVING FOOD SECURITY ISSUES IN CHINA

Ke Bingsheng
Research Center for Rural Economy, Ministry of Agriculture, China

Abstract

For a country like China that is home to more than a billion people, the principal objective of agricultural policy must be to ensure food security. The foremost measure that can be taken in this regard is to increase domestic production. While this may seem to inevitably lead to an adversarial relationship pitting food security against the environment and available resources, it is nonetheless also a relationship that can be mutually supportive. This paper examines the ways in which the achievement of China's food security goals can be advanced by means of better practices in resource use and environmental protection.

Food security and the resource environment

As a large country, China must base its efforts to ensure food security on the supplies available through domestic production. Even though grain production has marked significant increases in each of the last two years, domestic production of grains, cotton and edible oil is still insufficient to meet domestic demand. There are thus latent dangers imperilling China's food security.

At the present moment, there are a number of areas in which supply is lacking. After two years of excellent harvests, there is still a grain supply shortfall of roughly 30 million tonnes.[1] China imported in 2005 some 26.59 million tonnes of soyabeans, an amount greater by 50% than domestic production. Cotton imports amounted to 2.57 million tonnes, or 45% more than was supplied through domestic sources. In addition, some 6.27 million tonnes of cereals were imported (wheat, barley, rice). In order for domestic production to make up the shortfall in these areas alone, China would have to place an additional 290 million *mu* (19.3 million ha) of land under cultivation.[2] This is an area equivalent to 13% of the land currently sown. China also imported 6.21 million tonnes of edible oil and 1.39 million tonnes of sugar in 2005.

Meanwhile, demand pressures are constantly increasing. In the foreseeable future, as China's population continues to increase and average income levels continue to rise, demands on the supply of agricultural produce and foodstuffs will continue to increase as well. Forecasts by the responsible authorities indicate that demand for foodstuffs will grow by an average of 1% per year. Rapid development in the textile industry has also driven double-digit growth in demand for cotton, wool and similar agricultural products in recent years. In addition, as fuel alcohol has been introduced throughout the Chinese

1. Editor's note: in China grains are defined as including cereals, beans (including soybeans) and tubers.

2. Editor's note: the *mu* is a traditional unit of land measure equal to 1/15 of ha (0.0667 ha).

provinces of Henan, Anhui and Dongbei and in parts of other provinces as well, a great amount of grain is needed as a raw material for its manufacture. Growth in the fuel alcohol industry will accelerate in step with rises in petroleum prices and contribute to ever-greater demand for grains. Indeed, this unrelenting growth in demand for grains, animal products and industrial raw materials is sustained and irreversible. Unless these demands are satisfactorily met, the impact on people's livelihoods, social stability and development of the industrial economy will be deep and far-reaching. This is the main reason why the new socialist countryside programme is of concern to more than just the agricultural industry, rural residents and the internal workings of their communities.

Constraints on resource availability are increasingly evident. Between 1998 and 2004, China lost some 100 million *mu* (6.7 million ha) of cultivated land, and this downward tendency will inevitably continue as long as industrialising, urbanising and modernising trends remain at the forefront of development in China. If policies to protect farmland established by the Party's Central Committee and the State Council are duly implemented, the extent of this downward trend may possibly be controlled to a certain degree. However, should these policies be in any way disregarded in various regions, then the loss of farmland can only accelerate. The highly productive farmland in the northeastern Dongbei region, for example, is at particular risk as the region is economically well developed and there is constant pressure on land resources. In addition, shortfalls in water resources are growing ever more acute. On the one hand, available supply is dwindling, particularly in those regions where deep artesian sources are tapped for irrigation and water tables are dropping every year, such use is unsustainable. On the other hand, demand for water for industrial and domestic use is constantly increasing, which inevitably eats into the resources available for agriculture.

China is already past the point where it can increase its draw on available natural resources to increase agricultural production, and has no other option than to increase the utilisation rates of the resources at hand. The 11th 5-Year Plan sets a grain production goal of 500 million tonnes in 2010, which indicates an increase of 16 million tonnes over the baseline production of 2005. On the basis of average unit yields in Chinese agriculture, this indicates an additional need for 50 million *mu* of cultivable land. The 11th 5-Year Plan also forecasts, however, that not only will no increase in cultivable land be possible in that period, but a reduction in excess of 30 million *mu* is to be expected instead. In fact, by the year 2020, the area of available cultivable land may well decline by as much as 80 to 100 million *mu*. There is thus no alternative path open: any increase in grain production in China is dependent upon an increase in unit yields.

No large country can rely upon imports to ensure food security. Due, however, to our lack of land resources, China must nonetheless import a certain quantity of grains, cotton and edible oil in order to satisfy domestic demand for these commodities. Imports of this type, however, can only be of a supplementary nature. With a population of 1.3 billion, China cannot rely upon international markets wholly to supply domestic demand for agricultural products and foodstuffs. Indeed, China's imports of soy beans and of cotton already account for fully one-third of international trade in these commodities. Ten per cent of China's demand for rice is equal to one-half of world exports. While the capacity of world markets is limited to begin with and these cannot possibly supply everything to China, it is also true that massive imports by China can only result in a surge in world commodity prices, which in turn would mitigate against imports. Further constraints that exist in the form of limitations in the carrying capacities of ports and internal transportation networks, as well as in transportation costs, also stand against massive imports. In addition, large scale imports by China could well incur the displeasure of other developing countries. The conclusion to be drawn is that China must increase its own domestic agricultural productivity so as to ensure that a high level of self-sufficiency is maintained, particularly with respect to grains.

From a static perspective, the pursuit of food security through self-sufficiency can appear to be in conflict with the environment and place a strain on natural resources. Increased agricultural production requires development of untapped land resources, and any increase in the use of land resources leads to greater demands for water. Such a situation might appear to be quite detrimental to the protection of resources and the natural environment.

From a perspective of dynamic development, however, if the resource environment does not receive adequate protection, agricultural resources will not be used in a sustainable way, and the long-term objectives of food security will not be attained. Within the framework of sustainable development, however, measures taken to encourage agricultural productivity can also have an active and positive effect with respect to protection of the resource environment. If this were not the case, sustainable development of agricultural production could not be achieved.

Thus, the critical question is as follows. How can appropriate policies be determined that will provide for increased productivity in the agricultural sector and satisfy the nation's food security goals, while at the same time maintaining the sustainable development and utilisation in the resource environment that can ensure this food security over the long term? Possible measures to this end are discussed below, from a variety of perspectives.

Measures to protect existing cultivable land

A resource base of cultivable land is the basic building block for agricultural production. The principal reason for uncertainties in grain production in recent years has been the decreases in the area of land available for cultivation. Compared to 1990, gross production of wheat in 2005 was lower by 1%, and of rice by 5%. During this same period, however, the unit yields of both these grains increased quite significantly, but not enough to offset the even more significant losses of land under cultivation, which in the case of rice amounted to 14% and of wheat a staggering 30%.

Acceleration in the pace of industrialisation and urbanisation has led to the occupation of more and more agricultural land, in a trend that can be neither resisted nor reversed. Furthermore, the land that is being lost tends to be level, fertile land of excellent quality providing multiple annual harvests with high unit yields. There is the potential to develop new land resources or, in the case of land that has been abandoned in the past, return previously abandoned land to cultivation. However, these areas are not likely to make up fully for the land being lost as they are almost invariably of poor quality and low fertility, capable of offering no more than a single harvest per year, and with low unit yields.

While the occupation of land for non-agricultural purposes may indeed be inevitable, the waste of such land can certainly be avoided. According to the rules of economics, the most important consideration is to drive up the cost of occupying land for non-agricultural purposes. With respect to certain industrial enterprises in particular, increasing the cost of occupying agricultural land forces them to make the most efficient use of it, which means doing their utmost to use land in the most economical way possible. As the current system for transferring agricultural land to non-agricultural purposes through simple requisition scarcely satisfies this criterion, reforms are clearly called for. And the direction of such reforms involves allowing farmers who hold land collectively to enter the market directly, whether by outright sale or through participatory share-holding. Such approaches would go a long way to encouraging the most economical use of land by industrial enterprises. In addition, the State can also consider levying land-value increment taxes on agricultural land that is transferred to non-agricultural purposes. This would also serve to curtail excessive demand for land in these conditions.

Meanwhile, while many existing laws and regulations also provide protection to agricultural land, the critical issue is in how they are implemented and enforced. Unless the system for land expropriation is changed, requisitions of agricultural land by local governments can be driven by strong profit motives, and it is this potential profitability that inhibits the effective enforcement of legal statutes. Furthermore, in the context of the new socialist countryside programme, many villages are implementing planning and rehabilitation measures, which makes them very reluctant to curtail occupations of land.

Measures for the protection of water resources

Limited water resources are another important constraining factor in the development of Chinese agriculture. There are three aspects to this problem. First, there is a basic insufficiency of water supply in China, particularly in the north. Second, as the economy develops, the agricultural sector must compete with growing demand for water from industry and for domestic purposes. The third issue is that there is acute pollution of water resources. Figures compiled by the Ministry of Water Resources and the State Environmental Protection Administration show that in seven major river systems of China roughly 40% of water is rated Level 5 or below, which means it is unsuited even for use in irrigation.

The most important measure that can be taken to address the shortfall in water resources is to economise use. In the agricultural sector, water conservation technologies must be developed and widely promoted so that wasteful, low-efficiency methods of irrigation can be consigned to the past and replaced by piping systems for sprinkler or drip irrigation, etc. In all areas where water is lacking, facilities must be put in place to collect and store rainwater. Price reforms with respect to water use would introduce market-type incentives to water conservation and also encourage the development and promotion of conservation-oriented equipments that would improve utilisation efficiency. Meanwhile, further controls should be introduced to clamp down on the use of non-renewable artesian sources of water for irrigation.

A great variety of measures are required to address the ongoing pollution of water resources. Sources of the pollution of both surface and groundwater are many and varied, ranging from industrial effluents to domestic wastes to agricultural runoff and wastes. Agricultural activity contributes to water pollution through the excessive use of pesticides and chemical fertilisers, and the release of untreated wastes from livestock and poultry breeding operations. Such policies as the current rural biogas initiative, upgrades to human-waste disposal as part of the new socialist countryside programme, treatment of wastes from intensive livestock feeding operations, the use of formula fertilisers, etc. all have useful roles to play in combating water pollution.

Measures to increase land productivity

The most important step that can be taken to buffer the relationship between agricultural development and environmental protection is to improve the productivity of agricultural land and increase unit yields. There are two critical aspects to this: innovations in and promotion of new agricultural technologies, and improvements in the basic farmland infrastructure.

Improving capacities for innovation in agricultural technology

The experience of developed countries shows that, from a long-term perspective, technical innovation is a basic dynamic element in developing agricultural production. The European Union, for example, has used a number of policy incentives very effectively to promote increased productivity. Since the 1990s, while a number of restrictions have been put in place to deal with problems related to over-production, technical improvements in the agricultural industry have resulted in increased production nonetheless. This is a clear illustration of the great importance of technology in this sector.

China's own agricultural technology has made significant advances over the past few decades but, in relation to the needs of the sector, these must still be considered insufficient. In practical terms, while China may lead in some areas, including various areas of research, overall it lags behind, such as in applying the fruits of the research.[3] China has many achievements in learning or copying from others, but has few in terms of indigenous innovation. Commendable results have been achieved in general commodities such as grains and cotton, but less impressive ones in commercial crops and value-added products (sugars, vegetables, peanuts, etc.). Furthermore, improvements in crops have not been matched by similar progress in livestock breeding.

Avenues towards improved capabilities in technical innovation include:

- Increasing government investment in agrotechnical research. Central government documents already stipulate that, "Investments in agrotechnology should hold priority in public funding allocations so that the proportion of the State's investments in technology that go to agriculture is increased." The key issue here is actually implementing the policy already in place.
- Building up and providing major funding for "national teams" of agricultural researchers.
- Improving decision-making mechanisms for major technical projects. Consideration should be given to founding a National Council of Agricultural Experts to determine strategies for agrotechnical development, and to set up and administer or delegate major research projects.
- Reforming and improving internal management of research institutions so as to inspire the most active participation of technical personnel.

3. Unit yields are a useful illustration: China's unit yields of wheat are 50% of those in Europe; of corn, 60% of those in the USA; and of rice, 70% of those achieved in Egypt and Australia.

Improvements in basic farmland infrastructure

What is most required in this domain is increased government funding, particularly from central government sources. More funding should be made available in this way to China's central and western regions because of the limited financial resources of local governments in those areas. Available figures indicate that in 2005, average per capita local fiscal revenue amounted to 1 903 Yuan in 11 eastern coastal provinces, 668 Yuan in 8 central provinces, and 685 Yuan in 12 western provinces. In addition to greater investment in water conservancy facilities of all types, steps should also be taken as appropriate for local conditions to protect and improve land quality. Repair and reconditioning work is particularly called for in the terraced fields of mountain areas, and the raised fields of low-lying saline/alkali areas. As such work in terraced fields has noteworthy beneficial effects both on improving productivity and protecting the environment, it serves to advance agricultural production while also combating soil erosion and water loss.

Government investments in these two areas have cumulative and long-term effects. Once new technologies are developed, they can be used forever. Examples of the positive influence of government investments in the initial stages of development and promotion of new technology can be seen in seed technologies, the scattering method for transplanting rice seedlings, precision sowing, plant protection, plastic covering of seed beds, zero-tillage and minimum tillage cultivation, greenhouse techniques, etc. The short period of government investment is followed by years of beneficial applications in the field, with no little need of further financial support. Investments in agriculture and agricultural infrastructure continue to exert positive effects over years and even decades. This is true of the construction of reservoirs, irrigation channels and similar water conservancy facilities, and even more so with respect to terraced fields. In Yunnan and Guangxi, for example, terraces built hundreds of years ago remain a thoroughly positive feature of local agriculture today.

Measures to improve utilisation efficiency of feeds

Improving production efficiency in the livestock industry and boosting feed conversion rates both contribute to economies in feed and resource use. At present, the demand for feeds by China's livestock and aquaculture industries accounts for roughly one-third of total grain consumption demand. If improvements in feeds and feeding techniques can raise feed conversion rates by 10%, the savings would be equal to a production increase of 15 million tonnes of feed grains, which is in fact the target for increased production over the next five years.

The traditional model in which every household raises a few animals can encourage feed economies insofar as agricultural by-products and domestic kitchen wastes are used. Once a slightly larger operation is undertaken, however, then commercial feeds enter the picture and the various constraints associated with smallholdings become apparent, and the ultimate feed conversion rates are invariably lower than those achieved in large commercial feedlots. The main reason for this is the higher level of technical inputs in large operations, which favour higher conversion rates.

The main path to improved feed conversion rates passes on the one hand through technical innovation in both feeds and feeding, and on the other through increasing the size of feed operations, using specialised feeds or setting up concentrated breeding areas, and improving management procedures throughout the production chain.

Measures to exploit new possibilities for resource utilisation

Another important approach to boosting agricultural production can be found in converting non-usable resources to usable resources. There are two possibilities in this respect. One is to convert what were once considered non-exploitable natural resources to agricultural use, for example converting barren or abandoned land to arable farmland, or upgrading wild fruit trees to trees producing a quality commodity. Another is to convert what were once discarded agricultural by-products to usable commodities, such as converting crop straws to feeds, or poultry manures to fish-feeds.

Steps taken along these lines can do more than simply remove potentially detrimental factors from the environment as they can often have a net positive effect. For example, it was once customary in many areas to deal with crop straws by burning them, which of course represents a source of atmospheric pollution. But when such straws are reprocessed as animal feeds, an extra resource is made available at the same time as the environment is spared an unnecessary burden. Similarly, converting barren or abandoned land to arable farmland can also improve the natural beauty of an area.

External criteria for improving resource utilisation efficiency

Whether or not resource utilisation efficiencies can be realised depends on two sets of factors. One is whether the rural producers and other resource users have an active and progressive attitude with respect to resource conservation; the other is whether there exist real and significant possibilities to conserve resources and boost utilisation efficiencies. In both of these areas, the government can play a positive role. In the first, it is through systemic innovation and policy reforms, and in the second, through the provision of support and the necessary services.

The three areas discussed below are of particular importance.

The first is a system to govern land and resource rights that is clear-cut and offers long-term assurances. Only when producers have long-term rights to use the land resources they exploit will they attach importance to protecting them in such a way as to ensure their long-term, sustainable use. The current system of lease holding by rural households, for example, can be used to better effect. The key issue is to prevent local governments from failing to enforce existing land laws and using excuses of all types to alter the lease holding rights of rural producers. At the same time, it is worth considering extending the length of land leases under this system beyond the current 30 years. Ideally, on the basis of the current system, the 30-year term could be extended indefinitely, thus giving farmers permanent land-lease rights.

The second is to give free rein to the fundamental functions of market mechanisms in resource distribution. The basic nature of markets is expressed in pricing mechanisms, and these reflect availability of resources. The more limited the supply is of a resource, the higher its price will rise; this simple function serves to encourage users to strive for the highest utilisation efficiencies. The principle applies equally well to water resources, and to both agricultural and non-agricultural land resources. The pricing approach must be extended to water use and the use of land for non-agricultural purposes so as to encourage conservation and cut down on waste.

The third area to be addressed is government investment in the agricultural sector, which must be increased, particularly with respect to technical innovation in production and in resource use, and basic agricultural infrastructure. Technical progress and improvements to basic infrastructure represent the way forward towards harmonising development in agricultural production and protection of the resource environment. If farmers are left only to their own initiative, and do not receive the support of appropriate technical inputs and basic infrastructure, it is very difficult to simultaneously achieve the twin goals of developing production while also protecting the environment. Government investment and positive incentives are needed in these areas to make up for the deficiencies of the market. Infrastructure development and technical innovation are of great public value and clearly contribute to society. But to depend solely on market mechanisms in these areas leaves many problems unresolved. Furthermore, government investment in these areas has cumulative and long-lasting effects. The investments of one year will continue to exert favourable effects for many years afterwards. The central government should therefore be exerting greater influence in these areas because the issues of food security and environmental sustainability are of an importance that weighs upon the national condition as a whole. They are most emphatically not simply problems of a local nature restricted to certain regions and however active local governments may be in addressing them, their efforts alone can not suffice to obtain effective and satisfactory long-term solutions.

Chapter 13.

MODELS AND STRATEGIES FOR THE DEVELOPMENT OF CIRCULAR AGRICULTURE IN CHINA

Tang Huajun and Yin Changbin
Institute of Agricultural Resources and Regional Planning,
Chinese Academy of Agricultural Sciences

Abstract

This paper proposes a systematic definition of circular agriculture and describes its basic features and implications, with reference to the circular economy models of the Dupont Corporation, the Kalundborg Industrial Symbiosis of Denmark, the DSD recycling system of Germany and the recycling society of Ayacho, Japan. It further categorises prevailing models for the development of circular agriculture in China and analyses typical characteristics. The paper concludes with policy proposals and suggested strategies for the development of a circular economy in China's agricultural sector.

Introduction

The development of a circular economy is a declared national strategy. The objective to "Energetically develop a circular economy and progressively put in place structures for resource-efficient production and consumption" was embraced at a Central Economic Work Conference on 3 December 2004.[1] The deliberations of the Fifth Plenum of the 16th CPC Central Committee held in Beijing on 11 October 2005 subsequently called for: "accelerating progress towards a resource-efficient, environmentally-friendly society, energetically developing a circular economy, bolstering efforts to protect the environment, ensuring practical protection of natural ecosystems; striving to overcome environmental problems affecting economic and social development[2] and putting in place models for improving resource efficiency and encouraging healthy consumption throughout society."

Circular agriculture involves applying the principles of the circular economy to agricultural production. The production chains of agricultural production and product life cycles are extended so as to reduce material and resource inputs as well as waste outputs, and achieve a circle that is both ecologically and economically benign.

1. In the English language literature, circular agriculture is sometimes referred to as closed system agriculture. The underlying idea is to maximise the beneficial economic and societal footprint of the sector while concurrently minimising the ecological footprint.

2. Particular emphasis will be paid to overcoming problems which jeopardise public health and safety.

Features and implications of circular agriculture

Strategies aiming to ensure sustainable development have become a worldwide phenomenon since the 1990s, on the basis of efforts aiming to achieve maximal resource utilisation while reducing the production of wastes and pollutants to a minimum. Various concepts such as clean production, comprehensive resource utilisation, ecological design and sustainable consumption have been integrated and blended to formulate systematic strategies for a circular economy.

The 'circular economy' is a model for economic development that "promotes coordination and harmony of man and nature". It requires utilisation to the greatest possible degree of materials and energy that enter the system so as to boost resource utilisation rates and achieve a parallel reduction in pollutants produced. The result is greater quality and efficiency of economic activity. The theory of the circular economy has opened up new perspectives for agricultural development in China.

Concepts of circular agriculture

Circular agriculture is an entirely new concept and set of strategies, an improved approach to agricultural economy that addresses the coordinated development of populations, resources and the environment. At its core are the principles of sustainable development, circular economy and extended production chains. There are many elements to this, including but not limited to: technological innovations and organisational reforms, optimisation of the internal production structures of agri-ecological systems, extension of complementary production chains, multifaceted circular energy utilisation, maximum utilisation of bio-mass energy resources, promotion of clean production and conservation-minded consumption, stringent control of harmful inputs and waste production, maximum reduction of pollution and ecosystem destruction. Concurrently, efforts are made to increase the value of each stage of production, utilise every material link in the production process and improve the quality of the human environment. This is all done with the underlying objective of ensuring that production processes and producers take their place in a benign agri-ecological circle conducive to the harmonious development of the agricultural industry and rural communities.

Characteristics of circular agriculture

The principal characteristics of circular agriculture are as follows:

- In keeping with the principles of resource conservation, it emphasises improving the production environment and protecting farmland biodiversity as the foundation of stable, sustainable development.
- The circular agriculture framework embraces an industrial approach to agriculture management, stressing clean production practices, technological improvements and the use of environmentally friendly 'green' agro-chemicals so as to minimise environmental impact.
- It applies technology to optimise the structural order of agricultural systems, organising production according to the feedback model of: resources → product → waste → recycled resources, thus obtaining the maximum utility from resources committed.

- It extends the eco-industrial chain of agriculture by encouraging the reutilisation of wastes, elemental coupling, etc. in conjunction with related industries, thereby fostering production networks and collaborative development.
- By encouraging 'clean' and conservation-minded lifestyles in rural communities, it promotes a beneficial modern culture.

In summary, the principal characteristics of circular agriculture involve the extension of production chains and resource conservation.

The production chain of circular agriculture extends from crop cultivation, the forestry, fishing and animal husbandry industries and their associated processing, trade and service industries to the end consumers of the various products. It also includes the exchange and reutilisation of wastes and by-products and elemental coupling and linkages with other industries to shape a collaborative, inter-connected and inter-reliant network for agricultural production on an industrialised model. The various industries are further connected through the exchange and interlinkage of intermediate and waste products, thus completing the make-up of what can be a closed network. In it, resources are allocated in an optimal manner, wastes are put to efficient use and environmental impacts are kept to a minimum.

Superseding conventional agricultural practices

As a low-input, high-recycling, high-efficiency, high technology and industrialised set of practices, circular agriculture is distinct from conventional or traditional agriculture and represents a major re-engineering in how the sector operates. [5]

It surpasses conventional agricultural practice in four major areas:

- In its theoretical orientation, circular agriculture borrows from modern industrial practice, applying the concepts of clean production and circular economy to agricultural production and management. It advocates control over the entire production process and product life cycle so as to forestall any occurrence of pollution or contamination. It also requires that producers abide by the "3 Rs" of the circular economy, in that its priorities are to Reduce, Reuse and Recycle, and minimise production of wastes.
- With respect to production methodology, circular agriculture incorporates the essential practices of conventional agriculture, yet increases the efficiency and effectiveness of both inputs and output. It also advocates comprehensive application of high technology to the agricultural sector, and aims to gradually replace high material inputs with technological alternatives while maintaining high productivity.
- With respect to industrial cooperation, and efforts to enhance the degree of industrialisation of agriculture, circular agriculture aims to put in place eco-production systems throughout the agricultural sector and its related industries. The key point is to optimise internal production structures, extending and broadening production chains so as to achieve multilevel, circular utilisation of resources at both the agri-ecological and regional levels, and ensure the continuance of benign ecological cycles.
- With respect to production and eco-efficiency, circular agriculture advocates multilevel circular utilisation of resources with moderation in external inputs. This can lead to improved productivity, higher quality and reduced production costs. Both economic and

ecological efficiency are thereby enhanced and agriculture can achieve sustainable development in the true sense of the term.

Circular economy models from abroad

At least four principal models for the circular economy can be observed on the international scene. They are represented here by the individual enterprise model of the Dupont Corporation, the industrial symbiosis model of the Kalundborg region in Denmark, Germany's DSD recycling system, and the recycling society active in Ayacho, Japan.

Dupont's corporate-internal circular economy

The Dupont Corporation began experimenting with the principles of the circular economy in the 1980s. By initiating circular approaches to the use of materials in all areas of its operations and extending its production chains, Dupont was able to reduce consumption of energy and materials while also restricting the production of wastes and the release of harmful pollutants. The company also used recyclable resources wherever possible and sought to ensure greater durability in its products. With the help of some creative thinking, Dupont incorporated the "3 Rs" principle into its manufacturing processes, abandoning the use of certain harmful chemicals, reducing the quantities used of others, and developing new techniques for recycling the company's own products. By 1994, Dupont had cut down its production of plastic wastes by 25%, and emissions of atmospheric pollutants by 70%. [15]

Symbiosis at the Kalundborg Industrial Park

The industrial community at Kalundborg in Denmark is built around a core of four major enterprises – a power station, an oil refinery, a pharmaceutical plant and a plasterboard factory. Interacting on a trading basis, these various entities use the wastes and by-products of their neighbours as raw materials in their own production processes. This cuts down on processing costs and waste and, with its considerable economic efficiencies, results in a benign circle of economic development and environmental protection. The Park boasts a very talented management team that handles the coordination, organisation, account settlement, and oversight tasks for the four major plants and collaborating enterprises in the area. The team also coordinates funding and technical support for new waste recycling projects and, by facilitating logistical, energy and information flows, it achieves orderly and efficient production based on the circular model.

Germany's DSD collection and recycling system

The *Duales System Deutschland* (DSD) is a non-profit organisation tasked to manage the recycling of packaging waste. It was established by a group of 95 packaging producers, garbage recyclers and other businesses in 1995 and now counts some 16 000 members. These entities constitute a network that channels recyclable materials – identified by a green dot - to recycling plants for processing. The basic principle of the green-dot programme is that of producer responsibility – that the production of waste incurs a cost. Member enterprises thus pay fees that support the collection, cleaning, sorting and eventual recycling of discarded packaging materials.

Organic farming and resource recycling in Ayacho, Japan

This project began on a small scale with the collection of drainage sludge, poultry droppings and organic industrial wastes as fermentation materials for the production of methane to be used to generate electricity. Solid remainders from this process were composted and dried for use as fertilisers, while liquids were re-processed or discarded, insofar as they did not represent an environmental threat. The result of the programme was to achieve a high degree of waste-to-resource recycling and remove potential environmental threats. Sixteen years after it began, the programme has now developed into a considerable circular organic agriculture industry that has wrought great improvements to the eco-environment in Ayacho Town and Miyazaki Prefecture, while concurrently enabling steady economic development, greatly improving farmers' livelihoods and contributing to social order and stability. [8]

These models of circular economy from outside China all feature the "3 Rs" in that they emphasise reductions in resource use and waste production, and multiple re-use or recycling of waste materials. They also result in more economical resource use, cleaner production processes, eco-transformations of production chains, recycling of wastes and eco-friendly 'green' consumer practices in society at large. Indeed, the close interaction of agriculture with the natural environment is an inseparable condition that predisposes the agricultural economy to harmonious inclusion in the recycling processes of natural ecosystems, and this provides a model for the development of a circular agricultural economy.

Categories and characteristics of circular agriculture in China

Models based on integration of eco-agricultural patterns

The production process in this approach is permeated by the underlying idea of achieving concurrent benefits for the economy, the environment and society. It represents the bare essence of the eco-agricultural model. It stresses the overall ecological effects of agricultural development, and based on establishing circular mechanisms linking "resources to products, to reuse, to re-production", achieves harmony between economic development and ecological equilibrium. Low levels of resource consumption and waste production and maximised utilisation of energy and materials are its basic characteristics. Through farmers' associations or similar organisations, it brings widely dispersed farmers under centralised management and broadens the scope of production, resulting in the integration of cultivation, animal husbandry and product processing in one production model.

Farmers' associations in Linquan County of Anhui Province, for example, have developed what amounts to a classic model for circular agriculture in China. The production model developed by local farmer Wang Shouhong, which features linkage of forest products, grasses, livestock, fungi, biogas and fertilisers, complies with the principles of circular agriculture and results in greater economic and social benefits. It has proven particularly valuable in the handling of cattle manure, for which three downstream channels have been found, in the production of mushrooms, methane and foodstuffs respectively. His model has extended and broadened an originally simple eco-agricultural production chain to obtain greater economic efficiencies. [10]

Models targeting the multilevel utilisation of agricultural wastes

In conventional models of agricultural economy, bio-based industry using agricultural wastes as a resource has not been integrated into the circular loops of the overall production system. Given the degree of global interest in the exploitation of bio-energy resources and the growing maturity of bio-energy transformation technology, however, the industry merits close attention. Indeed, bio-energy resources are of particular interest for their potential in alleviating the global energy crisis. [11] The prime characteristic of the model as it applies here is that it integrates bio-industry as an important subsystem of the overall agricultural industry. Wastes are sought for their resource potential particularly in the processing stages where waste water, gases and solid residues are channelled for further utilisation. Throughout the whole materials circulation system, in fact, there is no such concept as 'waste', but only of 'resources'.

The model developed in the Eastern Xihu district of Wuhan, for example, achieves standardised, large-scale production of raw materials by integrating "industry, local community and farmers" in a system that brings in around RMB 60 million annually to local cattle farmers. At the same time, manures and liquid animal wastes are processed in a composting facility that uses modern fermentation and odour-reduction technology to produce high-quality organic fertilisers that bring in a further RMB 10 million of income.

Production park-based total circular agriculture

The successful experiences of Ecological Industrial Parks in North America are the inspiration for this model that aims to achieve total circular production. Four subsystems – cultivation, animal husbandry, product processing and bio-based production – are incorporated into a closed system for circular production. The flow of materials operates in two major loops. One is external, and includes the processes of transformation from production to consumption. The other is internal, and involves the recycling and reutilisation of waste resources. A set of symbiotic and interdependent relations is thus created that links the material flows and value streams of the different production systems and achieves the greatest possible extension of production chains.

An example is the Ecological Industrial Park model developed in the Sujiatun district of Shenyang, which has an agro-tourism concept at its core. The external framework consists of the various systems governing grape cultivation, wine production, anti-oxidant bio-engineering, bio-fertilisers production and environmental protection, such that the respective functions of the grape, manufacturing and agro-tourism industries are combined in a single entity that manifests itself in a closed-loop eco-economic network.

There are two principal production chains. The first is the closed loop of grape cultivation, product processing and fertilisation. Its noteworthy feature is the bioengineering reprocessing of agricultural wastes as organic fertilisers that are returned to the soil. The second loop includes the eco-residential area, household-waste processing, pond-based aquaculture and return of fish manures to fields as fertiliser. The prime feature of this loop is that the clean resources and organic fertilisers produced can be reutilised in the commercial and agricultural sectors as well. [14]

Directions and approaches for the development of circular agriculture in China

Directions for the development of circular agriculture

The overall progress of social and economic development is to be led according to scientific principles to promote a conservation-minded society and socialist development in rural communities. The various principles of harmlessness, low emissions, zero damage, high benefits, sustainability and environmental beautification apply to ensuring sustainability in both agricultural production and rural consumption, and to promoting coordinated development of the rural economy. Agriculture and rural industry, production and living conditions, rural infrastructure development and urbanisation will be the object of comprehensive planning, with the promotion of conservation-oriented development, clean production processes, utilisation of wastes as resources and clean consumption as starting points. Finally, further reforms in technological paradigms and methods of organisation will favour the development of circular economic systems in rural communities.

With respect to resource utilisation, conservation will be paramount, with first priority to be given to achieving increased efficiency of water and land usage, as well as of investments. To these ends, new concepts of cost-limitation must be applied to conventional practices, and every effort made to economise in the use of water, land, energy, fertilisers, chemicals and labour, as well as to increase the recycling of agricultural resources and bolster the capacity for sustainable development.

With respect to waste processing, the objective is to encourage reutilisation of wastes as resources, so as to achieve total reutilisation of bio-resources that accumulate during crop production, cut waste output in poultry and livestock raising and convert animal manure wastes for resource purposes. In particular, efforts are to be made to find alternative uses for the straw by-products of cultivation, moving beyond the familiar production of methane to investigate their potential in other bio-energy or microbial resource applications, while the most economical uses are to be found for manures. These various measures will promote circular utilisation and resource-oriented development.

With respect to the production chains of the agricultural industry, initiatives include promoting clean production, utilising the wastes of one link as resources in another, achieving increases in the value chain and expanding the scope of industrial approaches to agriculture. A major focus in this respect will be achieving internal extensions and interaction of various productive elements, and consolidating the concepts, directions and models in use to extend the domain of industrial-type management. Such extension is the wellspring for sustained growth in the value-added activity in the rural economy. It will strengthen regional economies and contribute to increases in farmer income, and as such will be instrumental in the development of circular agriculture.

With respect to infrastructure development in rural areas, efforts are to focus on developing circular communities with access to methane and solar power as energy resources so as to reduce reliance on outside inputs. "Clean consumption" will be encouraged with local treatment of refuse, while, with a view to building new socialist communities, practical steps will be taken to operate infrastructure improvements in rural areas in order to improve the people's living environment.

Strategies for development of circular agriculture

In the aim of building a harmonious socialist society, developing a circular economy in the agricultural sector and achieving a new socialist order in rural areas, work is presently required in the areas described below. [2]

Planning for the development of circular agriculture

With respect to provisions for agricultural development to be included in the 11th 5-Year Plan, plans should be drawn up to encourage circular agriculture, with particular focus on conservation, high-efficiency resource utilisation, recycling of wastes, clean production as production chains are extended, and 'clean' infrastructure improvements in rural communities. Meanwhile, directions, objectives and usable models of particular relevance to the next 5 years should be proposed, including engineering measures, support for important areas of activity and necessary safeguards. Specific planning is required to address economies in water, land and fertiliser use, as well as in labour and costs in general, and the comprehensive utilisation of agricultural resources. Similarly, major areas of focus should be identified, and development targets and policy measures drawn up.

Structural adjustments and optimising the agricultural economy

Regional agricultural development should be directed in accordance with the principles of the circular economy. With reference to the particular available resources and structural features of the agricultural industry in each area of the country, rational adjustments can be made to regional patterns of agricultural activity so as to put in place region-wide systems for circular agriculture. Specialisations of production can be promoted along regional lines, while the optimum-scale management schemes and industrial clusters required for effective circular agriculture should be put in place. Meanwhile, 'raw material' bases should be developed to support the extension of production chains and the industrial processing of wastes. Finally, infrastructure development in rural communities and smaller towns is required to ensure proper processing of refuse and lay the foundations for a pleasant living environment.

Technical integration systems to promote circular agriculture

Agricultural development should be supported by the application of advanced skills and high technology to restructure and improve upon traditional agricultural practices, particularly through the development of "factory farming" and "manufacturing agriculture." The technical component of agricultural products should be constantly increased and greater space found in the industry for technical integration and innovation so that due technological support is available to the agricultural economy.

Early efforts in this direction should emphasise: the technical linkages inherent in clean production; green production technologies and multilevel resource transformations; technical support to resource conservation through efficient utilisation and waste recycling; formulating technical standards for circular agriculture; eco-urbanisation technologies for developing rural communities; green consumption technologies for rural inhabitants. In addition, research should be conducted to establish new systems to support technological innovation and promote technical models of relevance to circular agriculture.

Pluralistic investment mechanisms and model programmes

In keeping with the policy of building a new socialist countryside, and guided by the principles of the circular economy, new, pluralistic investment mechanisms led by government inputs should be established during the 11th 5-Year Plan. In this respect, a few model programmes should be established on a nationwide basis. These models should address important aspects of the circular agricultural economy such as resource recycling, conversion of crop straws and animal wastes to resources through non-harmful processing, 'clean' infrastructure development in rural communities, development of bio-energy potential, circular utilisation of microbiological resources, etc. In addition, new development models for circular agriculture and new methods of resource utilisation adapted to the resources available in differing regions should be developed, so as to ingrain the principles of circular agriculture in the daily life of rural households and communities.

Structural innovation to create a favourable policy environment

There are at present several areas in which vigorous intervention by the authorities responsible for promoting innovations would be beneficial to the development of circular agriculture. These include bringing adjustments to state policy and to the legal system. Specific measures that would be conducive to creating the favourable environment needed would include increasing financial support to the agricultural sector, implementing market-style reforms in rural communities, establishing organisations dedicated to the promotion of circular agriculture, and improving basic agricultural infrastructure and environmental management. At the same time, social support service systems should be developed for rural inhabitants that align with national legislation concerning the circular economy, as there is some urgency to putting in place the legal safeguards and corresponding policy guarantees to support the development of circular agriculture.

The preliminary legislative work should begin as soon as possible and favourable policies put in place to address tax relief, financial guarantees and subsidies. In addition, practical and effective measures are required to promote infrastructural development in the countryside. Further measures urgently required include the drafting of regulations to govern clean production in the agricultural sector, as well as standards in this respect and for environmental cleanliness in rural areas. In this way, the development of circular agriculture and a conservation-minded rural society can be brought within the ambit of legally prescribed and standardised administrative management.

REFERENCES

1. Xinhua (2005), *Guanyu Zhiding Shiyiwu Guihua Jianyi de Shuoming* (On Making Suggestions for the 11th 5-Year Plan), 19 Oct.

2. Tang Huajun (2006), *Xunhuan Nongye Neihan, Fazhan Tujing yu Zhengce Jianyi* (Implication, Directions and Policy Suggestions for Circular Agriculture), China Agricultural Resources and Regional Planning, Vol. 27, No. 1, pp. 4-8.

3. Yin Changbin (2006), *Xunhuan Jingji Linian, Xunhuan Nongye Neihan yu Zhanlue Tujing* (Principles of Circular Economy, Implications and Strategic Directions of Circular Agriculture), Compilation of the Chinese Academy of Agricultural Sciences, p. 88.

4. Chen Demin (2002), *Xunhuan Nongye. Zhongguo Weilai Nongye de Fazhan Moshi.* (Circular Agriculture – A Model for the Future of Agriculture in China), Jingjishi, No. 11, p. 134.

5. Zhou Zhenfeng *et al., Guanyu Fazhan Xunhuanxing Nongye de Sikao* (Thoughts on Developing Circular Agriculture), Studies on Agriculture Modernization, Vol. 25, No. 5, pp. 349-351.

6. Jie Zhenhua (2003), *Guanyu Xunhuan Jingji Lilun yu Zhengce de Jidian Sikao* (Some Thoughts on the Theories and Policies of Circular Economy), Guangming Daily, 03 Nov.

7. Zhongguo Zixun Bao (2005), *Dangjin Shijie Xunhuan Jingji Sizhong Moshi* (4 Modern Models of Circular Economy), 29 Apr.

8. Wang Jun, Zhou Yan *et al.* (2005), *Qianlun Nongye Kechixu Fazhan de Xin Moshi* (Remarks on Sustainable Development in Agriculture), *Ziyuan Xunhuanxing Nongye* (Circular Resource Agriculture). Environmental Protection Science, 31 (129).

9. Zhou Yin *et al.* (2006), *Woguo Xunhuan Nongye Fazhan Moshi Fenlei Yanjiu* (Studies in Classifying Models of Circular Agriculture in China), Compilation of the Chinese Academy of Agricultural Sciences, p. 88.

10. Circular Economy Web, Linquan County, www.xhwww.com/index.asp.

11. Wang Xianhua and Chen Hanping (2004), *Guowai Shengwuzhineng Fazhan Zhanlue dui Woguo de Qishi* (Lessons for China in Developments in Bio-Energy Abroad), Papers of the 2004 National Conference on Bio-Energy Technologies and Sustainable Development.

12. People's Government of Eastern Xihu District, Wuhan, *Shishi Shengtai Liqu Zhanlue* (Implementing Ecological Strategies), *Gouzhu Hankou Shengtai Xincheng* (A new ecological city in Hankou), Hubei Environmental Protection Web: www.hbepb.gov.cn/

13. Deng Nansheng and Wu Feng (2001), *Guowai Shengtai Gongyeyuan Yanjiu Gaikuang* (A Survey of Ecological Industrial Parks Abroad), Journal of Safety and Environmental Studies, 1 (4): 25～26.

14. Ji Kunsen (2004), *Xunhuan Jingji Yuanli yu Yingyong* (Principles and Applications of Circular Economy), Anhui Scientific Publishing House, pp. 169-170.

15. Ma Rong (2005), *Deguo Xunhuan Jingji de Fazhan Gaikuang* (Development of Circular Economy in Germany), Zhongguo Huanbao Chanye, 5.

Chapter 14.

THE CROP PROTECTION INDUSTRY ROLE IN SUPPORTING SUSTAINABLE AGRICULTURE DEVELOPMENT IN CHINA

Jeff Au
Syngenta China

Abstract

Syngenta China is the subsidiary of Syngenta AG Switzerland. The company works closely with key stakeholders in developing and promoting sustainable agriculture in its daily business operations. Joint research projects with Chinese partners involve soil conservation in orchards and other commercial crops on sloped land, and no tillage and minimum tillage cultivation techniques for wheat, canola, rice and corn. This work has been conducted in cooperation with various agricultural research institutes since the mid-1980s. The research has demonstrated the benefits of soil, water and fertility conservation and cost saving techniques both to the environment and to farmers.

The company is also involved in transferring knowledge regarding conservation-oriented agricultural practices to farmers in collaborative work with extension agents at all levels. A good collaborative model is fundamental to develop and promote sustainable agriculture under current rural conditions. We find that the speed of adoption by farmers is directly related to the economic benefits they receive as a result of adoption.

The safe and responsible use of crop protection products is critical to the production of safe and healthy food. Our experience is that training to leading farmers, retailers and grass root level extension workers is an effective way to transfer this knowledge and technology and to support sustainable agricultural practices. The outcome is not only economic and health benefits to farmers, but also environmental and social benefits to the public. We discuss the key challenges to promoting sustainable agriculture in this paper.

Introduction

Syngenta was formed as a result of the merger of ICI, Ciba & Sandoz. Each of these companies has had over 100 years of business experience in China. Based in Basel, Switzerland, Syngenta is a global agribusiness company focusing on Seeds, Crop Protection Product (CPP) and Biotechnology. Its subsidiary, Syngenta China, has invested more than USD 150 million in China on Seeds and CPP and in the production and marketing of these products in both domestic and export markets.

China has 7% of the total global arable land but has to feed more than 20% of the global population. Further, arable land is declining due to the rapid urbanisation and industrialisation. Thus, China needs technology to increase the unit output to deal with this challenge. Sustainable agricultural practices are critical to meeting this daunting challenge. Improved seeds, CPP, fertiliser and good farm management of irrigation, machinery and cultivation method are elements of this advanced technology.

All stakeholders engaged in agriculture understand the importance of managing such advanced technologies in a sustainable way to support sustainable agriculture and achieve the correct balance between the economic, environment and social benefits to society. However, the majority of China's 700 million farmers have a relatively low education level, low income and have very small land holdings. It is, therefore, an enormous challenge to any industry or stakeholder engaged in agriculture to develop and transfer the new technology to this unique community in order to support and encourage sustainable agricultural practices. This paper reports what Syngenta China is doing to demonstrate the role that the industry can play.

The collaboration model

The key factor for success in business is to understand end-users' needs (*i.e.* farmers' needs) and provide a solution to their problems. It is important to remember that their need relates not just to the product you are offering, but the complete solution. Thus, the product we offer must be combined closely with appropriate farming techniques. It must provide additional benefits to both the grower and the public. The method or technique for applying the new technology must be suitable or appropriate to the local circumstances. For this reason, we are intensively involved in field trials to search and test for locally appropriate applications of our technology. Conducting this activity at the local level has proven to be critical to both successful adoption and to successful business operations.

New technology development

China has invested fairly heavily in research resulting in significant achievements. But the rate to which China has successfully applied, commercialised and extended its research is relatively low. Since 1980, Syngenta has worked with academic and research institutions to identify new techniques which have commercial potential. Syngenta also collaborates with the extension authorities at all levels to extend these techniques to commercial use by farmers. In this process, the funding is from the company but the basic motivation and research results have a strong commercial orientation. Between Syngenta and its collaborating research partners and extension institutions, strong commercially-oriented links are made and significant efficiencies and synergies are realised in jointly generating ideas, verifying and developing the new technology so that it is ready for extension and commercial adoption. Since such initiatives are initiated and funded by the company, all risks are absorbed by the company, which hopes for returns from future sales of its advanced products and services.

Long term investment

The Syngenta Agricultural Research, Education & Rural Community Development Fund has been established and co-managed by the Ministry of Agriculture's (MOA) Foreign Affair Department and Syngenta China. The first five years from 2001 to 2005 focused on scholarships and small research projects. These proved to be valuable model for agricultural development. Based on this success, the second five year plan started in 2006. Scholarships are now offered at ten agricultural universities for both B.Sc. and M.Sc. studies. Key areas sponsored include conservation agriculture research projects, rural community service by agriculture students, and food safety. Under this umbrella, there is more interaction and collaboration with and among students and university staff regarding new technological developments, potential collaborative projects and campus recruitment. PhD study and research is supported under a separate fund managed directly by our research institute in the United Kingdom.

Progress of conservation agriculture technology development

The company business mission is not just meeting the farmer's need but also includes benefits to the public. For this reason, when we develop new technologies, we also place an emphasis on supporting sustainable agricultural development. We engage in a lot of conservation agriculture technologies and rural development activities, with some results that are beneficial to farmer use, and some that have broader social and environmental benefits, as follows.

Success in research and good farmer adoption rates [13]

No-tillage cultivation of wheat and canola in Sichuan [1, 3, 4, 5, 7, 8]

The original idea to develop this cultivation method arose from a need to overcome the hurdle of field management during rainy autumn so as to improve drainage and cope with muddy field conditions for wheat farming. But the test on the no-tillage practice failed in the early 1980s due to a problem with weeds that could not be managed through conventional means. Only after the introduction of the contact kill herbicide, Gramoxone, was this approach successful in the mid 1980s. The benefits of no-tillage cultivation methods are not only overcoming this field management problem, but also generate significant labour and cost savings and a modest yield increase. Thus, the speed of adoption by farmer accelerated and reached 6 million *mu* (1 *mu* = 1/15 ha) in 5 to 10 years time. The key drivers of this farming technique are cost saving (labour, cattle and fuel) and more convenience in farming. The overall benefits are summarised as follows:

- Cost saving [1.4]: No need to plough the field with either cattle or tractor and the saving is 4 to 6 man-days per *mu*. The overall saving per *mu* in term of value is estimated at RMB 60-80. (Gross income per *mu* is estimated at RMB 400-500).

- The soil is not tilled. This significantly reduces the risk of soil erosion. Concurrently, moisture content at different levels of the top soil in no-tillage fields is noticeably higher than in conventionally tilled fields. This enhances the capability of the crop against any likely stress *e.g.* drought [3].

- Higher yield: A deep ditch is dug around the field for the no tillage practice. The ditch plays a drainage role during the wet season or in wet areas while, in the dry area/season, the no tillage field retains higher soil moisture. This kind of regulation function contributes to a higher yield and has resulted in sustained production in years after it was adopted, as follows: Wheat - 21-48 kg/*mu*; Oil seed rape – 23-44 kg/*mu* [3].

- The soil structure has been improved with higher organic matter content. [1]

No-tillage cultivation of cotton/rice and canola/wheat along the lower reaches of the Yangtze River

The crop pattern along Yangtze River is more or less similar to Sichuan. That is, wheat and canola in winter, rotated with cotton or rice in the summer. The success of no tillage practice in Sichuan further expands this technique into the cotton, rice and canola/wheat no tillage in Anhui, Jiangsu, Hubei, and Hunan.

After the harvest of wheat or canola, the cotton seedling is transplanted directly onto the field without ploughing the land. A spray of non-selective herbicide like Glyphosate or Gramoxone will then be applied if there are any weeds established before the transplant. The obvious benefit is the saving from not having to plough the field at the cost of RMB 30-50 per *mu* and the time associated with doing so. When the field is not ploughed, soil erosion and water evaporation are also minimised. The area of the no-tillage of Canola – Cotton – Canola/Wheat is estimated at 5.5 million *mu* and the area of no-tillage of Rice – Canola/Wheat is 14 million *mu* in 2004 (Syngenta internal survey).

No-tillage cultivation of rice in Sichuan and Southern China [1, 2, 6, 7, 9, 10, 11, 12]

Due to the earlier success of the no tillage technique for winter crops, farmers in Sichuan expanded this technique into the summer crop of no tillage rice because they are reaped significant economic benefits through cost and labour savings. In rice no tillage cultivation, there is an additional benefit of the water saving. The experience from Meishan indicates that there is 100 cubic meter of water saving during the early stages of field preparation and transplanting. The water savings also improved the local rural society's social harmony by lowering tension during the season when water demands and competition for water are high.

Other benefits of no-tillage [8]

The economic, environment and social benefits are well recognised by farmers and drive the adoption of no tillage cultivation. However, there are other less easily discerned benefits of crop residue treatment, as follows:

- Farmers usually burn the straws of wheat and rice, causing significant air pollution and disturbing air traffic in and around the Chengdu airport. It is now the case that the straw is returned to the field and eventually contributes to higher organic matter content in the field.
- The wheat straw is buried in the drainage of no-tillage paddy field. When soaked with the water, the decomposition process is sped up.
- The rice straw is laid on the no-tillage wheat soil bed after seed sowing. This inhibits the re-germination of weeds and reduces the weeds problem during the growing season, thereby virtually eliminating the need for other herbicides in field management.

The use of this no till technique has now been extended into South China, to the provinces of Zhejiang, Hunan, Jiangxi, Guangdong and Guangxi.

No or minimum tillage cultivation of corn in the Yellow River delta

Corn inputs are very low and, due to labour migration to other sectors and locations, Gramoxone is often adopted for mid season weed control. Due to its fast action of weed control, it is also a good tool to clean up any established weeds that arise from the use no till cultivation practices on corn. This avoids any competition between the growth of corn and weeds. The mixture of Gramoxone with pre-emergency herbicide can provide comprehensive seasonal weed control in no till fields. In these cases, the field is left unploughed, with the cut straw of the preceding wheat crop left on the top soil to minimise the evaporation of soil moisture and retard erosion. This practice has economic benefits for the farmer, but also results in environmental and societal benefits. The use of such herbicides has accelerated the adoption of no tillage cultivation techniques for corn in the Yellow River basin. It is estimated that between 60 and 70% of the basin is now under no till or minimum tillage.

Success in research but lagging behind in farmer adoption

Soil conservation on citrus plantation on sloped land in Zhejiang province. This project was undertaken in cooperation with the Environmental Resources and Soil Fertiliser Institute of Zhejiang Academy of Agriculture Sciences, Hangzhou 310021, P. R. China [14]

The application of herbicides in the hilly red soil orchards can efficiently reduce water and soil loss by 30 to 47% and increase fertility (OM & NPK) by 10 to 14% in comparison with traditional cultivation and hand weeding.[1] The recommended weed management programme for sloped orchard plantations not only prevents soil erosion and improves soil fertility but also results in a slight yield increase of between 4 and 8%. In this weed control programme, there are two herbicide choices, one alternative is long lasting Glyphosate, which has a higher soil erosion risk. The other alternative is quick acting effect Gramoxone, which has a lower soil erosion risk. Results to date suggest that farmers prefer the longer lasting weed control result to the quick acting alternative, even though both have a similar unit cost. This suggests that farmers may be less concerned with environmental benefits than with economic benefits, particularly if the latter are sustained.

Soil conservation on commercial plantations upstream in the Yangtze River Basin in Sichuan. This project was undertaken in cooperation with the Sichuan Academy of Forestry, Chengdu 610066, P. R. China [15]

After three years of research, a good weed management programme for commercial orchard and tea plantations on sloped land has been identified. The research shows that the recommended programme results in the prevention of water and soil erosion and positively contributes to the retention of soil fertility. Erosion can be reduced significantly under different herbicide treatment programmes, in comparison with traditional cultivation and hand weeding approaches in peach orchards with annual erosion of 777.3t/km^2.yr and in the tea plantation of 106.9t/km^2.yr at different level of slope. However, because the economic benefits between the different types of herbicide is not very significant, farmers still prefer the method with higher level of erosion risk but more

1. The annual soil erosion is 167.8/t•hm-2.

sustained weed control, while the alternative with a lower risk of erosion but less weed control was not the preferred choice for farmers. For anyone interested in extending new technologies to farmers, this suggests that farmers will first choose the option with the highest expected economic return.

Ongoing research with high potential for commercially adoption

No tillage cultivation of corn on sloped land in South West China. This project is undertaken in cooperation with Guangxi Maize Research Institute, Yining, Wuyu, Guangxi 530227 [16]

Preliminary results suggest that the yield of maize for no-tillage cultivation with weeds controlled by Gramoxone is similar to the traditional hand-weeding cultivation approach. However, the results also suggest that soil erosion under the Gramoxone no-tillage treatment is 43% less than under the traditional cultivation treatment. The economic analysis showed that the cost of Gramoxone no-tillage maize is reduced by 46% when compared with the traditional cultivation method and that total income is increased by RMB 825/ha (USD 100/ha).

The study indicates very obvious economic, environment and social benefits to farmer and the society. But most farmers are very low income and resource poor and may lack the motivation and means to improve efficiency. Because of this, it takes a relatively long time to promote at the farmer level for the adoption. The extension work is still ongoing and preliminary results are encouraging.

No tillage cultivation of corn in North West China

We have had a corn conservation tillage project in China's North West in cooperation with the Ministry of Agriculture for sometime. We found that it requires CPP to overcome problems with weeds. The new project with CPP kicked off in 2006. We are hopeful it will generate some practical insights that will be useful in enhancing this technique in anticipation of future extension to farmers. The outcome of this research and farming approach has the potential to reduce the incidence and magnitude of sandstorms in China.

High Yield Rice (HYR) with Integrated Pest Management (IPM)

Hybrid high yield rice research is the important government project with the goal of increasing China's total output of rice. Due to the change of physiology and morphology of this rice variety, it requires special fertiliser and CPP to secure the high yield. This project begins in 2006 and will run for five years. It will search for the "best practice" in the overall IPM programme. This crop protection programme will be a key element to secure the performance of new high yield rice varieties intended to boost the rice production.

Safe vegetable and fruit production for domestic and export market

China is number one in the world in terms of production and export of fruits and vegetables because these are labour intensive horticultural crops and need less land. However, the critical element for success is to control the residue level of pesticide and other chemicals and keep them within allowable limits. The best way to reduce the residual risk is at the farming stage. If the Good Agricultural Practice (GAP) can be well implemented at the farmer level, the farmers will not only benefit economically but it will contribute to better health and environmental protection in rural areas as well as assuring safer and healthier food for consumers.

Under China's existing rural conditions, the direct training to key farmer, grass root level extension worker and pesticide retailers has proven to be an effective model. The average farmer is influenced by those who have been trained and often follow their lead. The objective of this training is to equip farmers with the skills necessary to produce safe food. A major focus will be placed on the IPM and through integrated crop management, strides will be taken to achieve sustainable agriculture objectives. Syngenta has been working together with the National Agri-Technology Extension Service Center within MOA for many years on this project. Directly or indirectly, several hundreds of thousands of farmers benefit from this project every year.

This approach to training was also assessed by South West Agriculture University in 2004. The training events were conducted in Pangzhou county (Sichuan), targeting key farmers, CPP retailer and grass root level extension workers. In all, the study covered 80% of the townships and villages in the county. The university acted as a third party and conducted a training effectiveness assessment. The result indicated that farmers noticeably changed their behaviour in the application of CPP after receiving training. The trained "key" farmer could then further influence 24 other farmers on average. The trained input retailer could influence as many as 83 farmers and the trained extension workers could influence as many as 96 farmers. In the year under study, the market availability of vegetables in Pangzhou county was slightly higher than in the preceding year. CPP related incidents, however, were somewhat lower in comparison with the previous year.

The key challenge is how to scale up of this kind of campaign and gain more support from all stakeholders - including commercial companies, industry associations, universities, research institutes, government and non government bodies, farmers and the general public.

Challenges

- Farmers are mostly driven by economic returns when they consider adopting any new technology. Thus the economic benefit is the key element of any new idea, research project, or new technology - even if the new practice has very strong environmental or other intangible benefits. Commercial benefits to farmers are a critical consideration to widespread adoption, especially during the early stages of extension work.
- It is a very long process from idea to research, from research to development, from development to extension, and from extension to commercial production and use. The industry has to pay most of the cost and bear much of the risk of investment return throughout the whole process.
- Any new technology will bring with it new issues and challenges with the passage of time. Therefore, ongoing study is warranted to minimise the incidence and magnitude of potential adverse impacts. For example, the no tillage practice of wheat and rice and the crop residue in the field might change the pest occurrence pattern, and it requires further study to anticipate and manage such potential new problems.

- The large scale promotion of any new technology to encourage sustainable agriculture requires lot of extension effort, resources and new kinds of partnerships. New models for collaboration need to be identified, tested and piloted. It is not just the technology itself but also the model of extension that needs study and possible adjustment or fine-tuning.
- Market discipline. Farmers are provided with high quality branded products to ensure that their farming activities are in line with sustainable agriculture. However, the IPR awareness in the distribution channel is very low. There is an enormous number of poor quality or even fake products imitating quality branded products in the market place. The initial technology developing industry suffers as a result of these fakes, imitations and pirated products because these sub-standard substitutes make it difficult to recover their investments. Farmers also suffer because there is a risk that the products and inputs they purchase are not what they think they are and that they will under-perform or perform differently than expected. This adds to risks and costs and lowers productivity. Such sub-standard products can also contribute to risks to human health and safety – to farmers, to food consumers, to the public in general, and possibly to export destinations.

Conclusions

We have described what Syngenta has been doing in China to support the development of sustainable agriculture and what we have learned from this experience. It needs a high level of partnership with all related stakeholders from research, development to the extension for farmer adoption. It requires a long term commitment and more effort from all stakeholders, especially from industry and government. Once the market becomes more disciplined, commercial firms will invest more funds, time and effort because imitation products will be less able to erode their profits. Under such conditions, we expect that more engagement and commitment from industry will be forthcoming to support sustainable agriculture development in China.

REFERENCES

1. Crop Institute bulletin of Sichuan Agriculture Academy Science, 1998
    1. Du Jinquan: Study on rice no & minimum tillage-production effect and future.
    2. Du Jinquan: Technique for rice no-tillage high yield production.
    3. Du Jinquan: Study on mechanism of rice no & minimum tillage high yield production.
    4. Rui Nongtin (Meishan county government): No tillage is the reform of rice cropping technique.
2. Huang Qin *et al.*, Guangdong Agriculture Academic Science, 1997 (3)
    1. Preliminary study on rice no-tillage & broadcasting seedling in no-tillage.
    2. Preliminary study on seedling broadcasting in double crops rice no-tillage.
    3. Guangdong Agriculture Academic Science, 2000 (5)
       Special publication of study on rice no-tillage farming technique.
3. General station of Agriculture extension of Sichuan, 1996
   Study on wheat no tillage (project with Sichuan Agriculture Academic Science).
4. General station of Agriculture extension of Sichuan, 1996
   Farming technique of wheat no-tillage in Sichuan.
5. Yuen Wudon: Plant Protection Society of Sichuan, 1997
   The yield increase mechanism of wheat no-tillage.
6. Feng Shuan: Agriculture department of Meishan county of Sichuan, 1988
   The technical evaluation & economic assessment of rice no-tillage.
7. General station of Plant Protection of Sichuan, 1990
   Introduction of Gramoxone use in no-tillage.
8. Wen Chenjing (director of Sichuan Agriculture Bureau) Sichuan Science & Technology News 15th Oct 1999
   Excellent extension of double no-tillage technique.
9. Lu Jincheng: Plant Protection station of Zenshan county of Zhejiang, 2000
   The practice of direct seed planting of rice in no-tillage of early season crop.
10. Anjie county extension station, Zhejiang, Dec 2000
    The demonstration of direct seed rice in no-tillage in single crop rice.
11. Zhou Yinbin *et al.,* Rice Research Institute of Hunan Agriculture University, Aug 2000
    Study on the use of Gramoxone in double crop of rice no-tillage.
12. Danzhou Agriculture Extension Centre of Hainan, Dec 2000
    Preliminary study on rice seedling broadcasting in no-tillage.
13. Au, Jeff, The 18th Asian-Pacific Weed Science Society Conference Beijing PR China, June 2001
    The No-tillae technique with Gramoxone in rice & wheat production in the water saving & conservation agriculture.
14. Shui, Jian-Guo *et al.,* 13th International Soil Conservation Organization Conference, Brisbane, Australia, July 2004
    Effects of different natural vegetation management measure on red soil erosion in hilly orchards.
15. Min, An Min, *et al.,* To be published in the International Symposium on Soil Erosion & Dryland Farming, Yangning, Shaanxi China, Oct 2006
    Study on effect of herbicide application to soil and water conservation on slope garden.
16. Chen, Tian Yuan *et al.,* 20th Asian-Pacific Weed Science Society Conference Ho Chi Ming, 2005
    Study on the environment & economic benefits of the no-tillage cultivation practice of maize on slope with Gramoxone.

Chapter 15.

DOES CROP INSURANCE INFLUENCE AGROCHEMICAL USE IN THE CURRENT CHINESE SITUATION? A CASE STUDY IN THE MANASI WATERSHED, XINJIANG[1]

Zhong Funing, Ning Manxiu
Nanjing Agricultural University, Nanjing, China
and
Xing Li
Chinese Academy of Agricultural Sciences, Beijing, China

Abstract

Government subsidy to crop insurance has been advocated as a policy alternative to support growth of agricultural production and farmers' income in China since the country joining the WTO. However, cautions have been raised as the crop insurance programme may impact the environment negatively. This study tries to explore farmers' behaviour with regard to agrochemical uses with household data applied to simultaneous- equation system consisted of disaggregated input models. We find that decisions regarding fertiliser, pesticide and agro-film applications do have different impacts on crop insurance participation, and are in turn influenced by the latter in different ways. It is also implied that encouraging farmers' participation in crop insurance under current low-premium and low indemnity terms may not bring significantly negative impact on the environment.

Introduction

The vulnerability of China's agricultural sector due to lack of economy of scale, especially in production of bulk commodities, has become a hot topic after the country joined the WTO. Restricted by accession commitments that prohibit China from supporting agriculture with price and export subsidies, alternative policy measures have been sought by the government to comply with WTO rules. As one of the "Green Box" measures, the crop insurance with a modest budgetary subsidy has been advocated as a means of managing production risks while concurrently providing more stable and perhaps slightly higher income to farmers. However, some previous studies have shown that crop insurance has the potential to encourage applications of agrochemicals and, hence, bring negative impact on the environment and future growth of agriculture. If this is the case, the outcome of subsidising crop insurance may contradict China's policy goals and production targets in the long run. It is obvious that the potential environmental

1. An earlier version of this paper was first submitted to *Agricultural Economics* and will be published on behalf of the International Association of Agricultural Economists by Blackwell Publishing in 2007. The research for this study is sponsored by the Economy and Environment Program for Southeast Asia and the China National Natural Science Fund. The authors are grateful for valuable suggestions and comments from David Glover, Vic Adamowicz, Ted Horbulyk, and Stephan von Cramon-Taubadel.

impact of crop insurance depends on farmers' decisions with regard to agrochemical applications under existing social, economic, and environment conditions, as well as the terms stipulated in the insurance policies. This study explores whether farmers' decision of crop insurance participation is simultaneously made with that of agrochemical applications and, if so, to what extent it may impact the environment.

China has been able to supply enough food for its large and fast-growing population from a limited land endowment. This accomplishment has been achieved primarily by increasing use of modern inputs and advances in agricultural technology. However, the intensification of agriculture may have created serious problems of environmental degradation at the same time. Leaching of nitrate, pesticides and other agrochemicals into groundwater, surface water pollution from soil erosion and nutrients and pesticides in runoff, as well as other environmental problems all have been reported frequently in China, and are related to intensified farming practice in many cases.

Concerns with the potential impact of agricultural production on environmental quality have become prominent in policy discussions, and the environmental and economic literatures. A number of theoretical and empirical studies have been conducted to analyse the impact of crop insurance on input use (See, for instance, Quiggin, 1992; Ramaswami, 1993; Horowitz and Lichtenberg, 1993; Smith and Goodwin, 1996; and Babcock and Hennessy, 1996). Several empirical studies have estimated the intensive-margin effects of crop insurance on input use but reached contradictory conclusions. The main reason is that production conditions (for example, climate, rainfall) and the parameters of crop insurance programme (*i.e.* coverage, indemnity and premium levels) are heterogeneous across regions and crops, and the assumptions of farmers' decision-making process are different, *i.e.*, either simultaneous or recursive.

Smith and Goodwin (1996), for example, in an econometric analysis for Kansas's wheat farms in which the insurance and input decisions are determined simultaneously, conclude that nitrogen fertiliser expenditures decrease by USD 5/ac with crop insured. Likewise, Quiggin, Karagiannis and Stanton (1993) conclude that Midwest corn and soybean farmers who purchase crop insurance decrease chemical applications by about 10%.

In contrast, Horowitz and Lichtenberg (1993) assume that: farmers are risk averse; increased applications of fertiliser and pesticides increase the variability of yield and hence the probability of both high and low yields, and that the crop insurance decision has to be made ex ante - before any inputs are actually applied. Using a recursive structure in which the crop insurance decision influences input use but input use does not influence the crop insurance decision, Horowitz and Lichtenberg find that the purchase of crop insurance induces the Midwestern corn farmers to increase their fertiliser applications by approximately 19% and pesticide expenditures by 21%. The model developed by, among others, Horowitz and Lichtenberg argues that fertiliser and other chemical inputs have two distinct effects on yield distributions. In particular, increased application of chemical inputs raises expected yields as well as the variance of yields. To the extent that the effect of the increased variance may be large enough to offset the increase in expected yields, additional chemical inputs may actually raise the probability of low yields. If the losses due to low yields could be compensated by insurance indemnity, farmers may wish to increase application of chemical inputs. In this case, they will enjoy higher income when yields are high and get indemnity from participating in crop insurance when the yields turn to be low. Theoretically, therefore, insurance (which reduces the exposure to risk) has an ambiguous effect on chemical application rates, including fertiliser, pesticide and agro-films, depending on the actual characteristics of the insurance programmes.

However, it is widely accepted that pesticides do not increase yield potential, they only affect yields when damaging agents are present (Lichtenberg and Zilberman, 1986). Thus, increased application of pesticides should result in a decreased probability of low yields, which suggests that a farmer who insures against low yields should decrease, not increase, pesticide use (Babcock and Hennessy, 1996).

Babcock and Hennessy (1996), in a Monte Carlo analysis of crop insurance for Iowa farmers, find that crop insurance schemes will likely lead to relatively minor reductions in the applications of nitrogen fertiliser if coverage levels are at or below 70% of mean yield. If the coverage level is 90%, a farmer with high-risk aversion will reduce the fertiliser application rate by 10%. Babcock and Hennessy's results imply that not only the risk attitudes, but also the level of coverage of crop insurance, would influence the average per acre chemical use dramatically.

A careful comparison of those previous studies suggests that the relationship between crop insurance participation and agrochemical input applications depends on farmers' decision-making behaviour, production conditions, the type of agrochemicals, and the terms actually set in crop insurance programmes. Therefore, whether any specific crop insurance programme may or may not have negative environment impact in specific locations is a problem requiring empirical investigation.

The objective of this study is to increase the understanding of farmers' input decisions on chemical fertilisers, pesticides, and agro-film[2] when provided with crop insurance, in order to assess if the proposed government subsidy to crop insurance is acceptable in terms of environment concerns. This study seeks answers to the following questions:

- Does farmers' crop insurance purchase affect their decision on chemical use by kind, and how significant are the effects?
- What factors influence farmers' crop insurance purchase and their decisions on agrochemical inputs?
- What are the environmental and political implications of the effects of crop insurance?

Analytical framework

The typical framework employed to evaluate the impact of crop insurance purchase decisions on cropping patterns and the agro-chemicals usage utilises the standard assumption that farmers maximise expected utility of agricultural production profit. They are assumed to select among the factors of production such as fertiliser, pesticides, etc., as well as crop insurance, subject to physical and technical constraints (Wu and Adams, 2001; Babcock and Hennessy, 1996; Horowitz and Lichtenberg, 1993; Quiggin, 1992). If the crop is insured, the farmer may adopt different farming practices that increase the expected total return after considering crop insurance premiums and indemnities. It is also likely that the opportunity to adjust farming practices will affect the decision on whether insure the crop (Smith and Goodwin, 1996).

---

2. Agro-film is a kind of plastic film, widely used by farmers in draught and cold areas, to cover the field before sowing and/or during the growing season. It may keep soil moisture and raise soil temperature.

A simple model of crop insurance and inputs use can be derived from profit maximisation function. Let the production technology be given by

$$y = f(x, \omega) \tag{1}$$

where vector x denotes n inputs, vector ω denotes k random factors, and $y = f(\bullet)$ is the expected output. Intuitively, ω may be associated with inputs that are beyond the control of the farmer (properties of land, weather, etc) as well as potentially damaging events (hails, flood, pests and diseases, etc.) that affect production.

Let p be a random price per unit of output, while input prices denoted by the n-dimensional vector w>0 are presumed non-random. Let $\pi_1 = py - wx$ denote the state-contingent farm profit in absence of insurance. Suppose the farmer has an insured yield level, y^*, and that farmer can purchase this level of insurance at a fixed premium δ, where y^* and δ are assumed to be determined exogenously. If the actual yield, y, is less than y^*, an indemnity in the amount $I[\delta,(y^* - y)]$ is paid. The net revenue under crop insurance should be $\pi_2 = py + I[\delta,(y^* - y)] - \delta - wx$. For a particular realisation of ω, call it the trigger state denoted by ω^*, the insurer's expected payout is determined by the farmers' choice of x.

If all farmers are risk averse, they choose x to maximise the expected utility of profits:

$$\text{EU} = \int_{y_0}^{y^0} U(\pi)\ dG(y) \tag{2}$$

where $[y_0, y^0]$ is the bounded support of $G(y)$, which is the cumulative distribution function of y.

Equation (2) highlights the connection between farmers' expected utility of profit and their input decisions including that on crop insurance purchase.

From the analysis above, the relationship between farmer's use of x and their insurance purchase can be expressed as

$$x = h(p, \omega, \omega^*, w, \delta) \tag{3}$$

Equation (3) indicates that x is influenced by the output price p, the factors affecting the output ω, especially the trigger state under insurance ω^*, the input price w, and the insurance premium δ.

However, the factor ω is also influenced by applications of other inputs:

$$\omega = v(y, x) \tag{4}$$

In conclusion, the farmer's choice of a particular input, x, and their insurance purchase could interact simultaneously. Inputs may have different properties with regard to maximum utility of profit. Some inputs, such as fertiliser, are able to increase the expected yield. However, their application may well increase yield variation as well. In addition to input and output prices, the equilibrium level of application of such inputs is determined by expected yield and its variance. Some other inputs, such as pesticides, are able to reduce yield losses in case of serious infection of pests or diseases. As such, they are likely to reduce yield variation from the low-yield side. Crop insurance is likely to have the same property with regard to profit maximisation as it provided insurers with an indemnity when the actual yield falls below a pre-fixed trigger level. Whether material inputs and crop insurance interact simultaneously depends on many factors, but their functions in profit maximisation are key to understanding farmers' behaviour.

If the application of one input will increase the possibility of the crop being exposed to natural disasters, a farmer who is going to increase that input is more likely to participate in crop insurance, in order to get some compensation when an extreme adverse outcome occurs. If increased application of one input leads to higher expected yield as well as greater yield, and the existing crop insurance programme promises to provide indemnity to cover at least a part of the additional losses due to increased variation in yield. Farmers may take advantage of this to increase application of the input from previous equilibrium level and purchase crop insurance at the same time. However, if the trigger level falls below the range of yields that are plausible even though yield variation has increased, the farmer may not have the incentive to increase the input. The same would hold if the expected indemnity a farmer might receive is smaller than the total costs (actual costs plus increased risk). In this case, the farmer does not have the incentive to increase the input in question. In this type of circumstance, farmers may participate in crop insurance against extreme cases, but do not increase other inputs after purchasing insurance policies.

All above hypothesis can be tested with empirical estimation framework outlined below.

Model and data

A simultaneous equation system may be derived from Equation 2 in order to estimate farmers' demand for agrochemicals and crop insurance along with other exogenous variables as follows:

$$y_{1t} = \alpha_1 y_{2t}^i + \beta_1 X_{1t} + \mu_{1t}, \quad (5)$$

$$y_{2t}^i = \alpha_2 y_{1t} + \beta_2 X_{2t} + \mu_{2t}^i \quad (6)$$

where i = 1, 2, 3;

y_{1t} represents crop insurance purchase, modelled as a dichotomous-choice taking the value 1 if the farmers are voluntarily insured, and 0 if the farmer does not purchase insurance;

y_{2t}^i represents the ith chemical input use;

X_t is a vector of exogenous variables relevant to insurance purchases and agrochemicals use;

μ_{1t}, μ_{2t}^i are unobserved disturbances that are assumed to be normally distributed with constant variances; and

α_1, α_2 *and* β_1, β_2 are parameter vectors to be estimated.

The distribution of y_{1t} is discrete such that

$$y_{1t} = \begin{cases} 1 & \text{if } y_1^* > 0 \text{ (farmer purchased crop insurance)} \\ 0 & \text{Otherwise (farmer did not purchase insurance)} \end{cases}$$

For the identification purpose, it is assumed that exogenous variables included in X_{1t} and X_{2t} are allocated to either X_{1t} or X_{2t} but not both. The simultaneous-equation system of crop insurance and the agrochemical uses is estimated by a two-stage procedure (Maddala, pp. 246)[3]. In the first stage, the reduced form equations for the insurance decision and agrochemical use decisions can be written as:

$$y_{1t} = Z_t^{'}\Pi_1 + v_{1t}, \tag{7}$$

$$y_{2t}^{i} = Z_t^{'}\Pi_2 + v_{2t} \tag{8}$$

where Z_t is an appropriately defined vector.

Equation (7) will be estimated by the MLE Probit method, and equation (8) will be estimated by the ordinary least squares method (OLS). In the second stage, equation (5) will be estimated by MLE Probit method after substituting $Z_t^{'}\hat{\Pi}_2$ for y_{2it}, and equation (6) will be estimated by the OLS method after substituting $Z_t^{'}\hat{\Pi}_1$ for y_{1t}.

This two-stage procedure gives consistent estimates of model coefficients (Maddala, pp. 244, 1983). But the estimates of variance of the coefficients may be inconsistent because predicted values of endogenous variables are used in the second stage of estimation. Maddala (pp. 243-47) points out that the appropriate covariance matrix for a structural model with more than two discrete or censored endogenous variables is difficult to derive. So, bootstrap methods (Efron, 1979, 1987) are used to derive consistent estimates of variances in this analysis. Under the bootstrapping approach, a large number of pseudo samples equal in size to the number of observations in the original data is obtained by repeatedly drawing from the original data with replacement. Thus, it is possible that one observation is drawn several times. For each pseudo sample, Nelson and Olson's two–step procedures (Nelson and Olson, 1978) applied to generate a distribution of the consistently estimated structural parameter. Variances of model parameters are then consistently estimated by using the distribution.

Once valid estimates of parameters of the structural model and their respective covariance matrices have been obtained, a Wu-Hausman (Hausman, 1978; Wu, 1973) specification test will be performed. This will test the null hypotheses that (a) crop insurance purchase decisions are exogenous in chemical input uses, and (b) chemical input uses are exogenous in crop insurance purchase decisions. These estimates can be compared to those obtained by standard OLS and Probit estimates that ignore simultaneity. Under the null hypothesis that standard OLS and MLE Probit estimates yield correct specification. Hausman (1978) shows that

$$q = (\beta_0 - \beta_1)[V(\beta_0) - V(\beta_1)]^{-1}(\beta_0 - \beta_1)' \tag{9}$$

3. A similar method to estimate equations (5) and (6) is specified as "endogenous treatment effect" models, which also obtain consistent estimators by instrument variables approaches (Wooldridge, 2002).

has a χ^2 distribution with a degree of freedom equal to the number of coefficients being evaluated, where β_0 and β_1 are standard OLS/Probit and instrumental variables parameters estimates, and $V(\beta_0)$ and $V(\beta_1)$ are their respective covariance matrices.

The China United Property Insurance Corporation (CUPIC) started its crop insurance programmes in 1986, first inside the Xinjiang Production and Construction Co. (XJPCC)[4] as its subsidiary and by dint of administration force. During the last two decades, crop insurance offered by the CUPIC has extended to non-XJPCC farmers in Xinjiang, and even to other provinces in recent years. The participation rate of crop insurance inside XJPCC increased from 6.65 % in 1986 to 83.56% in 2003[5].

The existing insurance is designed to cover a part of the material costs in order to re-start production after a bad year. It is not designed to compensate for yield losses: the insured farmers may receive insurance benefit up to 60% of the average material cost when the yield drops to a level below 50% of normal one.

Farmers cannot select coverage level and a guaranteed price because the premium is fixed by the insurance policy:

$$\delta = \delta(\tilde{E}(c), \tilde{E}(y), \omega) = RMB\ 20 \text{ per } mu \tag{10}$$

Where, $\tilde{E}(c)$, $\tilde{E}(y)$ represent the proportional actual production history (APH) of costs and yields respectively, and ω denotes the damaging events that affect production, δ is fixed at RMB 20 per *mu* (area unit equal to 1/15 ha) in recent two years and is the same among different areas. In other words, a payable loss occurs if the actual yield is below a trigger level set as a certain per cent of historical record of average yield ($\tilde{E}(y)$), and the maximum indemnity a farmer may receive is a certain per cent of historical record of average cost ($\tilde{E}(c)$).

Under current terms, the trigger yield, y*, is set as 50% of historical average yield and the maximum indemnity is set as 60% of material costs (fixed at RMB 250 in recent years). If a farmer's realised yield, y, falls below the trigger yield, y*, he will receive a payment proportional to the difference between his actual yield and the trigger level up to RMB 250:

$$I = I[\delta, (y^* - y)] \leq RMB\ 250 \text{ per mu} \tag{11}$$

As discussed before, the procedures for estimating the relationship between insurance purchase and input use are derived from the profit maximisation function. In our study, pesticides, agro-films and fertilisers may have different risk properties and are

4. The XJPCC was established in the 1950s as a semi-military unit, performing multi-roles such as organising agricultural and industrial production, commerce and service, infrastructure construction and maintenance, providing government administration, education, health care, and etc., as well as supporting formal armed force in guarding the border. Its internal administration system was vertical under the planning economy. Following the overall trend of reform, the XJPCC is gradually shifted to an economic organisation with decreasing role in government function, and the decision making power has been slowly decentralised as well.

5. The insurance coverage is relatively low outside the XJPCC, resulted in 45% coverage for cotton production and roughly 25-30% for all crops in whole Xinjiang Autonomous Region.

disaggregated in the model. Therefore, a simultaneous equation system is derived from Equation 2 with all variables described in Table 1[6].

To obtain consistent estimators, all estimation methods must impose restrictions or identification conditions on the exogenous variables in the simultaneous models. Therefore, it is of interest to impose as many *a priori* restrictions as are theoretically reasonable and determine the validity of these restrictions. On theoretic grounds, those restrictions which influence cotton yield variability (or expectation of variability) inter-temporally and in a large surrounding area may affect farmers' crop insurance demand but not chemical and other inputs decisions. In contrast, those restrictions which influence yield level (but not yield variability) on their own farms may have an impact on farmers' chemical input decisions.

The cotton acreage (CA) is employed to reflect any scale effect. A positive correlation between cotton acreage and participation in crop insurance might exist because larger cotton producers' losses suffered from extreme weather are likely to be greater. This variable is not used in chemical input equations because the chemical inputs are measured by average amount per unit of sown area rather than total application.

The variable used in this study to measure yield fluctuation is the coefficient of standard deviation (CV) of cotton yields at the county level for the 1980-2002 time period[7]. A similar specification was employed by Goodwin (1993), Smith and Goodwin (1996), and Goodwin (2001). It is believed that farmers tend to look at the average yield variation in a large surrounding area instead of that in their own small farms to help in their assessment and managing of risk. This variable reflects the general trend in a large area that each farmer would suffer from natural disasters, and may not have strong impact on farmers' chemical input decisions on their own land during the current cropping season.

6. Crop insurance is not a condition for obtaining credit in China. As such, the potential linkage with credit is not considered in modeling.

7. Under APH insurance program, coverage levels and premium rates are computed based on the insured expected yield. The expected yield is usually calculated as the average yield at the county level over the preceding 10 years to avoid adverse selection.

Table 1. Model specification and variable definitions

Model specification

Crop Insurance purchase= *f* (FERTILISER, PESTICIDES, AGRO-FILM, CV, FTF, DISR, CA, EDU, FEXPER, RISKATT)

Fertiliser usage= *f* (COTTINS, EDU, FEXPER, DENSITY, DISEASE, SHRLIVE, LC, AVGCY, RISKATT)

Pesticides usage= *f* (COTTINS, EDU, FEXPER, DENSITY, DISEASE, SHRLIVE, LC, AVGCY, RISKATT)

Agro-film usage= *f* (COTTINS, EDU, FEXPER, DENSITY, DISEASE, SHRLIVE, LC, AVGCY, RISKATT)

Variable description

Variable	Description
COTTINS	A zero-one discrete variable indicating whether farmers purchased crop insurance in 2003 (1= purchased insurance, 0= no insurance purchased)
FERTILISER	Expenditures of aggregate fertiliser per *mu* including base fertiliser and late fertiliser in cotton production in 2003 (RMB per *mu*)
PESTICIDES	Expenditures of pesticides per *mu* in cotton production in 2003 (RMB per *mu*)
AGRO-FILM	Expenditures of agro-film per *mu* in cotton production in 2003 (RMB per *mu*)
DENSITY	Cotton planting density per *mu* on cotton acreages in 2003 (1 000 individual plants per *mu*)
DISESAE	The degree of losses caused by pests and diseases in the recent 4 years (1= average loss above 80% of normal yield, 5= average loss below 20% of normal yield)
CV	Coefficient of standard deviation of average cotton yield at county level from 1980-2002 (%)
AVGCY	Average cotton yields during recent two years (kg per *mu*)
CA	Acreage sown to cotton by one farm household (*mu*)
LC	Land capability (1= the best, 3= the worst)
SHRLIVE	Share of off-farm income and livestock sales in total net income (%)
DISR	Whether farmer received government disaster relief in the recent 4 years (1= yes, 0= no)
RISKATT	The operator's degree of risk preference for rural medicare which 1 denotes risk loving and 5 denotes risk hating
FEXPER	Number of years that farmer has occupied to agriculture in a village (year)
FTF	Whether farmer is in full-time farming (1= yes, 0= no)
EDU	Number of years that farmer has received the education (year)

The effect of government disaster relief programme (DISR) on purchasing crop insurance is ambiguous. It might be taken as a substitution to crop insurance on one hand. However, on the other hand, farmers who have received such assistance may think of being more exposed to extreme natural disasters, and crop insurance might be viewed as a complementary measure to government disaster relief programme in this case.

Other factors that would affect insurance purchase decisions include whether or not a farmer works on-farm full time. Full-time farmers (FTF) may be more willing to insure their crop because they do not have alternative revenue sources. However, as pointed out by Goodwin (1993), a negative effect may also be possible because full-time farmers may have a higher degree of specialised expertise in production practices than part-time farmers and thus may have a lower demand for crop insurance.

The share of total net income derived from livestock sales and off-farm income (SHRLIVE) is used to reflect farmers' budget constraints on chemical inputs. This variable also reflects farm diversification and thus may affect the demand for crop insurance. However, it is reasonable for using farmer's employment characteristics (FTF) rather than SHRLIVE to reflect the effect of farm diversification on crop insurance demand for satisfying identifiability criterion for simultaneous equations.

Cotton plant density (DENSITY) may influence agro-chemical application. It is expected that farmers who plant cotton in higher density may apply more pesticides, more agro-film and less fertiliser, because higher density will lead to higher insect populations and higher potential losses due to insect infection.

The losses of cotton caused by insects and diseases in the most recent four years (DISEASE) reflect farmer's individual yield losses under local environmental conditions in specific region. This has no obvious relationship between the yield variation in a large surrounding area. Therefore, the variable DISEASE may only influence farmers' chemical input decisions

Land quality (LC) and the average cotton yield in previous two years (AVGCY) also directly influence farmers' chemical input decisions. Land quality may have stable impact on farmers' cotton yield in the long term, and it is not obviously correlated with yield fluctuation. For the same reason, the average cotton yield in previous two years is the outcome of chemical inputs in the past. This reflects the production relationship between output and chemical inputs; this lagged yield may affect current chemical inputs decisions but not necessarily crop insurance decision.

Farmers' attitude towards risk certainly influences their decision on crop insurance and chemical inputs decisions. However, it is hard to measure their risk attitude directly. As a proxy, their stated preference in valuing health insurance is recorded in a 5-level scale and used in our estimation.

The demographic characteristics of farm operators including farming experience (FEXPER) and education level (EDU), which may affect adoption of production technology and the demand for insurance, are also included in the simultaneous model. All these variables may have positive or negative effect on crop insurance purchase and each chemical use.

To statistically examine the validity of our two sets of instrumental variables (one for crop insurance purchase structural equation and the other for chemical inputs structural equations), we conduct: a likelihood ratio test (LR) (Bollen, Guilkey and Mroz, 1995; Wooldridge, 2002) for equation (5)10; a Hausman over-identification restriction test (Wooldridge, 2002) for Eq. (6)8.

The primary data used in this study are collected through face-to-face interviews with farmer households in the Manasi Watershed, a major cotton producing area in Xinjiang. The sample is selected in 3 steps: 1) Five counties are chosen based on cotton sown acreages; 2) Four villages are selected from each county based on equal-interval in cotton yield; and 3) About 20 farmer households are randomly chosen in each village.

Among the 450 households interviewed, 340 respondents provided sufficiently complete information for inclusion in the analysis. Among the 340 farmers, 113 of them (33.23%) voluntarily purchased crop insurance while the remaining 227 (66.77%) did not purchase crop insurance in 2003. They all utilised fertiliser, pesticides and agro-film in cotton production.

---

8. The Hausman test is a Lagrange multiplier test (Hausman, 1983). The chi-square distributed test statistic with k–1 degrees of freedom, where k is the number of IVs, is $N \times R^2$, where N is the number of observations, and R^2 is the measure of goodness of fit of the regression of the residuals from the second stage equations on the variables, which are exogenous to the system. Under the null hypothesis that exclusion restrictions are valid, a statistically insignificant test statistic indicates that the instruments can be safely excluded from the chemicals input equations.

The values of the mean and standard deviation of each variable are presented in Table 2. In our sample, one-third of farmers participated in crop insurance programme, and the average expenditures, of all farmers, on chemical fertilisers, pesticides and agro-film applications are 94.13, 26.25 and RMB 31.51 per *mu*, respectively.

Table 2. Summary statistics

Variable	Mean	Std. Dev.
COTTINS	0.3324	0.4718
FERTILISER	94.0903	37.1337
PESTICIDES	26.2502	18.8074
AGROFILM	31.5155	8.1560
EDU	7.7412	2.6189
FEXPER	18.6206	11.3228
FTF	0.6265	0.4845
DISR	0.5382	0.4993
CV	16.5813	7.2850
CA	68.6176	75.9404
RISKATT	3.3559	1.3236
DENSITY	15.7342	6.4713
DISEASE	2.5147	1.6428
LC	2.0353	0.7634
AVGCY	206.4651	53.2625
SHRLIVE	8.4543	17.1037

Source: Field survey, July 2004.

Empirical results and discussions

The test of over-identifying restrictions produces likelihood ratio statistic of 6.84 for the crop insurance structural equation. Lagrange multiplier statistics of 5.54, 5.88 and 2.58 were yielded for agro-film input, fertiliser input and pesticides input application structural equations respectively. These do not exceed the Chi-square distributed critical value of 7.61 with four degrees of freedom for crop insurance equation and 6.25 with three degrees of freedom for chemical inputs equations at 10% level of significance. Therefore, we do not reject the null hypothesis that the instruments are uncorrelated with the error term. Four statistically insignificant test statistics indicate that the two sets of instruments can be safely excluded from the crop insurance and chemical inputs equations respectively.

As stated earlier, a Wu-Hausman specification test is performed to test the null hypotheses that (a) chemical input uses are exogenous in crop insurance purchase decisions; and (b) crop insurance purchase decisions are exogenous in chemical input uses before further application of econometric models. The estimated Wu-Hausman χ^2 Statistics are reported in Table 3.

Table 3. Results of Wu-Hausman specification test

Null hypothesis	Wu-Hausman χ^2 test statistics	P-value for the statistics
Exogeneity of chemical inputs in crop insurance discrete choice model	33.37	0.0000***
Exogeneity of fertiliser input in crop insurance discrete choice model	32.48	0.0000***
Exogeneity of agrofilm input in crop insurance discrete choice model	4.09	0.043**
Exogeneity of pesticides input in crop insurance discrete choice model	2.97	0.085*
Exogeneity of crop insurance discrete choice in fertiliser input model	0.09	0.76
Exogeneity of crop insurance discrete choice in agro-plastic film input model	0.59	0.441
Exogeneity of crop insurance discrete choice in pesticides input model	3.11	0.078*

Note: *, ** and *** indicate statistical significance at 10%, 5% and 1% levels, respectively.
Source: Estimated by this study.

The P-values of the estimated χ^2 statistics indicate that the exogeneity hypothesis is rejected in the crop insurance purchase model at 1% level of significance for FERTILISER, AGRO-FILM and PESTICIDES as a whole, and at 1%, 5% and 10%, respectively, for the three inputs individually. The same hypothesis is also rejected for the variable COTTINS in the pesticides input equation at 10% level of significance, but not so in the fertiliser and agro-plastic film input equations.

The results of the Wu-Hausman specification test suggest that farmers' decisions on crop insurance participation are endogenously made with that on agrochemical inputs; however, their decisions on agrochemical inputs, except pesticides, may not be endogenously made with that on crop insurance participation.

Nevertheless, it is still worthwhile to further test the hypothesis with a simultaneous equation system. The bootstrapped parameter estimates and implied marginal probability effects for the simultaneous equation Probit model of the discrete cotton insurance purchase decision are presented in Table 4. The standard errors of the coefficients are estimated from bootstrap method with 1 000 replications. The last column of Table 4 shows the changes in the probability of purchasing the crop insurance given one unit change in the explanatory variables and are computed at the means of all explanatory variables. The whole model highly fits the sample observations, as the correctly predicted percentage of purchasing crop insurance is close to 99%.

Table 4. Estimates of the probit model of cotton insurance decisions

(bootstrap method with 1 000 replications)

Explanatory variable	Coefficient[a]	Standard error	Marginal probability[b] (dy/dx)
CONSTANT	-119.3245**	33.7683	n.a
AGROFILM	1.2775**	0.3741	0.2637
PESTICIDES	-0.4682**	0.1609	-0.0966
FERTILISER	0.2165**	0.0797	0.0581
FEXPER	-0.1486**	0.0525	-0.0307
RISKATT	-0.1004	0.2321	-0.0207
DISR	0.6319	0.9272	0.1346
FTF	3.1369*	1.3954	0.7829
EDU	-0.6652*	0.2886	-0.1373
CV	4.1384**	1.2736	0.8543
CA	0.0488**	0.0163	0.0100
LR χ^2 (9)= 401.78	Number of observation= 340		
Prob > χ^2 = 0.0000	Pseudo R^2= 0.9293		
Per cent correctly predicted=98.82			

a. * and ** indicate statistical significance at the 5% and 1% level respectively.

b. dy/dx is for discrete change of dummy variable from 0 to 1.

Source: Estimated by this study.

The coefficients on variables FEXPER and EDU are negative and statistically significant at 1% and 5% level, respectively. It suggests that farmers who have more farming experience and with higher level of education tend to purchase crop insurance less frequently. One explanation for this is that those farmers have, or believe they have, better risk management skills, so need less protection from crop insurance programmes. The coefficients on cotton acreage (CA) and whether a farmer is full-time working in agriculture (FTF) are positive and statistically significant at 1% and 5% levels, respectively, as expected. Full-time farmers are more likely to participate in crop insurance because they don't have alternative sources of income to disperse risks, and the farmers with larger cotton production are more inclined to purchase crop insurance against severe natural disasters. As expected, the coefficient on standard deviation of average yield at county level (CV) is positive and highly significant, suggesting that farmers facing higher yield variation are also more likely to purchase crop insurance.

However, the regression coefficients for the variables used to represent a farmer' attitude towards risk, RISKATT, and receipt of disaster assistance, DISR, are not statistically significant. It is likely that a farmer's stated preference towards health insurance is not a good proxy for that towards crop insurance, or that the actual behaviour is somewhat different from the stated preference. It is also likely that farmers do not expect natural disaster to repeat or to apply more frequently on their own small farms.

The most interesting result concerns the effects of agricultural chemical inputs on crop insurance purchase decisions. As shown in Table 4, both AGRO-FILM and FERTILISER have positive coefficients significant at 1% level, while that of PESTICIDES is negative and significant at 1% level. This suggests that the more fertiliser and agro-film a farmer applies, for any reason, the more likely he or she will purchase crop insurance. On the contrary, the more pesticides a farmer applies, the less likely he or she is to participate in crop insurance. One plausible explanation is that the increasing application of chemical fertilisers and agro-film leads to stronger growth of individual cotton plants, resulted in higher possibility of being exposed to natural disaster and/or higher probability of lower (below-expected-average) yields. Therefore, farmers tend to purchase crop insurance against increased risk associated with more intensive application of chemical fertiliser and agro-films. On the contrary, as pesticides could prevent low yield to occur when damaging agents are present, farmers may participate less in crop insurance if they apply more pesticides.

The results of this research are consistent with the conventional wisdom that fertiliser is a risk-increasing input while pesticides are risk-reducing. It is also demonstrated that agro-film may have similar properties to fertiliser in this regard. The empirical estimates show that, *ceteris paribus*, each additional RMB spent on fertiliser and agro-film increases the probability of insurance purchase by 5.81% and 26.37% respectively. However, each additional RMB spent on pesticides lowers the probability of insurance purchase by 9.66%.

Parameter estimates for the structural equations of agricultural chemical inputs are presented in Table 5. Standard errors of these parameters are estimated from Bootstrap method with 1 000 replications.

Table 5. OLS and bootstrap estimates of chemical inputs

Explanatory variable	Explained variable					
	Pesticides		Fertiliser		Agrofilm	
	Coefficient[a]	St. Error	Coefficient	St. Error	Coefficient	St. Error
CONSTANT	18.3145**	7.8097	61.1593***	14.0200	24.2183***	2.5357
COTTINS	-5.0825**	2.6764	2.6714	6.3768	5.8695***	1.0898
EDU	-0.0803	0.3321	2.5650***	0.7777	-0.0157	0.1746
LC	2.1939*	1.4367	4.2422*	2.7169	0.3700	0.5144
SHRLIVE	0.0278	0.0523	0.0343	0.1308	0.0462	2.2116
DENSITY	0.2978**	0.1728	-0.1634	0.2527	0.0097	0.0600
DISEASE	-1.7042***	0.5351	1.9309*	1.2648	0.3794*	0.2609
FEXPER	-0.1767**	0.0968	0.2069	0.1948	0.0027	0.0437
AVGCY	0.0469**	0.0211	-0.0026	0.0486	0.0177**	0.0083
RISKATT	-0.5926	0.8495	-0.6676	1.6139	-0.0269	0.2857
Adj. R^2	0.0325		0.0188		0.1704	

[a] *, ** and *** indicate statistical significant at 10%, 5% and 1% levels, respectively; sample size= 340.
Source: Estimated by this study.

The coefficients on cotton density (DENSITY), as we expected, are positive in the PESTICIDES and AGRO-FILM equations, and negative in the FERTILISER equation, but only statistically significant in the PESTICIDES equation. This result suggests that the higher the cotton density a farmer chooses, the more pesticides he will apply as the possibility of insects and diseases occurrences is increasing with plant density.

The parameter estimates for DISEASE are negative and significant at 1% level in the PESTICIDES equation, and positive and significant at 10% level in both the FERTILISER and AGRO-FILM equations. They indicate that the less losses caused by the insects and diseases in recent four years, the less pesticides, and the more fertiliser and agro-film a farmer may apply. This result is consistent with the intuition that intensive application of fertiliser and agro-film has a positive impact on the infestation of pests and diseases.

The performance of soil characteristics, LC, has the expected sign in all three equations and is significant at 10% level in both PESTICIDES and FERTILISER equations, confirming that farmers with lower land productivity tend to apply more pesticides and fertiliser. The positive effect of the average cotton yields in recent two years (AVGCY) on application of pesticides and agro-film, at 5% level of significance, suggests that stronger growth of cotton plants requires better protection against pests and diseases, as well as drought. The coefficient on the share of off-farm income and net income derived from livestock (SHRLIVE), as one might expect, is positive but not significant for all three equations.

Again, the most interesting result of the chemical input equations is the coefficient on COTTINS, or crop insurance purchase. As shown in Table 5, the coefficient of COTTINS is negative and significant at 5% level for PESTICIDES equation, and positive for both FERTILISER and AGRO-FILM equations, only significant at 1% level for AGRO-FILM equation but insignificant for FERTILISER equation. The results show that the greater the probability that a farmer purchases crop insurance, the less he or she will spend on pesticides, and the more he might spend on both agro-film and fertiliser. This confirms the Horowitz and Lichtenberg's presumption that fertiliser and agro-film are risk-increasing inputs while pesticides are risk-reducing inputs.

In reality, rainfall is severely inadequate, land fertility is fairly low and the occurrence of aphid pests is relatively frequent in Xinjiang, located in northwest of China. This means that increases in fertiliser and agro-film may lead to subsequent and substantial increases in insect populations and an increase in the potential for large losses. There have been recent reports that pests are becoming an increasingly serious problem in Xinjiang, as opposed to its reputation of low pest infestations in the past. Under such conditions, increased application of fertiliser and agro-film chemical inputs raises expected yields but also increases the danger of being exposed to greater disasters. Under such circumstances, additional fertiliser and agro-film chemical inputs may actually increase the probability of low yields while increased pesticides should result in a decreased probability of low yields. This means that a farmer who insures against yield shortfalls is likely to increase fertiliser and agro-film use, and decrease pesticide use.

The empirical results indicate that, on average, farmers with insurance apply RMB 5.87 more of agro-film and RMB 2.67 more of fertiliser respectively per *mu* (about 20.2% and 2.9% higher in percentage terms) than the quantities applied by farmers who do not participate in crop insurance. At the same time, farmers with crop insurance tend to apply RMB 5.08 less of pesticides per *mu*, or 18.99% below the level observed on farms without insurance.

Demographic characteristics of farmers are also included in the analysis. The variable EDU has a strong positive impact, suggesting that farmers with more education spend more on fertiliser. This is consistent with the expectation that better educated operators are more adept at acquiring and processing information from various sources, and then adopting and implementing recommendations and solutions relevant to their specific problems (Mishra, Nimon and El-Osta, 2005). The variable FEXPER has significant and negative effect on the applications of pesticides, suggesting that pesticides might have been over-applied and that more experienced farmers may wisely reduce their application level.

Conclusions and recommendations

The methodology adopted in this study is similar to that of Smith and Goodwin (1996), but the empirical results are different. The results of this study strongly indicate that crop insurance purchase decision depends on farmers' production behaviour: those who apply more chemical fertilisers and agro-film are more likely to participate in crop insurance programmes while those who apply more pesticides will do the opposite. The results also indicate that farmers' agrochemical input application decisions are influenced by that on crop insurance participation differently: pesticides are likely to be applied less if cotton production is insured, while agro-film and chemical fertilisers are likely to be applied more, although the latter case is not statistically significant. In contrast, Smith and Goodwin find that farmers with crop insurance will use chemical input less intensively, and they claim that this must be because the expected return to crop insurance declines with input use.

The difference between our results and that found by Smith and Goodwin might be explained by several factors. First, our estimation is disaggregated for each of the agrochemical inputs and this may have resulted in different findings compared with a study based on aggregate data. Since the output impacts of pesticides, agro-film and fertilisers are not identical, they may have different risk properties. This is born out by our empirical results which suggest that the effect of crop insurance on each of the inputs is not identical.

Second, the current crop insurance programme in China is designed to compensate for only a part of the material costs incurred at the presence of severe yield losses. The indemnity farmers expect to receive is much lower than what might be received from a typical crop insurance programme in the US, which provides compensation equal to a relatively larger portion of yield losses. In fact, the trigger level of yield in China is two standard deviations below the average one, and the maximum indemnity for null harvest is only 60% of the average material cost. Hence, the incentive for increasing fertiliser and agro-film inputs in order to raise expected yield and taking advantage of obtaining indemnity in case of resulted higher yield variation is very low under current conditions. Therefore, moral hazard is not likely to be a big issue in China today under current terms stipulated in crop insurance policies. This means that farmers insure their crop against natural disasters and catastrophes but not the additional – but much less substantial - yield variation resulted from increased agrochemical applications.

This research has confirmed that fertiliser and agro-film are risk-increasing inputs while pesticides are risk-reducing inputs. So, they influence farmers' decision on crop insurance participation in different ways. This research has also found that, under current terms of low premium and low indemnity, crop insurance may not bring serious impact on the environment in Xinjiang, with the exception of the accumulation of small pieces of broken agro-film. It should be stated, however, that the results may be different in other agronomic and other ecological zones. The increase in fertiliser application is relatively small compared with current average level and it is not statistically significant, but the reduction of pesticides application is significant.

If China is to encourage farmers' participation in crop insurance with subsidy to premium in Xinjiang, bringing the participation rate up to 80% from 44.84% at present, the total pesticides application is likely to be reduced by about 2%, while the total application of agro-film is likely to be increased by 8.38%. At the same time, if the insignificance is ignored, the total application of fertilisers is likely to be increased by 2.3% in the whole region assuming no changes in cotton sown areas.

If, however, the government chooses to subsidise crop insurance such that indemnity increase while the premium paid by farmers remains the same, changes in the pattern of agrochemical inputs may be induced. Our estimates indicate that, an increase of indemnity up to 80% of the material costs from current 60% level may induce farmers to apply an additional RMB 1.17 of agrofilm, and RMB 0.53 of fertiliser per *mu*, and reduce expenditures on pesticides by RMB 1.02 per *mu*. This implies a total increase of agro-film application by 3.7% and a total increase of fertiliser application of less than 0.56%, as well as a total decrease of pesticides by 3.89% in Xinjiang as a whole.

It should be noted that farmers in Xinjiang grow one crop in a year while their counterparts in most other regions grow two, or even more, crops. As agrochemical application rates are based on crop bases, the chemical residuals in the soil are likely to be much lower in Xinjiang. It also should be noted that high application rate of agro-film in Xinjiang is associated with the extreme climate there. As it is used to protect soil moisture from evaporation and to raise soil temperature in the early growing season, its importance and application rates will be reduced in other regions, especially in central and south China.

Taking all of the above into consideration, if the current terms are not to be changed dramatically, crop insurance subsidies appear to be an acceptable policy alternative for supporting agriculture and farmers' income under WTO rules. Farmers' welfare will be increased with reduced premiums and/or increased indemnities, while output would concurrently be stimulated by carefully designed, government supported, crop insurance programmes.

In order to remedy potential environment threats with accumulation of agro-film in the soil, such crop insurance policies and subsidies might be better implemented in agro-ecological areas and applied to crops where agro-film is not a necessary input. At the same time, development and adoption of easy-pickup agro-film, bio-degradable film or special machines to clear the soil at low cost should also be considered.

Further studies should be conducted to investigate explicitly the relationship between farmers' decision on crop insurance and the terms stipulated in the insurance policy, such as trigger and indemnity levels, as they clearly influence whether it is desirable to increase expected yield and variations in yield at the same time. Also, more detailed studies are needed to explicitly explore if different fertilisers and pesticides may have different risk properties. Welfare analysis should be extended further, to determine whether the expansion and enrichment of government supported crop insurance will bring meaningful benefits to farmers and Chinese society in general and to explore whether and what type of approach might be practical for China.

REFERENCES

Babcock, B. and Hennessy, "Input Demand under Yield and Revenue Insurance", *Amer. J. Agri. Econ.* 78 (1996): 416-427.

Bollen, K.A., D.K.Guilkey and T.A. Mroz, "Binary Outcomes and Endogenous Explanatory Variables: Tests and Solutions with an Application to the Demand for Contraceptive Use in Tunisia", *Demography*, Vol. 32 (February 1995):111-131.

Efron, B., "Bootstrap Methods: Another Look at the Jacknife", *Ann. of Statist.*7 (January 1979): 1-26.

Efron, B., "Better Bootstrap Confidence Intervals", *J. Amer. Statist. Assoc.*82 (March 1987): 171-185.

Hausman, J.A., "Specification Tests in Econometrics", *Econometrica* 46 (November 1978):1251-1272.

Horowitz J. and E. Lichtenberg. "Risk-Reducing and Risk-Increasing Effects of Pesticides", *J. of Agri. Econ.* 45 (1) (1994): 82-89.

Horowitz, J. and E. Lichtenberg, "Insurance, Moral Hazard, and Chemical Use in Agriculture", *Amer. J. Agr. Econ.* 75 (November 1993): 926-35.

Lichtenberg, E. and D. Zilberman, "The Econometrics of Damage Control: Why Specification Matters*", Amer. J. Agr. Econ.* 68 (February 1986): 261-273.

Loehman, E. and C. Nelson, "Optimal Risk Management, Risk Aversion, and Production Function Properties", *J. of Agr and Res. Econ.* 2, 17 (1992): 219-231.

Maddala, G.S., *Limited Dependent and Qualitive Variables in Econometrics*, Cambridge University Press, 1983.

Mishra, A.K., R.W. Nimon and H.S. El-Osta, "Is Moral Hazard Good for The Environment? Revenue Insurance and Chemical Input Use", *Journal of Environmental Management* 74 (2005): 11-20.

Nelson, C. and L. Olson, "Specification and Estimation of a Simultaneous Equation Model with Limited Dependent Variables", *International Economic Review* 19 (1978): 695-709.

Pope, R. and R. Kramer, "Production Uncertainty and Factor Demands for the Competitive Firm", *South Econ. J.* 46 (1979): 489-501.

Quiggin, J., "Some Observations on Insurance, Bankruptcy and Input Demand", *J. of Econ. Behavior and Organization* 18 (1992): 101-110.

Quiggin, J., G. Karagiannis and J. Stanton, "Crop Insurance and Crop Production: An Empirical Study of Moral Hazard and Adverse Selection", *Austr. J. Agr. Econ* 37, 2 (August 1993): 95-113.

Ramaswami, B. "Supply Response to Agricultural Insurance: Risk Reduction and Moral Hazard Effects", *Amer. J. Agr. Econ.* 75 (November 1993): 914-925.

Smith, H.V. and A.E. Baquet, "The Demand for Multiple Peril Crop Insurance: Evidence from Montana Wheat Farms", *Amer. J. Agr. Econ.* 78 (February 1996): 189-201.

Smith Vincent and Barry Goodwin, "Crop Insurance, Moral Hazard, and Agricultural Chemical Use", *Amer. J. Agr. Econ.* 78 (1996): 428-438.

Wooldridge, Jeffery, *Econometric Analysis of Cross-Section and Panel Data., MIT* Press, 2002.

Wu, D., "Alternative Tests of the Independence Between Stochastic Regressors and Disturbances", *Econometrica* 41 (July 1973): 733-750.

Wu, J., "Crop Insurance, Acreage Decisions, and Non-point Source Pollution", *Amer. J. Agr. Econ.* 81(1999): 305-320.

Wu, J. and R. M. Adams, "Production Risks, Acreage Decisions and Implications for Revenue Insurance Programs", *Can. J. Agr. Econ.*49 (2001): 19-35.

Chapter 16.

NON-POINT SOURCE AGRICULTURAL POLLUTION: ISSUES AND IMPLICATIONS

Huang Jikun, Hu Ruifa, Cao Jianmin and Scott Rozelle

Abstract

Fertiliser application rates have doubled since 1980 and pesticide use has increased almost three-fold over the same period. While chemicals have played an important role in increasing agricultural production, they can also increase production costs, increase the risk of certain food quality and food safety problems, and contribute to environmental pollution. Chemical fertilisers are now over-applied at rates between 20 and 50%. For pesticides, the over application rate is even higher, falling between 40 and 55%. There is some circumstantial evidence that tenure and migration issues play a role in this pattern of excess application of commercial inputs as migrant workers apply inputs "all at once" because the time they have during their home visits is limited. But there is even more evidence that the government, scientific community, plant breeders, extension agents, and input suppliers have convinced farmers that "if a little bit is good, a lot is better". While these findings are tentative, they suggest that incentives within and among the existing research community, extension education system, and agricultural input suppliers need examination.

Introduction

China's use of commercial fertilisers and plant protection products has increased remarkably over the past few decades. For example, nitrogen fertiliser use in China has increased more than 40 fold since 1960 (Figure 1). The intensity of China's fertiliser use is three times higher than the world average (Figure 2) and is now the fourth highest in the world. Pesticide use has trebled for rice and wheat and nearly quadrupled for maize since 1980. The use of pesticides for commercial crops like cotton, tomatoes, and apples is even higher than for staple food grains (Figure 3).

Figure 1. N-Fertiliser consumption in the world and China, 1960-1999

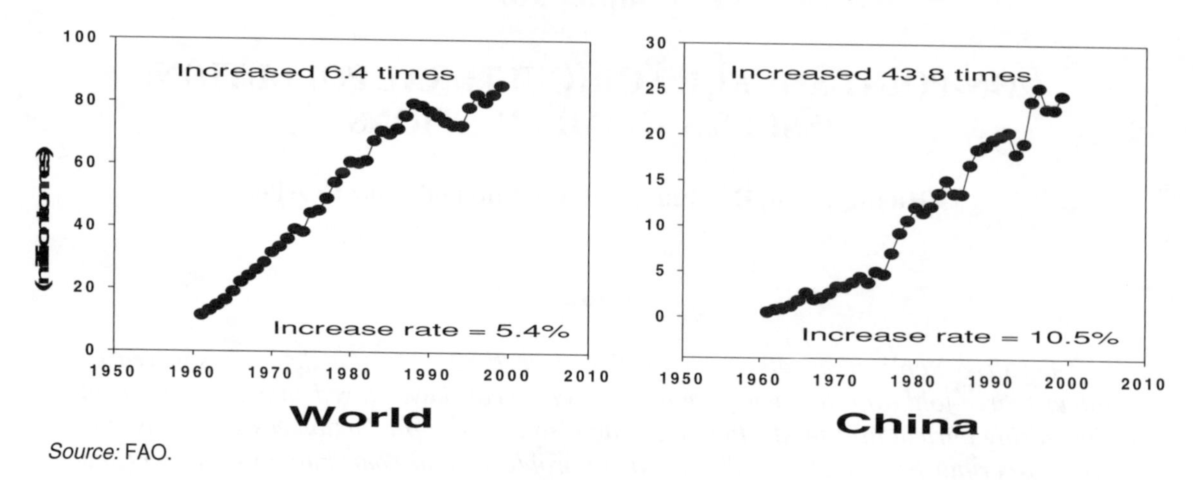

Source: FAO.

Figure 2. The intensity of fertiliser use in the world and China, 1965-2004

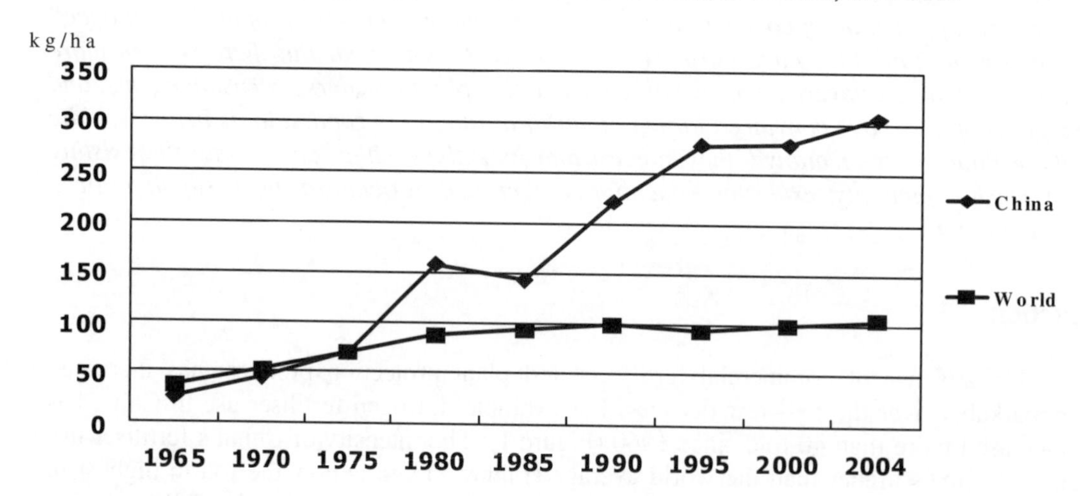

Source: FAO.

Figure 3. Pesticide use in crop production in China, 1980-2001

(RMB per hectare in 1995 prices)

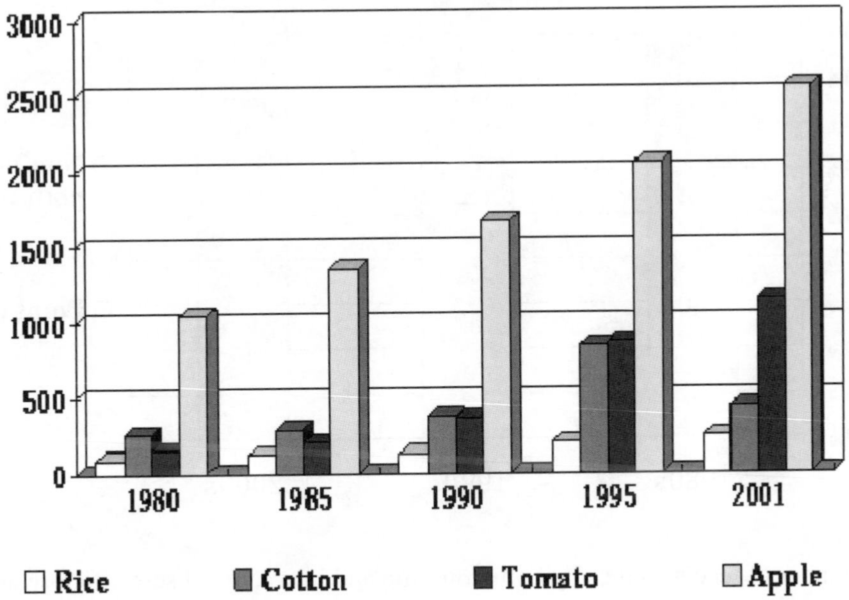

The relationship between increased input use, farmer income and the environment

The increased use of such commercial inputs has contributed to significant gains in agricultural production. However, there is growing evidence of the environmental costs of these high levels of fertiliser and pesticide applications as they have contributed to pollution in the form of higher nutrient loading in river basins, watercourses, lakes and streams and chemical residues in the soil. For instance, recent surveys suggest that non-point source agricultural pollution is now the main source of pollution in China's waterways (Figure 4). In addition to the environmental effects of high input applications, these inputs also raise the cost of production, lower farmers' net income, and can exacerbate food safety and quality problems - particularly when chemical residue levels exceed acceptable human food safety levels.

Figure 4. Agricultural NPP has become the main source of pollution

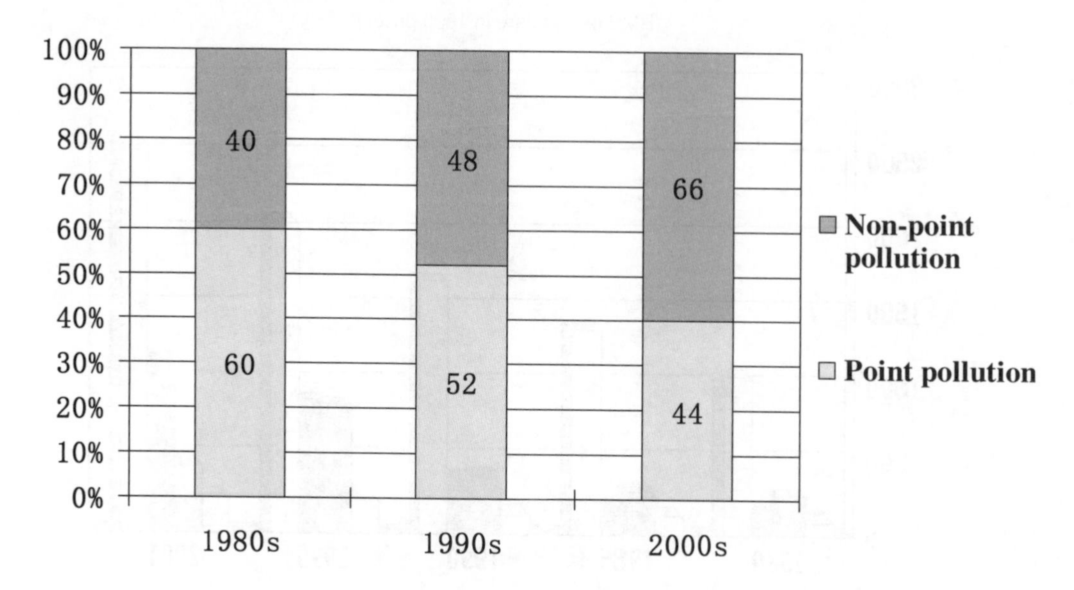

A number of recent studies by both domestic and international scientific organisations indicate that both fertilisers and pesticides are over applied in China. Recent studies by the Chinese Academy of Agricultural Sciences, the Chinese Academy of Sciences, the Centre for Chinese Agricultural Policy, and the International Rice Research Institute suggest that fertilisers are over applied by between 20 and 50%. Similar work focusing on pesticides suggests that they are over applied by 40 to 50%.

Ongoing agronomic research and survey work suggests farmers are misinformed

Using a rapid rural appraisal (survey) approach (RRA), work is now underway at the Centre for Chinese Agricultural Policy that supports earlier findings: commercial fertiliser and pesticide applications can be reduced by 25 to 35% without reducing yield in any significant way but without incurring any input costs. Not only do such practices contribute to serious environmental problems, they also directly reduce farmers' incomes because they add expense without adding to yield or revenue. They may even reduce revenue due to lower prices and reduced demand arising from heightened consumer concerns about chemical residue levels in the food that they eat.

As the over application of fertilisers and pesticides effectively amounts to literally throwing money down natures drain, we have to ask whether farmers are aware of what they are doing. It may be that farmers over use such inputs in order to manage risks. There is some circumstantial evidence that tenure and migration issues may play a role in the pattern of excess application of commercial inputs. When migrant workers return home, they often apply inputs "all at once" rather than in optimal amounts at critical times in the growing cycle because the time they have during their home visit is limited. The chapters by Zhong Funing and his colleagues and Jeff Au elsewhere in this volume echo our observations regarding this phenomenon, although they do not explore it in detail. Research undertaken by China's Ministry of Water Resources has also made similar findings.

But the main finding of our preliminary research and survey results is that most farmers have absolutely no clue that 25% or more of these inputs were wasted and that their incomes and the environment suffered unnecessarily as a result. Most farmers were absolutely stunned to find out how many resources were wasted. How did such a situation arise? Although these survey results are preliminary in nature, there is sufficient and substantial evidence to suggest that the government, scientific community, plant breeders, extension agents, and input suppliers have all convinced farmers that "if a little bit is good, a lot is better". This is the case even in the face of strong evidence to the contrary.

Closing remarks

The results of our research to date indicate that incentives and motives within and among the existing government, scientific community, plant breeders, extension agents, and agricultural input suppliers need to be re-examined. Is the information farmers are getting credible? Do their sources of information have vested interests in ensuring higher input use and in reaching sales or output targets that are not necessarily to the benefit of farmers or the environment?

In the coming few years, efforts will be made to further explore this issue and to educate a broader group of farmers about the merits of reducing input use, both in terms of their own income and in terms of environmental damage.

ANNEX.

AGENDA AND LIST OF PARTICIPANTS

Agenda

WORKSHOP ON ENVIRONMENT, RESOURCES AND AGRICULTURAL POLICIES IN CHINA

19-21 June 2006, Asia Hotel, Beijing, China

Monday, 19 June 2006

Opening of the meeting

Chair: Mr. CHEN Xiaohua, Director General, the Ministry of Agriculture (MoA), China

Keynote speech	Mr. FAN Xiaojian, Vice Minister of MoA, China
Opening address: Agri-environmental policy issues in OECD countries	Mr. Herwig SCHLÖGL, Deputy Secretary-General of OECD

Session 1. Agri-environmental situation and policies in China: practice and outcomes

This session will stocktake the key issues in agricultural resource management and policies in China with a focus on identifying pressures on the current and future use of agricultural resources arising from the competition for these resources by various users, both agricultural and non-agricultural, the long-term restructuring of agricultural production and climate change.

Chair: Mr. ZHANG Hongyu, Director General, the Ministry of Agriculture (MoA), China

Presentations

The new socialist countryside and its implications for China's agriculture and natural resources	Mr. TANG Renjian, Leading Group for Finance and Economy, CCCP, China
Selected aspects of water management in China: state, policy responses and future trends	Mr. Krzysztof MICHALAK, OECD
Effects of integrated ecosystem management on land degradation control and poverty reduction	Mr. HAN Jun, DRC of the State Council
Water resources and agricultural production in China: the present situation	Mr. MA Xiaohe, NDRC

Discussants: Mr. Achim FOCK, World Bank; Mr. WEN Tiejun, Renmin University.

General discussion

Session 2. Experiences in agricultural resource management and environmental protection in OECD countries

This session will provide examples of agri-environmental policies in OECD countries as well as an overview of decision support tools and policy tools which might be of relevance for China. Examples will show how market-based mechanisms can contribute to improvements in water use for agriculture and ecosystems and in the reduction in agricultural sources of water pollution.

Chair: Mr. Herwig SCHLÖGL, Deputy Secretary-General of OECD

Presentations

Agri-environmental policies in OECD countries and natural resource management	Mr. Wilfrid LEGG, OECD
Market mechanisms in water allocation in Australia	Mr. Seamus PARKER, Department of Natural Resources, Mines and Water, Queensland, Australia
The Dutch approach to water quality problems related to fertilisation and crop protection	Mr. Peter VAN BOHEEMEN, the Dutch Ministry of Agriculture
Policy issues regarding water availability and water quality in agriculture in the United States	Mr. Dennis WICHELNS, Rivers Institute at Hanover College, Indiana, USA
Decision support tools to aid policy design and implementation for sustainable resource use in agriculture	Mr. Kevin PARRIS, OECD

Discussants: Mr. QIN Fu, Chinese Academy of Agricultural Sciences; Mr. CHEN Zhijun, FAO.

General discussion

Tuesday, 20 June 2006

Session 3. Policy options for China

This session will provide an overview of various policy options which might be of relevance for China.

Chair: Mr. LU Xiaoping, Deputy Director General, MoA

Presentations

Fertiliser use in Chinese agriculture	Mr. CHEN Mengshan, MoA
Conserving agricultural biodiversity through water markets in China: Lessons from the Millennium Ecosystem Assessment	Mr. Jeff MCNEELY, IUCN-The World Conservation Union
A resource utilisation approach to resolving food security issues in China	Mr. KE Bingsheng, RCRE, MoA
Models and strategies for the development of circular agriculture in China	Mr. TANG Huajun, Chinese Academy of Agricultural Sciences
The crop protection industry role in supporting sustainable agriculture development in China	Mr. Roy RU, Syngenta-China
Does crop insurance influence agrochemical use in the current Chinese situation? A case study in the Manasi Watershed, Xinjiang	Mr. ZHONG Funing, Nanjing Agriculture University

Discussants: Mr. CHEN Zhijun, FAO; Mr. Brad GILMOUR, Agriculture and Agri-Food Canada.

General discussion

Session 4. General conclusions

This session will wrap-up the meeting and will focus on concrete policy suggestions as resulting from the workshop.

Chair: Mr. CHEN Xiaohua, the MoA, China

Panel discussion

Panellists will be asked to provide answers to the following three key questions:

- How to find a balance between the objectives of expanding food production, raising rural welfare, opening up the domestic market to international trade, and protecting the environment (policy coherence between agricultural, environmental, energy, rural development and trade policies)?
- How to shift policy measures towards market-based policy instruments to ensure that the polluter pays and providers of environmental benefits are paid?
- What would be the best policy mix in China to involve a combination of taxes and payments, regulations, and voluntary and stakeholder participation in watershed management?

Participants of the panel:

Mr. HUANG Shouhong, the State Council, China;

Mr. CHEN Zhijun, FAO;

Mr. Bryan LOHMAR, ERS/USDA;

Mr. ZHANG Xiaoshan, CASS, China;

Mr. HUANG Jikun , CAS, China;

Mr. Jeff MCNEELY, IUCN-The World Conservation Union;

Mr. Krzysztof MICHALAK, OECD.

General discussion

Summaries by: Mr. ZHANG Hongyu, MoA and Mr. Wilfrid LEGG, OECD.

Closing remarks by: Mr. Herwig SCHLÖGL, Deputy Secretary-General of OECD and Mr. FAN Xiaojian, Vice Minister of MoA, China.

Close of discussion

Wednesday, 21 June 2006

One-day study tour organised by the Chinese MoA to illustrate water availability and water quality issues in rural China

List of Participants / Liste des Participants

Argentina / Argentine

Mr. Mariano RIPARI
Agricultural Counsellor
Embassy of Argentina
Sam Li Tum
East Road, Bldg II
Beijing, China

Tel: +86 10 6532 0789
Fax: +86 10 6532 0270
Email: cecilia@consejeria-china.org

Australia / Australie

Ms. Fiona BARTLETT
Manager
Australian Government Department of Agriculture, Fisheries and Forestry
Natural Resource Management Division
GPO Box 858
Canberra ACT 2601, Australia

Tel: +61 2 6272 5921
Fax: +61 2 6271 6448
Email: fiona.bartlett@daff.gov.au

Mr. Toby MCGRATH
Water Entitlements Specialist
Australian Department of Agriculture, Fisheries and Forestry
Water Entitlements and Trading Planning Project
Rm. 1319, Golden Land Building No. 32 Liang Maqiao Road
Chaoyang District, Beijing, China

Tel: +86 135 110 22446
Email: mcgrath_toby@yahoo.com.au

Mr. Robert SPEED
Australian Team Leader
Australian Department of Agriculture, Fisheries and Forestry
Water Entitlements and Trading Planning Project
Rm. 1319, Golden Land Building
No. 32 Liang Maqiao Road
Chaoyang District, Beijing, China

Tel: +86 135 110 22446
Email: Robert_a_speed@yahoo.com.au

Mr. Vincent HUDSON
Counsellor Agriculture (Policy)
Australian Embassy
Sanlitun
21 Dongzhimenwai Dajie
10060 Beijing, China

Tel: +86 10 5140 4212
Email: vincent.hudson@dfat.gov.au

Mr. Seamus PARKER
Project Director
South-East Queensland Water Management
GPO Box 2454
Brisbane, Queensland 4001, Australia

Tel: +61 7 3225 1903
Email: seamus.parker@nrm.qld.gov.au

Brazil / Brésil

Mr. Antonio MORAES
Trade Policy Coordinator
Ministry of Agriculture, Livestock and Food Supply
Secretary of Agricultural Policy
Esplanada dos Ministérios Bloco D Sala 511
70043-900 Brasilia, DF, Brazil

Tel: +55 61 3218 2365
Fax: +55 61 3225 4726
Email: moraes@agricultura.gov.br

Canada

Mr. Peter CHEN
Senior Trade Policy Analyst
Agriculture and Agri-Food Canada
1071, 930 Carling Ave.
Ottawa, Ontario K1A 0C5, Canada

Tel: +1 613 759 7655
Fax: +1 613 759 7503
Email: chenp@agr.gc.ca

Mr. Brad GILMOUR
Team Leader - Asia Pacific and Technical Issues
Agriculture and Agri-Food Canada
International Agri-Food Analysis
930 Carling Ave.
Ottawa, Ontario K1A OC5, Canada

Tel: +1 613 759 7404
Fax: +1 613 759 7036
Email: gilmourb@agr.gc.ca

China / Chine

Mr. Mengshan CHEN
Director-General
Ministry of Agriculture
Department of Crop Farming
Beijing, China

Mr. Xiaohua CHEN
Director-General
Ministry of Agriculture
Department of Policy and Legislation
Beijing, China

Mr. Xiaojian FAN
Vice Minister
Ministry of Agriculture
No. 11, Nongzhanguan Nanli
100026 Beijing, China

Mr. Jun HAN
Director-General
Development Research Center of the State Council
Research Department of Rural Economy
No. 225 Chaoyangmen Neidajie
100010 Beijing, China

Tel: +86 10 6523 7011
Fax: +86 10 6523 6060
Email: hanjun@drc.gov.cn

Mr. Jikun HUANG
Director
Chinese Academy of Sciences
Center for Chinese Agricultural Policy
917 Building, Anwai, Datun Road
100101 Beijing, China

Tel: +86 10 6217 6604
Fax: +86 10 6217 8579
Email: jikhuang@public.bta.net.cn

Mr. Shouhong HUANG
Director-General
Policy Research Center of the State Council
Rural Department
China

Mr. Bingsheng KE
Director-General
Ministry of Agriculture
Research Centre for Rural Economy (RCRE)
No. 56 Xi Si Zhuan Ta Hutong
100810 Beijing, China

Tel: +86 10 66 17 36 04
Email: ke@agri.gov.cn

Mr. Xiaoping LU
Deputy Director-General
Ministry of Agriculture
Department of International Cooperation
Beijing, China

Mr. Xiaohe MA
Vice President
National Development and Reform Commission
Macro Research Academy
China

Email: mxhe2002@x263.net

Mr. Fu QIN
Director-General
Chinese Academy of Agricultural Sciences
Institute of Agricultural Economy
China

Mr. Huajun TANG
Director-General
Chinese Academy of Agricultural Sciences
Institute of Agricultural Resources and Regional Planning
China

Mr. Renjian TANG
Vice Minister
Office of the Leading Group for Finance and Economy, CCCP
China

Mr. Tiejun WEN
Dean
Renmin University of China
College of Rural Development
China

Mr. Hongyu ZHANG
Director-General
Ministry of Agriculture
Department of Policy and Law
Beijing, China

Mr. Funing ZHONG
Professor
Nanjing Agriculture University
College of Economy
China

Denmark / Danemark

Ms. Anna TAUBY SOERENSEN
Agricultural Assistant
San Li Tun
Dong Wu Jie 1
100600 Beijing, China

Tel: +86 10 8532 9900
Fax: +86 10 8532 9999
Email: antaso@um.dk

France

Mr. Bruno VINDEL
Sub-Director of Evaluation, Future Studies and Guidance Division
Ministry of Agriculture
3 rue Barbet-de-Jouy
07 SP 75349 Paris, France

Tel: +33 1 49 55 42 10
Fax: +33 1 49 55 49 70
Email: bruno.vindel@agriculture.gouv.fr

Germany / Allemagne

Mr. Klaus J. SUPP
Counsellor, Food, Agriculture and Consumer Protection
Embassy of the Federal Republic of Germany
Dongzhimenwai Dajie 17
Chaoyang District
100600 Beijing, China

Tel: +86 10 8532 9306
Fax: +86 10 6532 5335
Email: wiss-10@peki.diplo.de

Japan / Japon

Mr. Hiroki SASAKI
Officer
Ministry of Agriculture, Forestry and Fisheries
Environment Policy Division
1-2-1, Kasumigaseki, Chiyoda-ku
Tokyo 100-8950, Japan

Tel: +81 3 3502 8056
Fax: +81 3 3591 6640
Email: hiroki_sasaki@nm.maff.go.jp

Mr. Mitsuaki SHINDO
First Secretary
Permanent Delegation
Agriculture, Forestry and Fisheries
11 avenue Hoche
75008 Paris, France

Tel: +33 1 53 76 61 93
Fax: +33 1 45 63 05 44
Email: shindo@deljp-ocde.fr

Korea / Corée

Mr. Bong-Soon CHOI
Deputy Director
Ministry of Agriculture and Forestry
Multilateral Cooperation
88, Gwnmunro
427-719 Gwacheon-si, Gyeonggi-do, Korea

Tel: +82 2 500 1713
Fax: +82 2 507 2095
Email: chbs@maf.go.kr

Ms. Ji-Yeon YANG
Assistant Director
Ministry of Agriculture and Forestry
88, Gwanmunro
Gwacheon-si, Gyeonggi-do, Korea

Tel: +82 2 500 1723
Fax: +82 2 504 6659
Email: yjy430@maf.go.kr

Mexico / Mexique

Mr. Jorge RUEDA SOUSA
Minister Counsellor, Agriculture Representation for EU
Mexican Embassy
94, Av. Franklin Roosevelt
1050 Bruxelles, Belgium

Tel: +32 2 644 1300
Fax: +32 2 644 12835
Email: jorgerueda@sagarpaue.be

Netherlands / Pays-Bas

Mr. Peter VAN BOHEEMEN
Staff member
Ministry of Agriculture, Nature and Food Quality
P.O. Box 20401
2500 EK Den Haag, Netherlands

Tel: +31 70 378 4915
Fax: +31 70 378 6156
Email: p.j.m.van.boheemen@minlnv.nl

New Zealand / Nouvelle-Zélande

Mr. Steven AINSWORTH
First Secretary (Agriculture)
New Zealand Embassy, Beijing
1 Ritan Dongerjie
Chaoyang District
100600 Beijing, China

Tel: +86 10 6532 2731, ext. 223
Fax: +86 10 6532 4317
Email: steven.ainsworth@mfat.govt.nz

United Kingdom / Royaume-Uni

Mr. Leo HORN
Environmental Economic Adviser
Department for International Development (DFID)
30th Fl., Kerry Centre South Tower
1, Guanghua Road
Chaoyang District
100020 Beijing, China

Tel: +86 10 8529 6882, ext. 2018
Fax: +86 10 8529 6003
Email: l-horn@dfid.gov.uk

Mr. John WARBURTON
Senior Environmental Adviser
Department for International Development (DFID)
30th Fl., Kerry Centre South Tower
1, Guanghua Road
Chaoyang District
100020 Beijing, China

Tel: +86 10 8529 6882, ext. 2024
Fax: +86 10 8529 6003
Email: j-warburton@dfid.gov.uk

United States / Etats-Unis

Ms. Christine BOYLE
Visiting Researcher at Chinese Center for Agricultural Policy (CCAP)
National Science Foundation
P.O. Box 27435
98165-2435 Seattle, WA, United States

Tel: +1 206 979 1002
Email: cboyle@email.unc.edu

Mr. Caleb O'KRAY
Agricultural Economist / U.S. Delegation Assistant
U.S. Foreign Agricultural Service, U.S. Embassy, Beijing
Qi Jia Yuan Diplomatic Compound
No. 5-2 Jianguomenwai
100600 Beijing, China

Tel: +86 10 6532 1953, ext. 302
Fax: +86 10 6532 2962
Email: Caleb.O'Kray@usda.gov

Mr. Bryan LOHMAR
Economist, Market and Trade Economics Division
U.S. Department of Agriculture
Economic Research Service
1800 M Street, NW
Washington, DC 20036, United States

Tel: +1 202 694 5226
Fax: +1 202 694 5793
Email: BLOHMAR@ers.usda.gov

Mr. Dennis WICHELNS
Professor of Economics and Executive Director
Rivers Institute at Hanover College
P.O. Box 48
Hanover, Indiana 47243, United States

Tel: +812 866 6846
Fax: +812 866 6828
Email: wichelns@hanover.edu

U.N. Food and Agricultural Organization (FAO) / Organisation des Nations Unies pour l'Alimentation et l'Agriculture (FAO)

Mr. Zhijun CHEN
Water Resources Development and Conservation Officer
FAO Regional Office for Asia and the Pacific
39, Phra Atit Road
Maliwan Mansion 10200, Thailand

Tel: +66 2 697 4329
Fax: +66 2 697 4445
Email: Zhijun.Chen@fao.org

World Bank / Banque mondiale

Mr. Achim FOCK
Senior Economist
The World Bank
No. 1 Jianguomenwai Avenue
Level 16, China World Tower 2
100004 Beijing, China

Tel: +86 10 5861 7681
Fax: +86 10 5861 7800
Email: afock@worldbank.org

Business and Industry Advisory Committee (BIAC) / Comité consultatif économique et industriel (BIAC)

Mr. Lihua CHEN
Regional Vice President for China of the International Fertilizer Industry Association (IFA)
International Fertilizer Industry Association (IFA)
China National Chemical Construction Corporation (CNCCC)
Kai Kang Mansion No. 15 Sanqu Anzhenxili
Chaoyang District
100029 Beijing, China

Tel: +86 10 64419836
Fax: +86 10 64419698
Email: cncccifa@cnccc.com.cn

Mr. Ximing HU
Agronomist
Unilever China
30/F Tower B, City Centre of Shanghai
100 Zunyi Road
Shanghai, China

Tel: +86 21 6237 0788
Fax: +86 21 6237 2295
Email: Ximing.Hu@Unilever.com

Ms. Laura HU
Government Affairs Manager
Syngenta (China) Investment Co., Ltd.
21F, Xin Mei Union Square, 999 Pudong South Road
200120 Shanghai, China

Tel: +86 021 6888 0077
Fax: +86 021 6888 2277
Email: Laura.Hu@syngenta.com

Mr. Herbert OBERHÄNSLI
Assistant to the Chairman for Economic Affairs
Nestlé S.A.
Economics and International Relations
Avenue Nestlé 55
1800 Vevey, Switzerland

Tel: +41 21 924 2357
Fax: +41 21 924 4582
Email: herbert.oberhaensli@nestle.com

Mr. Roy RU
Professional Product Development Manager
Syngenta (China) Investment Co., Ltd.
21/F Xin Mei Union Square
999 Pudong South Road
200120 Shanghai, China

Tel: +86 21 6888 0077
Fax: +86 21 6888 2277
Email: roy.ru@syngenta.com

Mr. Sihai WU
President of the International Fertilizer Industry Association (IFA)
International Fertilizer Industry Association (IFA)
Sino-Arab Chemical Fert. Co. Ltd (SACF)
East Section, Jianshe Road
066003 Qinhuangdao, China

Tel: +86 335 316 1088
Fax: +86 335 316 1303
Email: hach@sacf.com

Mr. Xiaomei ZHANG
Technical Manager of Agricultural Chemical
BASF (China) Co. Ltd.
15/F, Beijing Sunflower Tower
No. 37 Maizidian Street
Chaoyang District
100026 Beijing, China

Tel: +86 10 6591 8899
Fax: +86 10 8527 5617
Email: zhangxm@basf-china.com.cn

Mr. Alfred ZHOU
Public Affairs and Communication Manager
Syngenta (China) Investment Co., Ltd.
21F, Xin Mei Union Square, 999 Pudong South Road
200120 Shanghai, China

Tel: +86 021 6888 0077
Fax: +86 021 6888 2277
Email: Alfred.Zhou@syngenta.com

OECD / OCDE

2, rue André Pascal
75016 Paris, France

Mr. Herwig SCHLÖGL
Deputy Secretary-General
GENERAL SECRETARIAT

Tel: +33 1 45 24 80 38
Fax: +33 1 44 30 62 71
Email: Herwig.SCHLOGL@oecd.org

Mr. Wilfrid LEGG
Head of Policies and Environment Division
DIRECTORATE FOR FOOD, AGRICULTURE AND FISHERIES

Tel: +33 1 45 24 95 36
Fax: +33 1 44 30 61 02
Email: Wilfrid.LEGG@oecd.org

Mr. Kevin PARRIS
Senior Analyst
DIRECTORATE FOR FOOD, AGRICULTURE AND FISHERIES

Tel: +33 1 45 24 95 68
Fax: +33 1 45 24 18 90
Email: Kevin.PARRIS@oecd.org

Mr. Andrzej KWIECINSKI
Senior Analyst
DIRECTORATE FOR FOOD, AGRICULTURE AND FISHERIES

Tel: +33 1 45 24 95 08
Fax: +33 1 44 30 61 19
Email: Andrzej.KWIECINSKI@oecd.org

Mr. Krzysztof MICHALAK
Administrator
ENVIRONMENT DIRECTORATE

Tel: +33 1 45 24 96 00
Fax: +33 1 45 24 96 71
Email: Krzysztof.MICHALAK@oecd.org

IUCN - World Conservation Union

Mr. Jeffrey MCNEELY
Chief Scientist
IUCN - World Conservation Union
28, rue Mauverney
1196 Gland, Switzerland

Tel: +41 22 999 0284
Fax: +41 22 999 0025
Email: jam@iucn.org

Chinese Secretariat of the Workshop

Ms. Hui WANG
Director
Ministry of Agriculture
Department for Sectoral Policy and Law
11, Nongzhanguan Nanli
100026 Beijing, China

Mr. Wengsheng JIANG
Deputy Director
Ministry of Agriculture
Department for Sectoral Policy and Law
11, Nongzhanguan Nanli
100026 Beijing, China

Ms. Jiemei YANG
Ministry of Agriculture
Department for Sectoral Policy and Law
11, Nongzhanguan Nanli
100026 Beijing, China

Ms. Pan GONG
Institute of Agricultural Resources and Regional Planning of the Chinese Academy of Agricultural Sciences

Mr. Chunhua YANG
Director
Ministry of Agriculture
Department for Sectoral Policy and Law
11, Nongzhanguan Nanli
100026 Beijing, China

Ms. Na LI
Division Chief
Ministry of Agriculture
Department for Sectoral Policy and Law
11, Nongzhanguan Nanli
100026 Beijing, China

Tel: +86 10 6419 2782
Fax: +86 10 6419 2777
Email: nanali8830@sohu.com

Mr. Yinngbin HE
Institute of Agricultural Resources and Regional Planning of the Chinese Academy of Agricultural Sciences

OECD PUBLICATIONS, 2, rue André-Pascal, 75775 PARIS CEDEX 16
PRINTED IN FRANCE
(51 2006 10 1 P) ISBN 92-64-02846-3 – No. 55343 2006